SCHAUM'S OUTLINE OF

THEORY AND P[illegible]

OF

ANALOG and DIGITAL COMMUNICATIONS

•

HWEI P. HSU, Ph.D.
Professor and Chairman
Department of Electrical Engineering
Fairleigh Dickinson University

SCHAUM'S OUTLINE SERIES
McGRAW-HILL, INC.

New York St. Louis San Francisco Auckland Bogotá Caracas
Lisbon London Madrid Mexico City Milan Montreal
New Delhi San Juan Singapore Sydney Tokyo Toronto

HWEI P. HSU received his B.S. from National Taiwan University and M.S. and Ph.D. from Case Institute of Technology. He is presently Professor of Electrical Engineering and Chair of the Electrical Engineering Department, Fairleigh Dickinson University.

This book is printed on recycled paper containing a minimum of 50% total recycled fiber with 10% postconsumer de-inked fiber. Soybean based inks are used on the cover and text.

Schaum's Outline of Theory and Problems of
ANALOG AND DIGITAL COMMUNICATIONS

1 2 3 4 5 6 7 8 9 10 11 12 13 14 15 16 17 18 19 20 SHP SHP 9 8 7 6 5 4 3

ISBN 0-07-030636-2

Sponsoring Editor: John Aliano
Production Supervisor: Denise Puryear
Editing Supervisor: Maureen Walker
Cover design by Amy E. Becker.

Library of Congress Cataloging-in-Publication Data

Hsu, Hwei P. (Hwei Piao), date
Schaum's outline of theory and problems of analog and digital communications/Hwei P. Hsu.
p. cm.—(Schaum's outline series)
Includes index.
ISBN 0-07-030636-2
1. Signal theory (Telecommunication) 2. Modulation theory.
I. Title.
TK5102.5.H48 1993
621.382'2—dc20 92-38804
CIP

Preface

This book is intended as a supplement to all textbooks in communication theory for electrical engineers; it may be used as a self-contained textbook or for self-study. Each topic is introduced in a chapter with numerous solved problems. The solved problems constitute an integral part of the text.

Chapter 1 presents the mathematical groundwork of signal and linear system analysis based on Fourier methods. Based on this analysis, Chapters 2 and 3 treat noiseless modulation theory, including amplitude and angle modulation. Chapter 4 deals with the sampling and digital transmission of analog signals. Probability and random signals are introduced in Chapters 5 and 6. No previous knowledge of probability is assumed. Chapter 7 applies this analysis to the study of the effect of noise on communication systems. The last chapter, Chapter 8, presents an elementary treatment of information theory and coding.

I wish to express my thanks to Professor Gordon Silverman, Manhattan College, for his assistance, comments, and careful review of the manuscript. I also wish to acknowledge the assistance of Jiri Naxera and Wenchang Wu for proofreading many of the chapters in the manuscript and some of the drawings. Furthermore, I wish to thank Mr. David Burleigh for his assistance, Mr. John Aliano for his helpful comments and encouragement, and Ms. Maureen Walker for her great care in preparing this book. Finally, I am grateful to my wife, Daisy, whose understanding and constant support were necessary factors in the completion of this work.

HWEI P. HSU

Contents

Chapter 1

Signals and Systems

1.1 CLASSIFICATION OF SIGNALS

A signal is a function representing a physical quantity. Mathematically, a signal is represented as a function of an independent variable t. Usually t represents time. Thus a signal is denoted by $x(t)$.

A. Continuous-Time and Discrete-Time Signals:

A signal $x(t)$ is a *continuous-time* signal if t is a continuous variable. If t is a discrete variable, that is, $x(t)$ is defined at discrete times, then $x(t)$ is a *discrete-time* signal. Since a discrete-time signal is defined at discrete times, it is often identified as a *sequence* of numbers, denoted by $\{x(n)\}$ or $x[n]$, where $n =$ integer.

B. Analog and Digital Signals:

If a continuous-time signal $x(t)$ can take on any values in the continuous interval (a, b), where a may be $-\infty$ and b may be $+\infty$, then the continuous-time signal $x(t)$ is called an *analog* signal. If a discrete-time signal $x[n]$ can take only a finite number of distinct values, then we call this signal a *digital* signal.

Since a discrete-time signal $x[n]$ is often formed by sampling a continuous-time signal $x(t)$ such that $x[n] = x(nT_s)$, where T_s is the *sampling interval*, in the following we deal mainly with continuous-time signals.

C. Real and Complex Signals:

A signal $x(t)$ is a *real* signal if its value is a real number and is a *complex* signal if its value is a complex number.

D. Deterministic and Random Signals:

Deterministic signals are those signals whose values are completely specified for any given time. *Random* signals are those signals that take random values at any given time, and these must be characterized statistically. Random signals are discussed in Chap. 6.

E. Energy and Power Signals:

The *normalized energy content* E of a signal $x(t)$ is defined as

$$E = \int_{-\infty}^{\infty} |x(t)|^2 \, dt \tag{1.1}$$

The *normalized average power* P of a signal $x(t)$ is defined as

$$P = \lim_{T \to \infty} \frac{1}{T} \int_{-T/2}^{T/2} |x(t)|^2 \, dt \tag{1.2}$$

If $0 < E < \infty$, that is, if E is finite (so $P = 0$), then $x(t)$ is referred to as an *energy* signal. If $E = \infty$, but $0 < P < \infty$, that is, P is finite, then $x(t)$ is referred to as a *power* signal.

F. Periodic and Nonperiodic Signals:

A signal $x(t)$ is *periodic* if there is a positive number T_0 such that

$$x(t+T_0)=x(t) \tag{1.3}$$

The smallest positive number T_0 is called the *period*, and the reciprocal of the period is called the *fundamental frequency* f_0:

$$f_0=\frac{1}{T_0} \quad \text{hertz (Hz)} \tag{1.4}$$

From Eq. (*1.3*) it follows that

$$x(t+nT_0)=x(t) \tag{1.5}$$

Any signal for which there is no value of T_0 satisfying Eq. (*1.3*) is said to be *nonperiodic* or *aperiodic*.

A periodic signal is a power signal if its energy content per period is finite, and then the average power of the signal need only be calculated over a period.

1.2 SINGULARITY FUNCTIONS

An important subclass of nonperiodic signals in communication theory is the *singularity* functions (or, as they are sometimes called, the *generalized* functions). Here we deal with only two important singularity functions: the unit step function $u(t)$ and the unit impulse function $\delta(t)$.

A. Unit Step Function:

The *unit step function* $u(t)$ is defined as

$$u(t)=\begin{cases}1 & t>0\\ 0 & t<0\end{cases} \tag{1.6}$$

which is shown graphically in Fig. 1-1.

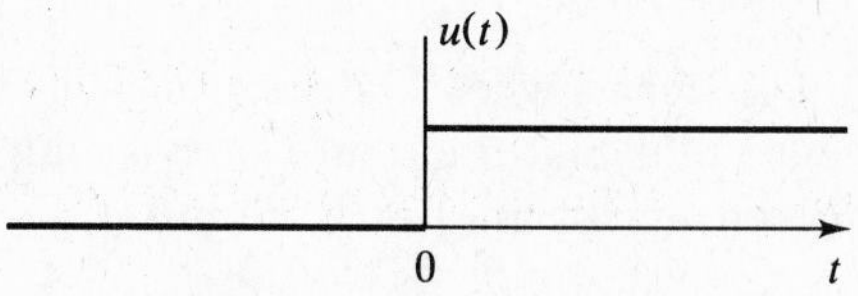

Fig. 1-1 Unit step function

Note that it is discontinuous at $t=0$ and that the value at $t=0$ is undefined.

B. Unit Impulse Function:

The *unit impulse function*, also known as the *Dirac delta function*, $\delta(t)$ is not an ordinary function and is defined in terms of the following process:

$$\int_{-\infty}^{\infty}\phi(t)\delta(t)\,dt=\phi(0) \tag{1.7}$$

where $\phi(t)$ is any test function continuous at $t=0$.

Some additional properties of $\delta(t)$ that can be derived from definition (*1.7*) are

$$\int_{-\infty}^{\infty} \phi(t)\delta(t-t_0)\,dt = \phi(t_0) \tag{1.8}$$

$$\delta(at) = \frac{1}{|a|}\delta(t) \tag{1.9}$$

$$\delta(-t) = \delta(t) \tag{1.10}$$

$$x(t)\delta(t) = x(0)\delta(t) \tag{1.11}$$

$$x(t)\delta(t-t_0) = x(t_0)\delta(t-t_0) \tag{1.12}$$

An alternate definition of $\delta(t)$ is provided by the following two conditions:

$$\int_{t_1}^{t_2} \delta(t-t_0)\,dt = 1 \qquad t_1 < t_0 < t_2 \tag{1.13}$$

$$\delta(t-t_0) = 0 \qquad t \neq t_0 \tag{1.14}$$

Conditions (*1.13*) and (*1.14*) correspond to the intuitive notion of a unit impulse as the limit of a suitably chosen conventional function having unity area in an infinitesimally small width. For convenience, $\delta(t)$ is shown schematically in Fig. 1-2.

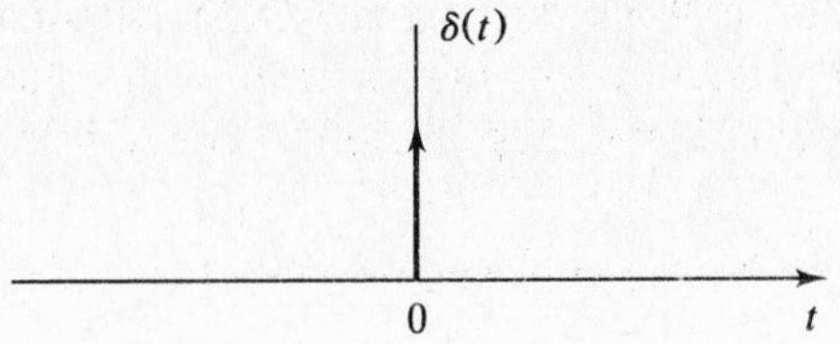

Fig. 1-2 Unit impulse function

C. Derivative of Singularity Functions:

If $g(t)$ is a generalized function, its derivative $g'(t)$ is defined by the following relation:

$$\int_{-\infty}^{\infty} g'(t)\phi(t)\,dt = -\int_{-\infty}^{\infty} g(t)\phi'(t)\,dt \tag{1.15}$$

where $\phi'(t)$ is the derivative of $\phi(t)$. By using Eq. (*1.15*), the derivative of $u(t)$ can be shown to be $\delta(t)$; that is,

$$\delta(t) = u'(t) = \frac{du(t)}{dt} \tag{1.16}$$

1.3 REVIEW OF FOURIER SERIES

A. Complex Exponential Fourier Series:

Let $x(t)$ be a periodic signal with period T_0. Then we define the complex exponential Fourier series of $x(t)$ as

$$x(t) = \sum_{n=-\infty}^{\infty} c_n e^{jn\omega_0 t} \tag{1.17}$$

where $\omega_0 = 2\pi/T_0 = 2\pi f_0$, which is called the *fundamental angular frequency*. The coefficients c_n

are called the *Fourier coefficients*, and they are given by

$$c_n = \frac{1}{T_0}\int_{t_0}^{t_0+T_0} x(t)e^{-jn\omega_0 t}\,dt \tag{1.18}$$

where t_0 is any value. Setting $t_0 = -T_0/2$, we have

$$c_n = \frac{1}{T_0}\int_{-T_0/2}^{T_0/2} x(t)e^{-jn\omega_0 t}\,dt \tag{1.19}$$

B. Frequency Spectra:

If the periodic signal $x(t)$ is real, then

$$c_n = |c_n|e^{j\theta_n} \qquad c_{-n} = c_n^* = |c_n|e^{-j\theta_n} \tag{1.20}$$

where $|c_n|$ is the amplitude, θ_n is the phase angle of c_n, and the asterisk indicates the complex conjugate. Note that

$$|c_{-n}| = |c_n| \qquad \theta_{-n} = -\theta_n \tag{1.21}$$

A plot of $|c_n|$ versus the angular frequency $\omega = 2\pi f$ is called the *amplitude spectrum* of the periodic signal $x(t)$. A plot of θ_n versus ω is called the *phase spectrum* of $x(t)$. These are referred to as *frequency spectra* of $x(t)$. Since the index n assumes only integers, the frequency spectra of a periodic signal exist only at the discrete frequencies $n\omega_0$. These are therefore referred to as *discrete frequency spectra* or *line spectra*. From Eq. (*1.21*) we see that the amplitude spectrum is an even function of ω and the phase spectrum is an odd function of ω.

C. Power Content of a Periodic Signal:

The *power content* of a periodic signal $x(t)$ with period T_0 is defined as the mean-square value over a period:

$$P = \frac{1}{T_0}\int_{-T_0/2}^{T_0/2} |x(t)|^2\,dt \tag{1.22}$$

D. Parseval's Theorem for the Fourier Series:

Parseval's theorem for the Fourier series states that if $x(t)$ is a periodic signal with period T_0, then

$$\frac{1}{T_0}\int_{-T_0/2}^{T_0/2} |x(t)|^2\,dt = \sum_{n=-\infty}^{\infty} |c_n|^2 \tag{1.23}$$

1.4 FOURIER TRANSFORM

To generalize the Fourier series representation (*1.17*) to a representation valid for nonperiodic signals in the frequency domain, we introduce the Fourier transform.

A. Definition:

Let $x(t)$ be a nonperiodic signal. Then the *Fourier transform* of $x(t)$, symbolized by $\mathscr{F}$, is defined by

$$X(\omega) = \mathscr{F}[x(t)] = \int_{-\infty}^{\infty} x(t)e^{-j\omega t}\,dt \tag{1.24}$$

The *inverse* Fourier transform of $X(\omega)$, symbolized by $\mathscr{F}^{-1}$, is defined by

$$x(t)=\mathscr{F}^{-1}[X(\omega)]=\frac{1}{2\pi}\int_{-\infty}^{\infty}X(\omega)e^{j\omega t}\,d\omega \tag{1.25}$$

Equations (*1.24*) and (*1.25*) are often called the *Fourier transform pair*. Writing $X(\omega)$ in terms of amplitude and phase as

$$X(\omega)=|X(\omega)|e^{j\theta(\omega)} \tag{1.26}$$

we can show, for real $x(t)$, that

$$X(-\omega)=X^*(\omega)=|X(\omega)|e^{-j\theta(\omega)} \tag{1.27}$$

or

$$|X(-\omega)|=|X(\omega)| \qquad \theta(-\omega)=-\theta(\omega) \tag{1.28}$$

Thus, just as for the complex Fourier series, the *amplitude spectrum* of $x(t)$, denoted by $|X(\omega)|$, is an even function of ω, and the *phase spectrum* $\theta(\omega)$ is an odd function of ω. These are referred to as *Fourier spectra* of $x(t)$. Equation (*1.27*) is the necessary and sufficient condition for $x(t)$ to be real (see Prob. 1.17).

B. Properties of the Fourier Transform:

We use the notation

$$x(t)\leftrightarrow X(\omega)$$

to denote the Fourier transform pair.

1. ***Linearity (Superposition):***

$$a_1x_1(t)+a_2x_2(t)\leftrightarrow a_1X_1(\omega)+a_2X_2(\omega) \tag{1.29}$$

where a_1 and a_2 are any constants.

2. ***Time Shifting:***

$$x(t-t_0)\leftrightarrow X(\omega)e^{-j\omega t_0} \tag{1.30}$$

3. ***Frequency Shifting:***

$$x(t)e^{j\omega_0 t}\leftrightarrow X(\omega-\omega_0) \tag{1.31}$$

4. ***Scaling:***

$$x(at)\leftrightarrow\frac{1}{|a|}X\left(\frac{\omega}{a}\right) \tag{1.32}$$

5. ***Time Reversal:***

$$x(-t)\leftrightarrow X(-\omega) \tag{1.33}$$

6. ***Duality:***

$$X(t)\leftrightarrow 2\pi x(-\omega) \tag{1.34}$$

7. ***Differentiation:***

Time differentiation:

$$x'(t)=\frac{d}{dt}x(t)\leftrightarrow j\omega X(\omega) \tag{1.35}$$

Frequency differentiation:

$$(-jt)x(t) \longleftrightarrow X'(\omega) = \frac{d}{d\omega}X(\omega) \tag{1.36}$$

8. Integration:

$$\int_{-\infty}^{t} x(\tau)\,d\tau \longleftrightarrow \frac{1}{j\omega}X(\omega) + \pi X(0)\delta(\omega) \tag{1.37}$$

C. Fourier Transforms of Some Useful Signals:

$$\delta(t) \longleftrightarrow 1 \tag{1.38}$$

$$\delta(t-t_0) \longleftrightarrow e^{-j\omega t_0} \tag{1.39}$$

$$1 \longleftrightarrow 2\pi\delta(\omega) \tag{1.40}$$

$$e^{j\omega_0 t} \longleftrightarrow 2\pi\delta(\omega-\omega_0) \tag{1.41}$$

$$\cos\omega_0 t \longleftrightarrow \pi\delta(\omega-\omega_0) + \pi\delta(\omega+\omega_0) \tag{1.42}$$

$$\sin\omega_0 t \longleftrightarrow -j\pi\delta(\omega-\omega_0) + j\pi\delta(\omega+\omega_0) \tag{1.43}$$

$$u(t) \longleftrightarrow \pi\delta(\omega) + \frac{1}{j\omega} \tag{1.44}$$

$$e^{-at}u(t) \qquad a>0 \longleftrightarrow \frac{1}{j\omega+a} \tag{1.45}$$

$$e^{-a|t|} \qquad a>0 \longleftrightarrow \frac{2a}{\omega^2+a^2} \tag{1.46}$$

1.5 CONVOLUTION

The *convolution* of two signals $x_1(t)$ and $x_2(t)$, denoted by $x_1(t) * x_2(t)$, is a new signal $x(t)$, defined by

$$x(t) = x_1(t) * x_2(t) = \int_{-\infty}^{\infty} x_1(\tau)x_2(t-\tau)\,d\tau \tag{1.47}$$

A. Properties of Convolution:

$$x_1(t) * x_2(t) = x_2(t) * x_1(t) \tag{1.48}$$

$$[x_1(t) * x_2(t)] * x_3(t) = x_1(t) * [x_2(t) * x_3(t)] \tag{1.49}$$

$$x_1(t) * [x_2(t) + x_3(t)] = x_1(t) * x_2(t) + x_1(t) * x_3(t) \tag{1.50}$$

B. Convolution with δ Functions:

$$x(t) * \delta(t) = x(t) \tag{1.51}$$

$$x(t) * \delta(t-t_0) = x(t-t_0) \tag{1.52}$$

C. Convolution Theorems:

Let $$x_1(t) \leftrightarrow X_1(\omega) \quad \text{and} \quad x_2(t) \leftrightarrow X_2(\omega)$$

Then
$$x_1(t) * x_2(t) \leftrightarrow X_1(\omega) X_2(\omega) \tag{1.53}$$

and
$$x_1(t) x_2(t) \leftrightarrow \frac{1}{2\pi} X_1(\omega) * X_2(\omega) \tag{1.54}$$

Equation (*1.53*) is referred to as the *time convolution theorem* and Eq. (*1.54*) as the *frequency convolution theorem*.

1.6 CORRELATION AND SPECTRAL DENSITY

A. Correlation of Energy Signals:

Let $x_1(t)$ and $x_2(t)$ be real-valued energy signals. Then the *cross-correlation* function $R_{12}(\tau)$ of $x_1(t)$ and $x_2(t)$ is defined by

$$R_{12}(\tau) = \int_{-\infty}^{\infty} x_1(t) x_2(t-\tau)\, dt \tag{1.55}$$

The *autocorrelation* function of $x_1(t)$ is defined as

$$R_{11}(\tau) = \int_{-\infty}^{\infty} x_1(t) x_1(t-\tau)\, dt \tag{1.56}$$

Properties of correlation functions:

$$R_{12}(\tau) = R_{21}(-\tau) \tag{1.57}$$

$$R_{11}(\tau) = R_{11}(-\tau) \tag{1.58}$$

$$R_{11}(0) = \int_{-\infty}^{\infty} [x_1(t)]^2\, dt = E \tag{1.59}$$

where E is the normalized energy content of $x_1(t)$.

B. Energy Spectral Density:

Let $R_{11}(\tau)$ be the autocorrelation function of $x_1(t)$. Then

$$S_{11}(\omega) = \mathscr{F}[R_{11}(\tau)] = \int_{-\infty}^{\infty} R_{11}(\tau) e^{-j\omega\tau}\, d\tau \tag{1.60}$$

is called the *energy spectral density* of $x_1(t)$. Now taking the inverse Fourier transform of Eq. (*1.60*), we have

$$R_{11}(\tau) = \mathscr{F}^{-1}[S_{11}(\omega)] = \frac{1}{2\pi} \int_{-\infty}^{\infty} S_{11}(\omega) e^{j\omega\tau}\, d\omega \tag{1.61}$$

If $x_1(t)$ is real, then we have

$$S_{11}(\omega) = \mathscr{F}[R_{11}(\tau)] = |X_1(\omega)|^2 \tag{1.62}$$

and
$$R_{11}(\tau) = \frac{1}{2\pi} \int_{-\infty}^{\infty} |X_1(\omega)|^2 e^{j\omega\tau}\, d\omega \tag{1.63}$$

Setting $\tau = 0$, we have

$$R_{11}(0) = \frac{1}{2\pi} \int_{-\infty}^{\infty} |X_1(\omega)|^2 \, d\omega \tag{1.64}$$

Thus, from Eq. (*1.59*),

$$E = \int_{-\infty}^{\infty} [x_1(t)]^2 \, dt = \frac{1}{2\pi} \int_{-\infty}^{\infty} |X_1(\omega)|^2 \, d\omega \tag{1.65}$$

This is the reason why $S_{11}(\omega) = |X_1(\omega)|^2$ is called the energy spectral density of $x_1(t)$. Equation (*1.65*) is also known as *Parseval's theorem* for the Fourier transform.

C. Correlation of Power Signals:

The *time-average* autocorrelation function $\bar{R}_{11}(\tau)$ of a real-valued power signal $x_1(t)$ is defined as

$$\bar{R}_{11}(\tau) = \lim_{T \to \infty} \frac{1}{T} \int_{-T/2}^{T/2} x_1(t) x_1(t - \tau) \, dt \tag{1.66}$$

Note that

$$\bar{R}_{11}(0) = \lim_{T \to \infty} \frac{1}{T} \int_{-T/2}^{T/2} [x_1(t)]^2 \, dt = P_1 \tag{1.67}$$

If $x(t)$ is periodic with period T_0, then

$$\bar{R}_{11}(\tau) = \frac{1}{T_0} \int_{-T_0/2}^{T_0/2} x_1(t) x_1(t - \tau) \, dt \tag{1.68}$$

D. Power Spectral Density:

The *power spectral density* of $x_1(t)$, denoted $\bar{S}_{11}(\omega)$, is defined as

$$\bar{S}_{11}(\omega) = \mathscr{F}\left[\bar{R}_{11}(\tau)\right] = \int_{-\infty}^{\infty} \bar{R}_{11}(\tau) e^{-j\omega\tau} \, d\tau \tag{1.69}$$

Then

$$\bar{R}_{11}(\tau) = \mathscr{F}^{-1}\left[\bar{S}_{11}(\omega)\right] = \frac{1}{2\pi} \int_{-\infty}^{\infty} \bar{S}_{11}(\omega) e^{j\omega\tau} \, d\omega \tag{1.70}$$

Setting $\tau = 0$, we get

$$\bar{R}_{11}(0) = \frac{1}{2\pi} \int_{-\infty}^{\infty} \bar{S}_{11}(\omega) \, d\omega \tag{1.71}$$

Thus, from Eq. (*1.67*),

$$P_1 = \lim_{T \to \infty} \frac{1}{T} \int_{-T/2}^{T/2} [x_1(t)]^2 \, dt = \frac{1}{2\pi} \int_{-\infty}^{\infty} \bar{S}_{11}(\omega) \, d\omega \tag{1.72}$$

This is the reason why $\bar{S}_{11}(\omega)$ is called the power spectral density of $x_1(t)$.

1.7 SYSTEM REPRESENTATION AND CLASSIFICATION

A. System Representation:

A system is a mathematical model of a physical process that relates the input signal (source or excitation signal) to the output signal (response signal).

Let $x(t)$ and $y(t)$ be the input and output signals, respectively, of a system. Then the system is viewed as a *mapping* of $x(t)$ into $y(t)$. Symbolically, this is expressed as

$$y(t) = \mathscr{T}[x(t)] \tag{1.73}$$

where $\mathscr{T}$ is the *operator* that produces output $y(t)$ from input $x(t)$, as illustrated in Fig. 1-3.

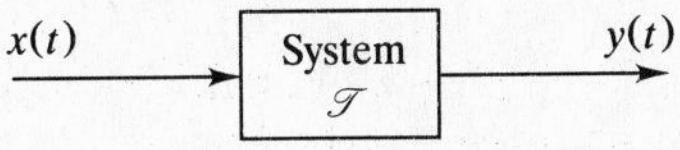

Fig. 1-3 Operator representation of a system

B. System Classification

1. Continuous-Time and Discrete-Time Systems:

If the input and output signals $x(t)$ and $y(t)$ are continuous-time signals, then the system is called a *continuous-time system*. If the input and output signals are discrete-time signals or sequences, then the system is called a *discrete-time system*. In the following, we deal mainly with the continuous-time system.

2. Linear Systems:

If the operator $\mathscr{T}$ in Eq. (*1.73*) satisfies the following two conditions, then $\mathscr{T}$ is called a *linear operator* and the system represented by $\mathscr{T}$ is called a *linear system*.

***a.* Additivity**

$$\mathscr{T}[x_1(t) + x_2(t)] = \mathscr{T}[x_1(t)] + \mathscr{T}[x_2(t)] = y_1(t) + y_2(t) \tag{1.74}$$

for all input signals $x_1(t)$ and $x_2(t)$.

***b.* Homogeneity**

$$\mathscr{T}[ax(t)] = a\mathscr{T}[x(t)] = ay(t) \tag{1.75}$$

for all input signals $x(t)$ and scalar a.

Any system that does not satisfy Eq. (*1.74*) and/or Eq. (*1.75*) is classified as a *nonlinear* system.

3. Time-Invariant Systems:

If the system satisfies the following condition, then the system is called a *time-invariant* or *fixed* system:

$$\mathscr{T}[x(t - t_0)] = y(t - t_0) \tag{1.76}$$

where t_0 is any real constant. Equation (*1.76*) indicates that the delayed input gives delayed output.
A system which does not satisfy Eq. (*1.76*) is called a *time-varying* system.

4. Linear Time-Invariant (LTI) Systems:

If the system is linear and time-invariant, then the system is called a *linear time-invariant* (LTI) system.

1.8 IMPULSE RESPONSE AND FREQUENCY RESPONSE

A. Impulse Response:

The *impulse response* $h(t)$ of an LTI system is defined to be the response of the system when the input is $\delta(t)$, that is,

$$h(t) = \mathscr{T}[\delta(t)] \tag{1.77}$$

The function $h(t)$ is arbitrary, and it need not be zero for $t < 0$. If

$$h(t) = 0 \qquad \text{for } t < 0 \tag{1.78}$$

then the system is called *causal*.

B. Response to an Arbitrary Input:

The response $y(t)$ of an LTI system to an arbitrary input $x(t)$ can be expressed as the convolution of $x(t)$ and the impulse response $h(t)$ of the system, that is,

$$y(t) = x(t) * h(t) = \int_{-\infty}^{\infty} x(\tau) h(t - \tau)\, d\tau \tag{1.79}$$

Since the convolution is commutative, we also can express the output as

$$y(t) = h(t) * x(t) = \int_{-\infty}^{\infty} h(\tau) x(t - \tau)\, d\tau \tag{1.80}$$

C. Response of Causal Systems:

From Eqs. (*1.78*) and (*1.79*) or (*1.80*), the response $y(t)$ of a causal LTI system is given by

$$y(t) = \int_{-\infty}^{t} x(\tau) h(t - \tau)\, d\tau = \int_{0}^{\infty} x(t - \tau) h(\tau)\, d\tau \tag{1.81}$$

A signal $x(t)$ is called *causal* if it has zero values for $t < 0$. Thus if the input $x(t)$ is also causal, then

$$y(t) = \int_{0}^{t} x(\tau) h(t - \tau)\, d\tau = \int_{0}^{t} x(t - \tau) h(\tau)\, d\tau \tag{1.82}$$

D. Frequency Response:

Applying the time convolution theorem of the Fourier transform (*1.53*) to Eq. (*1.79*), we obtain

$$Y(\omega) = X(\omega) H(\omega) \tag{1.83}$$

where $X(\omega) = \mathscr{F}[x(t)]$, $Y(\omega) = \mathscr{F}[y(t)]$, and $H(\omega) = \mathscr{F}[h(t)]$.

And $H(\omega)$ is referred to as the *frequency response* (or *transfer function*) of the system. Thus

$$H(\omega) = \mathscr{F}[h(t)] = \frac{Y(\omega)}{X(\omega)} \tag{1.84}$$

The relationships represented by Eqs. (*1.77*), (*1.79*), and (*1.83*) are depicted in Fig. 1-4.

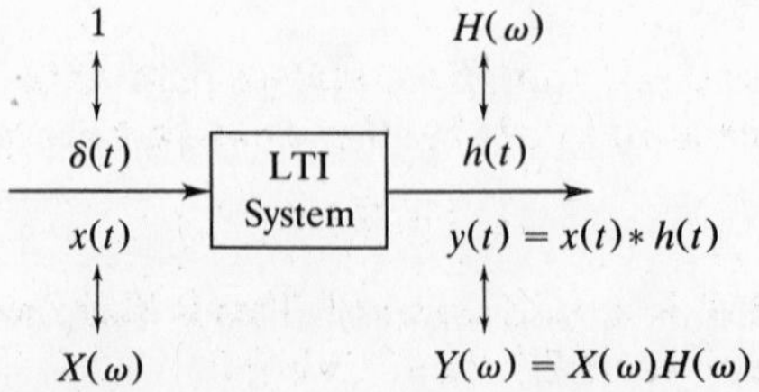

Fig. 1-4 Relationships between inputs and outputs in an LTI system

By taking the inverse Fourier transform of Eq. (*1.83*), the output becomes

$$y(t)=\frac{1}{2\pi}\int_{-\infty}^{\infty} X(\omega)H(\omega)e^{j\omega t}\,d\omega \tag{1.85}$$

Thus, we see that either the impulse response $h(t)$ or the frequency response $H(\omega)$ completely characterizes the LTI system.

1.9 FILTER CHARACTERISTICS OF LINEAR SYSTEMS

A. Frequency Spectra:

The frequency response $H(\omega)$ is a characteristic property of an LTI system. It is, in general, a complex quantity, that is,

$$H(\omega)=|H(\omega)|e^{j\theta_h(\omega)} \tag{1.86}$$

In the case of an LTI system with a real-valued impulse response $h(t)$, $H(\omega)$ exhibits conjugate symmetry [Eq. (*1.27*)], that is,

$$H(-\omega)=H^*(\omega) \tag{1.87}$$

which means that

$$|H(-\omega)|=|H(\omega)| \tag{1.88a}$$

$$\theta_h(-\omega)=-\theta_h(\omega) \tag{1.88b}$$

That is, the amplitude $|H(\omega)|$ is an even function of frequency, whereas the phase $\theta_h(\omega)$ is an odd function of frequency. Let

$$Y(\omega)=|Y(\omega)|e^{j\theta_y(\omega)} \qquad X(\omega)=|X(\omega)|e^{j\theta_x(\omega)}$$

Rewriting Eq. (*1.83*), we have

$$\begin{aligned}|Y(\omega)|e^{j\theta_y(\omega)}&=|X(\omega)|e^{j\theta_x(\omega)}|H(\omega)|e^{j\theta_h(\omega)}\\&=|X(\omega)||H(\omega)|e^{j[\theta_x(\omega)+\theta_h(\omega)]}\end{aligned} \tag{1.89}$$

Thus, we have

$$|Y(\omega)|=|X(\omega)||H(\omega)| \tag{1.90a}$$

$$\theta_y(\omega)=\theta_x(\omega)+\theta_h(\omega) \tag{1.90b}$$

Note that the amplitude spectrum of the output signal is given by the product of the amplitude spectrum of the input signal and the amplitude of the frequency response. The phase spectrum of the output is given by the sum of the phase spectrum of the input and the phase of the frequency response. Therefore, an LTI system acts as a filter on the input signal. Here the word *filter* is used to denote a system that exhibits some sort of frequency-selective behavior.

B. Distortionless Transmission:

For distortionless transmission through a system, we require that the exact input signal shape be reproduced at the output. Therefore, if $x(t)$ is the input signal, the required output is

$$y(t)=Kx(t-t_d) \tag{1.91}$$

where t_d is the *time delay* and K is a *gain constant*. This is illustrated in Fig. 1-5(*a*) and (*b*). Taking the Fourier transform of both sides of Eq. (*1.91*), we get

$$Y(\omega)=Ke^{-j\omega t_d}X(\omega) \tag{1.92}$$

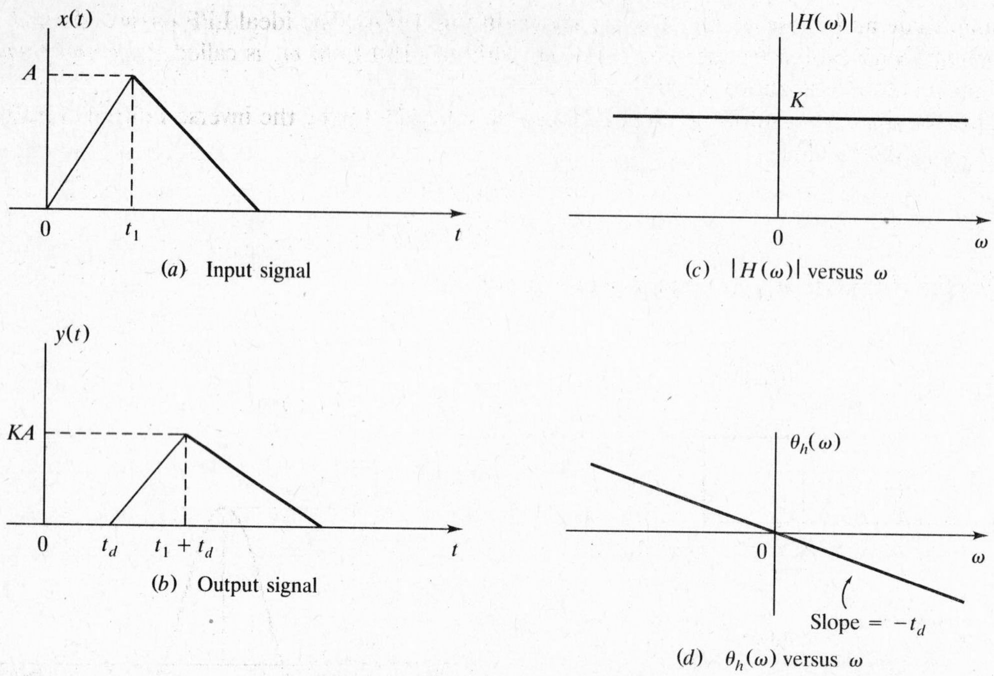

Fig. 1-5 Distortionless transmission

From Eq. (*1.83*), we see that for distortionless transmission, the system must have

$$H(\omega) = |H(\omega)|e^{j\theta_h(\omega)} = Ke^{-j\omega t_d} \tag{1.93}$$

That is, the amplitude of $H(\omega)$ must be constant over the entire frequency range, and the phase of $H(\omega)$ must be linear with frequency. This is illustrated in Fig. 1-5(*c*) and (*d*).

C. Amplitude Distortion and Phase Distortion:

When the amplitude spectrum $|H(\omega)|$ of the system is not constant within the frequency band of interest, the frequency components of the input signal are transmitted with different amounts of gain or attenuation. This effect is called *amplitude distortion*.

When the phase spectrum $\theta_h(\omega)$ of the system is not linear with frequency, the output signal has a different waveform from the input signal because of different delays in passing through the system for different frequency components of the input signal. This form of distortion is called *phase distortion*.

1.10 FILTERS

A filter is a system whose frequency response $H(\omega)$ takes significant values only in certain frequency bands. Filters are usually classified as low-pass, high-pass, bandpass, or bandstop.

A. Ideal Low-Pass Filter:

An ideal low-pass filter (LPF) is defined by

$$H_{\text{LPF}}(\omega) = \begin{cases} e^{-j\omega t_d} & \text{for } |\omega| \le \omega_c \\ 0 & \text{otherwise} \end{cases} \tag{1.94}$$

The amplitude and phase of $H_{\text{LPF}}(\omega)$ are shown in Fig. 1-6(a). The ideal LPF passes all input signal components with radian frequencies below ω_c without distortion; ω_c is called the *cutoff frequency*. All signal components above ω_c are rejected.

The impulse response of the ideal LPF can be found by taking the inverse Fourier transform of Eq. (*1.94*), which yields

$$h_{\text{LPF}}(t) = \frac{\sin \omega_c(t - t_d)}{\pi(t - t_d)} \tag{1.95}$$

The impulse response $h_{\text{LPF}}(t)$ is shown in Fig. 1-6(b).

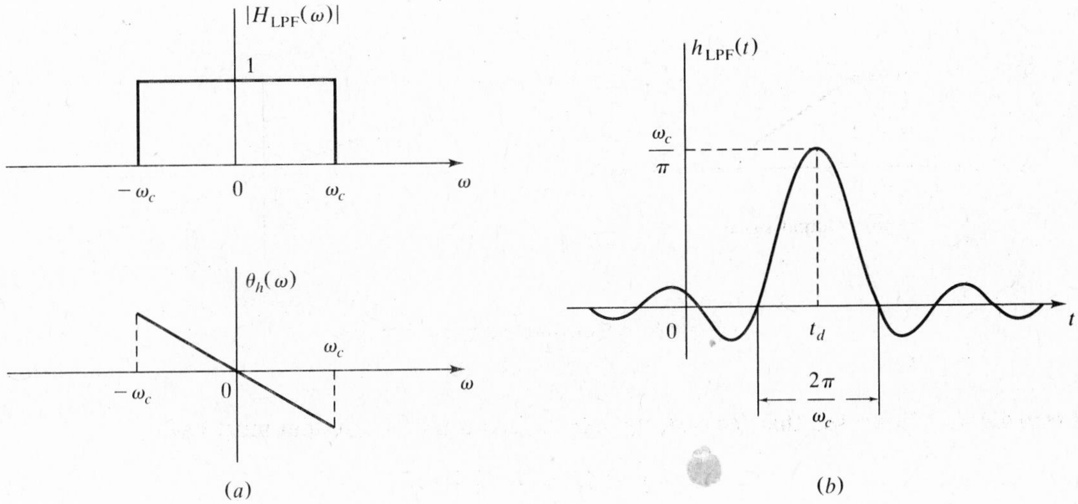

Fig. 1-6 Frequency response and impulse response of an ideal LPF

B. Ideal High-Pass Filter:

An ideal high-pass filter (HPF) is defined by

$$H_{\text{HPF}}(\omega) = \begin{cases} e^{-j\omega t_d} & \text{for } |\omega| \geq \omega_c \\ 0 & \text{otherwise} \end{cases} \tag{1.96}$$

The amplitude and phase of $H_{\text{HPF}}(\omega)$ are illustrated in Fig. 1-7. The ideal HPF rejects all input signal components at frequencies less than ω_c and passes all components above ω_c without distortion.

The frequency response of the ideal HPF can be expressed as

$$H_{\text{HPF}}(\omega) = e^{-j\omega t_d} - H_{\text{LPF}}(\omega) \tag{1.97}$$

where $H_{\text{LPF}}(\omega)$ is given by Eq. (*1.94*).

C. Ideal Bandpass Filter:

An ideal bandpass filter (BPF) is defined by

$$H_{\text{BPF}}(\omega) = \begin{cases} e^{-j\omega t_d} & \text{for } \omega_{c_1} \leq |\omega| \leq \omega_{c_2} \\ 0 & \text{otherwise} \end{cases} \tag{1.98}$$

The amplitude and phase of $H_{\text{BPF}}(\omega)$ are shown in Fig. 1-8.

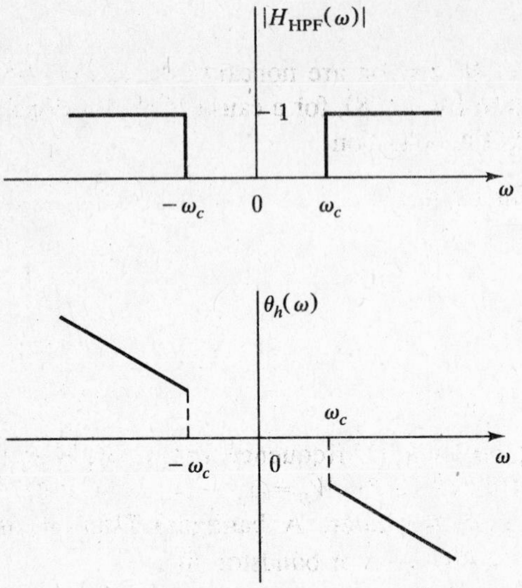

Fig. 1-7 Frequency response of an ideal HPF

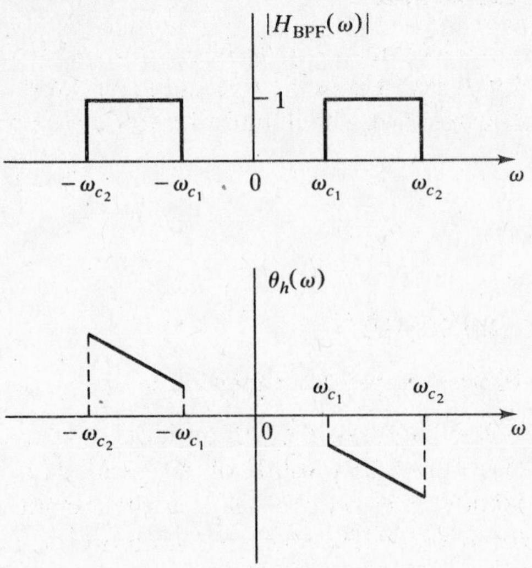

Fig. 1-8 Frequency response of an ideal BPF

D. Ideal Bandstop Filter:

An ideal bandstop filter (BSF) is designed to reject only those signals in a specified band of frequencies between ω_{c_1} and ω_{c_2} and pass the remaining components without distortion. The amplitude and phase of the frequency response of the ideal BSF are shown in Fig. 1-9.

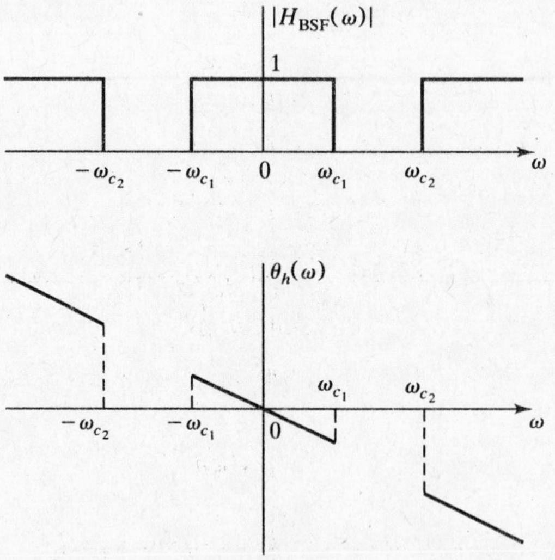

Fig. 1-9 Frequency response of an ideal BSF

The frequency response of the ideal BSF can be expressed as

$$H_{\mathrm{BSF}}(\omega) = e^{-j\omega t_d} - H_{\mathrm{BPF}}(\omega) \tag{1.99}$$

where $H_{\mathrm{BPF}}(\omega)$ is given by Eq. (*1.98*).

E. Causal Filter:

Notice that all ideal filters discussed in the preceding section are noncausal since $h(t) \neq 0$ for $t < 0$. It is not possible to build ideal filters. As shown in Eq. (*1.78*), for a causal filter (or physically realizable filter) its impulse response $h(t)$ must satisfy the condition

$$h(t) = 0 \qquad \text{for } t < 0$$

1.11 BANDWIDTH

A. Filter (or System) Bandwidth:

The bandwidth W_B of an ideal low-pass filter equals its cutoff frequency, that is, $W_B = \omega_c$ [Fig. 1-6(*a*)]. The bandwidth of an ideal bandpass filter is given by $W_B = \omega_{c_2} - \omega_{c_1}$ (Fig. 1-8). The midpoint $\omega_0 = \frac{1}{2}(\omega_{c_1} + \omega_{c_2})$ is the center frequency of the filter. A bandpass filter is called *narrowband* if $W_B \ll \omega_0$. No bandwidth is defined for a high-pass or bandstop filter.

For nonideal or practical filters, a common definition of filter (or system) bandwidth is the 3-dB bandwidth $W_{3\text{ dB}}$. In the case of a low-pass filter, $W_{3\text{ dB}}$ is defined as the positive frequency at which the amplitude spectrum $|H(\omega)|$ drops to a value equal to $|H(0)|/\sqrt{2}$, as illustrated in Fig. 1-10(*a*). In the case of a bandpass filter, $W_{3\text{ dB}}$ is defined as the difference between the frequencies at which $|H(\omega)|$ drops to a value equal to $1/\sqrt{2}$ times the peak value $|H(\omega_0)|$ at the filter's middle frequency ω_0 (called the midband frequency), as illustrated in Fig. 1-10(*b*). This definition is somewhat arbitrary and may become ambiguous and nonunique with multiple peak frequency responses, but it is a widely accepted criterion of measuring a system's bandwidth. Note that each of the preceding bandwidth definitions is defined along the positive frequency axis only and always defines positive-frequency, or one-sided, bandwidth only.

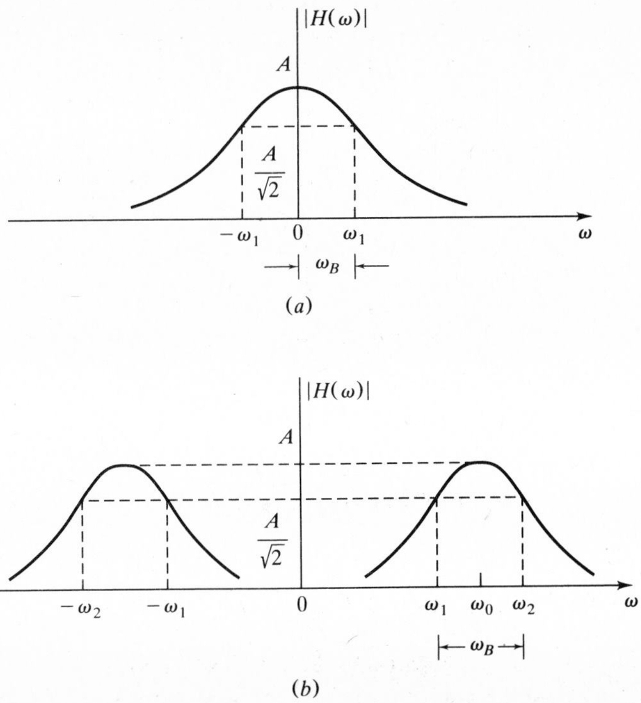

Fig. 1-10 Filter bandwidth

B. Signal Bandwidth:

The *bandwidth* of a signal can be defined as the range of frequencies in which "most" of the energy or power lies. This definition is rather ambiguous and is subject to various conventions. Thus there are numerous definitions for signal bandwidth. For example, the *power bandwidth* of a signal is defined as $\omega_2 - \omega_1$, where $\omega_1 < \omega < \omega_2$ defines the frequency band in which 99 percent of the total power resides.

The bandwidth of a signal $x(t)$ also can be defined on a similar basis as the filter bandwidth, such as the 3-dB bandwidth, by using the magnitude spectrum $|X(\omega)|$ of the signal. Indeed, if we replace $|H(\omega)|$ by $|X(\omega)|$ in Figs. 1-6, 1-7, and 1-8, we have frequency domain plots of low-pass, high-pass, and bandpass signals.

1.12 RELATIONSHIP BETWEEN INPUT AND OUTPUT SPECTRAL DENSITIES

Consider an LTI system with frequency response $H(\omega)$, input $x(t)$, and output $y(t)$. If $x(t)$ and $y(t)$ are energy signals, then by Eq. (*1.62*) their energy spectral densities are $S_{xx}(\omega) = |X(\omega)|^2$ and $S_{yy}(\omega) = |Y(\omega)|^2$, respectively. Since $Y(\omega) = H(\omega)X(\omega)$, it follows that

$$S_{yy}(\omega) = |H(\omega)|^2 S_{xx}(\omega) \tag{1.100}$$

A similar relationship holds for power signals and power spectral densities; that is,

$$\bar{S}_{yy}(\omega) = |H(\omega)|^2 \bar{S}_{xx}(\omega) \tag{1.101}$$

Solved Problems

CLASSIFICATION OF SIGNALS

1.1. Sketch the following signals, and determine whether the signals are power or energy signals or neither.

(*a*) $x(t) = A \sin t, \ -\infty < t < \infty$
(*b*) $x(t) = A[u(t+a) - u(t-a)], \ a > 0$
(*c*) $x(t) = e^{-a|t|}, \ a > 0$
(*d*) $x(t) = u(t)$
(*e*) $x(t) = tu(t)$

(*a*) $x(t) = A \sin t$ (see Fig. 1-11). Now $x(t)$ is periodic with $T = 2\pi$ and hence a power signal.

$$P = \frac{1}{T}\int_0^T [x(t)]^2\,dt = \frac{1}{2\pi}\int_0^{2\pi} A^2 \sin^2 t\,dt$$

$$= \frac{A^2}{2\pi}\int_0^{2\pi} \frac{1}{2}(1 - \cos 2t)\,dt = \frac{A^2}{2}$$

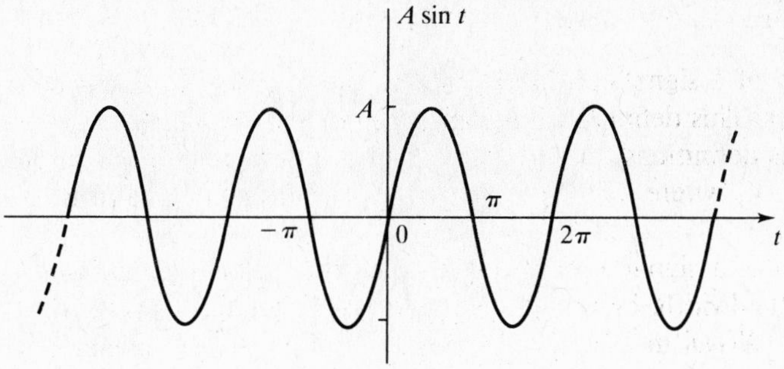

Fig. 1-11

(*b*) $x(t) = A[u(t+a) - u(t-a)]$, $a > 0$ (see Fig. 1-12). And $x(t)$ has a finite duration, thus $x(t)$ is an energy signal.

$$E = \int_{-a}^{a} A^2\, dt = 2aA^2$$

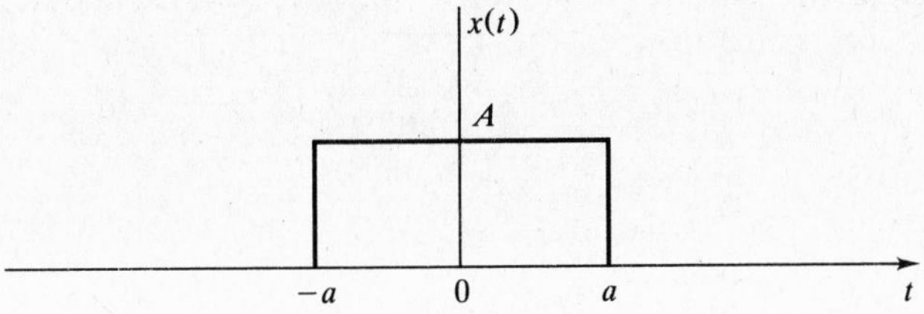

Fig. 1-12

(*c*) $x(t) = e^{-a|t|} = \begin{cases} e^{-at} & t > 0 \\ e^{at} & t < 0 \end{cases}$ (See Fig. 1-13.)

$$E = \int_{-\infty}^{\infty} [x(t)]^2\, dt = \int_{-\infty}^{\infty} e^{-2a|t|}\, dt$$

$$= 2\int_{0}^{\infty} e^{-2at}\, dt = \frac{1}{a} < \infty$$

Thus, $x(t)$ is an energy signal.

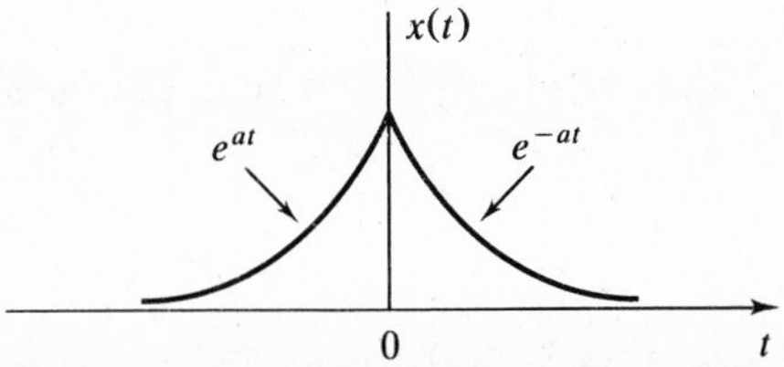

Fig. 1-13

(*d*) $x(t)=u(t)$ (See Fig. 1-14.)

$$E=\lim_{T\to\infty}\int_{-T/2}^{T/2}[x(t)]^2\,dt=\lim_{T\to\infty}\int_{0}^{T/2}1^2\,dt$$

$$=\lim_{T\to\infty}\frac{T}{2}=\infty$$

$$P=\lim_{T\to\infty}\frac{1}{T}\int_{-T/2}^{T/2}[x(t)]^2\,dt=\lim_{T\to\infty}\frac{1}{T}\int_{0}^{T/2}1^2\,dt$$

$$=\lim_{T\to\infty}\frac{1}{T}\frac{T}{2}=\frac{1}{2}$$

Thus $x(t)$ is a power signal.

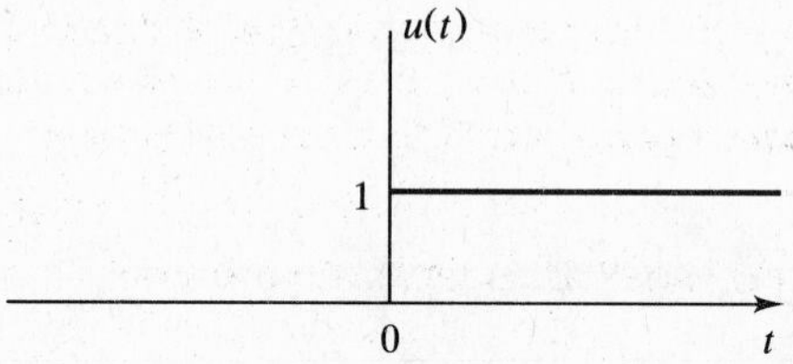

Fig. 1-14

(*e*) $x(t)=tu(t)$ (See Fig. 1-15.)

$$E=\lim_{T\to\infty}\int_{-T/2}^{T/2}[x(t)]^2\,dt=\lim_{T\to\infty}\int_{0}^{T/2}t^2\,dt$$

$$=\lim_{T\to\infty}\frac{(T/2)^3}{3}=\infty$$

$$P=\lim_{T\to\infty}\frac{1}{T}\int_{-T/2}^{T/2}[x(t)]^2\,dt=\lim_{T\to\infty}\frac{1}{T}\int_{0}^{T/2}t^2\,dt$$

$$=\lim_{T\to\infty}\frac{1}{T}\frac{(T/2)^3}{3}=\lim_{T\to\infty}\frac{T^2}{24}=\infty$$

Thus $x(t)$ is neither an energy signal nor a power signal.

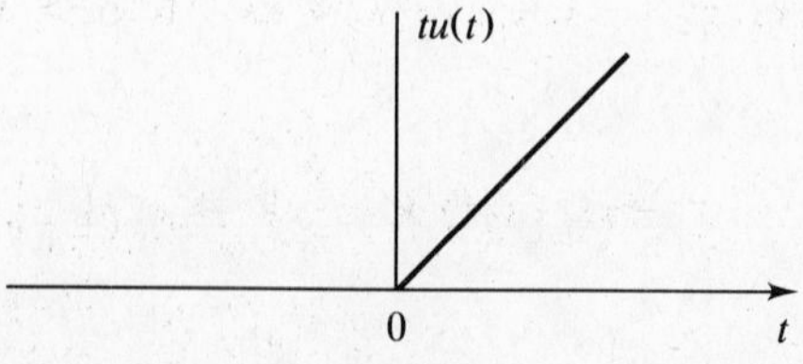

Fig. 1-15

1.2. Let $x_1(t)$ and $x_2(t)$ be periodic signals with periods T_1 and T_2, respectively. Under what conditions is the sum

$$x(t) = x_1(t) + x_2(t)$$

periodic, and what is the period of $x(t)$ if it is periodic?

From Eq. (*1.5*),

$$x_1(t) = x_1(t + T_1) = x_1(t + mT_1) \qquad m \text{ an integer}$$

$$x_2(t) = x_2(t + T_2) = x_2(t + nT_2) \qquad n \text{ an integer}$$

If, therefore, T_1 and T_2 are such that

$$mT_1 = nT_2 = T \tag{1.102}$$

then $$x(t+T) = x_1(t+T) + x_2(t+T) = x_1(t) + x_2(t) = x(t)$$

that is, $x(t)$ is periodic. Thus, the condition for $x(t)$ to be periodic is

$$\frac{T_1}{T_2} = \frac{n}{m} = \text{rational number}$$

The smallest common period is the least common multiple of T_1 and T_2 and is given by Eq. (*1.102*) if the integers m and n are relatively prime. If the ratio T_1/T_2 is an irrational number, then the signals $x_1(t)$ and $x_2(t)$ do not have a common period and $x(t)$ cannot be periodic.

1.3. Are the following signals periodic? If so, find their period.

(*a*) $x(t) = \cos \frac{1}{3}t + \sin \frac{1}{4}t$

(*b*) $x(t) = \cos t + 2\sin \sqrt{2}\, t$

(*a*) $\cos \frac{1}{3}t$ is periodic with period $T_1 = 6\pi$, and $\sin \frac{1}{4}t$ is periodic with period $T_2 = 8\pi$. Since $T_1/T_2 = 6\pi/8\pi = \frac{3}{4}$ is a rational number, $x(t)$ is periodic with period $T = 4T_1 = 3T_2 = 24\pi$.

(*b*) $\cos t$ is periodic with period $T_1 = 2\pi$, and $\sin \sqrt{2}\, t$ is periodic with period $T_2 = \sqrt{2}\pi$. Since $T_1/T_2 = 2\pi/(\sqrt{2}\pi) = \sqrt{2}$ is not a rational number, $x(t)$ is not periodic.

SINGULARITY FUNCTIONS

1.4. Verify properties (*1.9*) and (*1.10*):

(*a*) $\delta(at) = \dfrac{1}{|a|}\delta(t)$

(*b*) $\delta(-t) = \delta(t)$

The proof will be based on the following *equivalence property*:

Let $g_1(t)$ and $g_2(t)$ be generalized functions. Then the equivalence property states that $g_1(t) = g_2(t)$ if and only if

$$\int_{-\infty}^{\infty} g_1(t)\phi(t)\, dt = \int_{-\infty}^{\infty} g_2(t)\phi(t)\, dt \tag{1.103}$$

for all test functions $\phi(t)$.

(*a*) With a change of variable, $at = \tau$, hence $t = \tau/a$, $dt = (1/a)\, d\tau$, we obtain the following equations:

If $a > 0$,

$$\int_{-\infty}^{\infty} \delta(at)\phi(t)\, dt = \frac{1}{a}\int_{-\infty}^{\infty} \delta(\tau)\phi\left(\frac{\tau}{a}\right) dt$$

$$= \frac{1}{a}\phi\left(\frac{\tau}{a}\right)\bigg|_{\tau=0} = \frac{1}{|a|}\phi(0)$$

If $a < 0$,

$$\int_{-\infty}^{\infty} \delta(at)\phi(t)\,dt = \frac{1}{a}\int_{\infty}^{-\infty} \delta(\tau)\phi\left(\frac{\tau}{a}\right) dt$$

$$= -\frac{1}{a}\int_{-\infty}^{\infty} \delta(\tau)\phi\left(\frac{\tau}{a}\right) dt$$

$$= -\frac{1}{a}\phi\left(\frac{\tau}{a}\right)\Big|_{\tau\to 0} = \frac{1}{|a|}\phi(0)$$

Thus, for any a,

$$\int_{-\infty}^{\infty} \delta(at)\phi(t)\,dt = \frac{1}{|a|}\phi(0)$$

Now, using Eq. (*1.7*) for $\phi(0)$, we obtain

$$\int_{-\infty}^{\infty} \delta(at)\phi(t)\,dt = \frac{1}{|a|}\phi(0) = \frac{1}{|a|}\int_{-\infty}^{\infty} \delta(t)\phi(t)\,dt$$

$$= \int_{-\infty}^{\infty} \frac{1}{|a|}\delta(t)\phi(t)\,dt$$

for any $\phi(t)$. Then, applying the equivalence property (*1.103*), we get

$$\delta(at) = \frac{1}{|a|}\delta(t)$$

(*b*) Setting $a = -1$ in the above equation, we get

$$\delta(-t) = \frac{1}{|-1|}\delta(t) = \delta(t)$$

which shows that $\delta(t)$ is an even function.

1.5. Verify property (*1.12*):

$$x(t)\delta(t-t_0) = x(t_0)\delta(t-t_0)$$

If $x(t)$ is continuous at $t = t_0$, then

$$\int_{-\infty}^{\infty} [x(t)\delta(t-t_0)]\phi(t)\,dt = \int_{-\infty}^{\infty} \delta(t-t_0)[x(t)\phi(t)]\,dt$$

$$= x(t_0)\phi(t_0)$$

$$= x(t_0)\int_{-\infty}^{\infty} \delta(t-t_0)\phi(t)\,dt$$

$$= \int_{-\infty}^{\infty} [x(t_0)\delta(t-t_0)]\phi(t)\,dt$$

for all $\phi(t)$. Hence, by the equivalence property (*1.103*), we conclude that

$$x(t)\delta(t-t_0) = x(t_0)\delta(t-t_0)$$

1.6. Show that the following properties hold for the derivative of $\delta(t)$:

(*a*)
$$\int_{-\infty}^{\infty} \phi(t)\delta'(t)\,dt = -\phi'(0) \qquad (1.104)$$

(*b*)
$$t\delta'(t) = -\delta(t) \qquad (1.105)$$

(*a*) Using Eqs. (*1.15*) and (*1.7*), we have

$$\int_{-\infty}^{\infty} \delta'(t)\phi(t)\,dt = -\int_{-\infty}^{\infty} \delta(t)\phi'(t)\,dt = -\phi'(0)$$

(*b*) Again using Eqs. (*1.15*) and (*1.7*), we have

$$\begin{aligned}
\int_{-\infty}^{\infty} [t\delta'(t)]\phi(t)\,dt &= \int_{-\infty}^{\infty} \delta'(t)[t\phi(t)]\,dt \\
&= -\int_{-\infty}^{\infty} \delta(t)\frac{d}{dt}[t\phi(t)]\,dt \\
&= -\int_{-\infty}^{\infty} \delta(t)[\phi(t)+t\phi'(t)]\,dt \\
&= -[\phi(t)+t\phi'(t)]\big|_{t=0} \\
&= -\phi(0) \\
&= -\int_{-\infty}^{\infty} \delta(t)\phi(t)\,dt \\
&= \int_{-\infty}^{\infty} [-\delta(t)]\phi(t)\,dt
\end{aligned}$$

Thus, by the equivalence property (*1.103*), we conclude that

$$t\delta'(t) = -\delta(t)$$

1.7. Evaluate the following integrals;

(*a*) $\int_{-\infty}^{\infty} (t^2 + \cos \pi t)\,\delta(t-1)\,dt$

(*b*) $\int_{-\infty}^{\infty} e^{-t}\,\delta(2t-2)\,dt$

(*c*) $\int_{-\infty}^{\infty} e^{-2t}\,\delta'(t)\,dt$

(*a*)
$$\begin{aligned}
\int_{-\infty}^{\infty} (t^2 + \cos \pi t)\,\delta(t-1)\,dt &= (t^2 + \cos \pi t)\big|_{t=1} \\
&= 1 + \cos \pi = 1 - 1 = 0
\end{aligned}$$

(*b*)
$$\begin{aligned}
\int_{-\infty}^{\infty} e^{-t}\delta(2t-2)\,dt &= \int_{-\infty}^{\infty} e^{-t}\delta[2(t-1)]\,dt \\
&= \int_{-\infty}^{\infty} e^{-t}\frac{1}{|2|}\delta(t-1)\,dt \\
&= \frac{1}{2}e^{-t}\Big|_{t=1} = \frac{1}{2e}
\end{aligned}$$

(*c*)
$$\begin{aligned}
\int_{-\infty}^{\infty} e^{-2t}\delta'(t)\,dt &= -\int_{-\infty}^{\infty} \frac{d}{dt}(e^{-2t})\,\delta(t)\,dt \\
&= 2e^{-2t}\big|_{t=0} = 2
\end{aligned}$$

FOURIER SERIES

1.8. Find the complex Fourier series for the signal

$$x(t) = \cos \omega_0 t + \sin^2 \omega_0 t$$

We could use Eq. (*1.19*) to compute the Fourier coefficients, but using *Euler's* identities, we have

$$\begin{aligned} x(t) &= \frac{1}{2}(e^{j\omega_0 t} + e^{-j\omega_0 t}) + \left[\frac{1}{2j}(e^{j\omega_0 t} - e^{-j\omega_0 t})\right]^2 \\ &= \frac{1}{2}e^{j\omega_0 t} + \frac{1}{2}e^{-j\omega_0 t} - \frac{1}{4}(e^{j2\omega_0 t} - 2 + e^{-j2\omega_0 t}) \\ &= -\frac{1}{4}e^{-j2\omega_0 t} + \frac{1}{2}e^{-j\omega_0 t} + \frac{1}{2} + \frac{1}{2}e^{j\omega_0 t} - \frac{1}{4}e^{j2\omega_0 t} \\ &= \sum_{n=-\infty}^{\infty} c_n e^{jn\omega_0 t} \end{aligned}$$

Thus, we find that $c_0 = \frac{1}{2}$, $c_1 = c_{-1} = \frac{1}{2}$, $c_2 = c_{-2} = -\frac{1}{4}$, and all other values of c_n are equal to zero.

1.9. Find the complex Fourier series of the unit impulse train $\delta_T(t)$ shown in Fig. 1-16(*a*) and defined by

$$\delta_T(t) = \sum_{n=-\infty}^{\infty} \delta(t - nT) \tag{1.106}$$

Let

$$\delta_T(t) = \sum_{n=-\infty}^{\infty} c_n e^{jn\omega_0 t} \qquad \omega_0 = \frac{2\pi}{T}$$

From Eq. (*1.19*), the coefficients c_n are

$$c_n = \frac{1}{T}\int_{-T/2}^{T/2} \delta_T(t) e^{-jn\omega_0 t}\, dt = \frac{1}{T}\int_{-T/2}^{T/2} \delta(t) e^{-jn\omega_0 t}\, dt = \frac{1}{T}$$

Thus

$$\delta_T(t) = \sum_{n=-\infty}^{\infty} \frac{1}{T} e^{jn\omega_0 t} \tag{1.107}$$

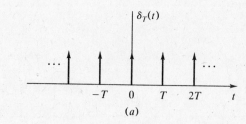

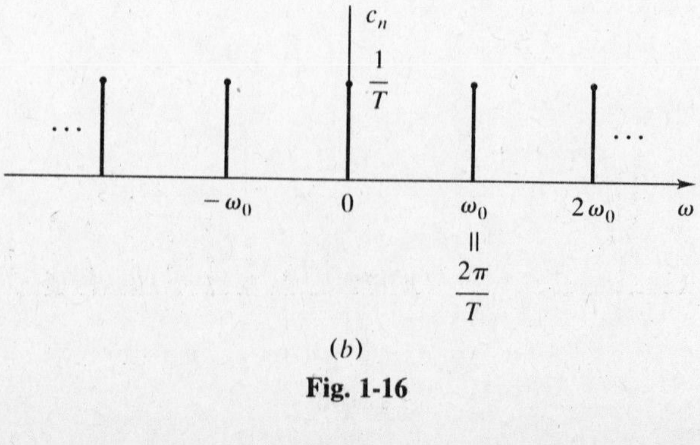

Fig. 1-16

1.10. Find and sketch the magnitude spectra for the periodic square pulse train signal $x(t)$ shown in Fig. 1-17(a) for

(a) $d = T/4$ and (b) $d = T/8$.

From Eq. (1.18) (setting $t_0 = 0$), with $\omega_0 = 2\pi/T$, we obtain

$$c_n = \frac{1}{T}\int_0^T x(t)e^{-jn\omega_0 t}\,dt = \frac{A}{T}\int_0^d e^{-jn\omega_0 t}\,dt$$

$$= \frac{A}{T}\frac{1}{-jn\omega_0}e^{-jn\omega_0 t}\Big|_0^d = \frac{A}{T}\frac{1}{jn\omega_0}\left(1 - e^{-jn\omega_0 d}\right)$$

$$= \frac{A}{T}\frac{1}{jn\omega_0}e^{-jn\omega_0 d/2}\left(e^{jn\omega_0 d/2} - e^{-jn\omega_0 d/2}\right)$$

$$= \frac{Ad}{T}\frac{\sin(n\omega_0 d/2)}{n\omega_0 d/2}e^{-jn\omega_0 d/2} \qquad (1.108)$$

Note that $|c_n| = 0$ whenever $n\omega_0 d/2 = m\pi$, that is,

$$n\omega_0 = \frac{m2\pi}{d} \qquad m = 0, \pm 1, \pm 2, \ldots$$

(a) $d = T/4$, $n\omega_0 d/2 = n\pi\, d/T = n\pi/4$

$$|c_n| = \frac{A}{4}\left|\frac{\sin(n\pi/4)}{n\pi/4}\right|$$

The magnitude spectrum for this case is shown in Fig. 1-17(b).

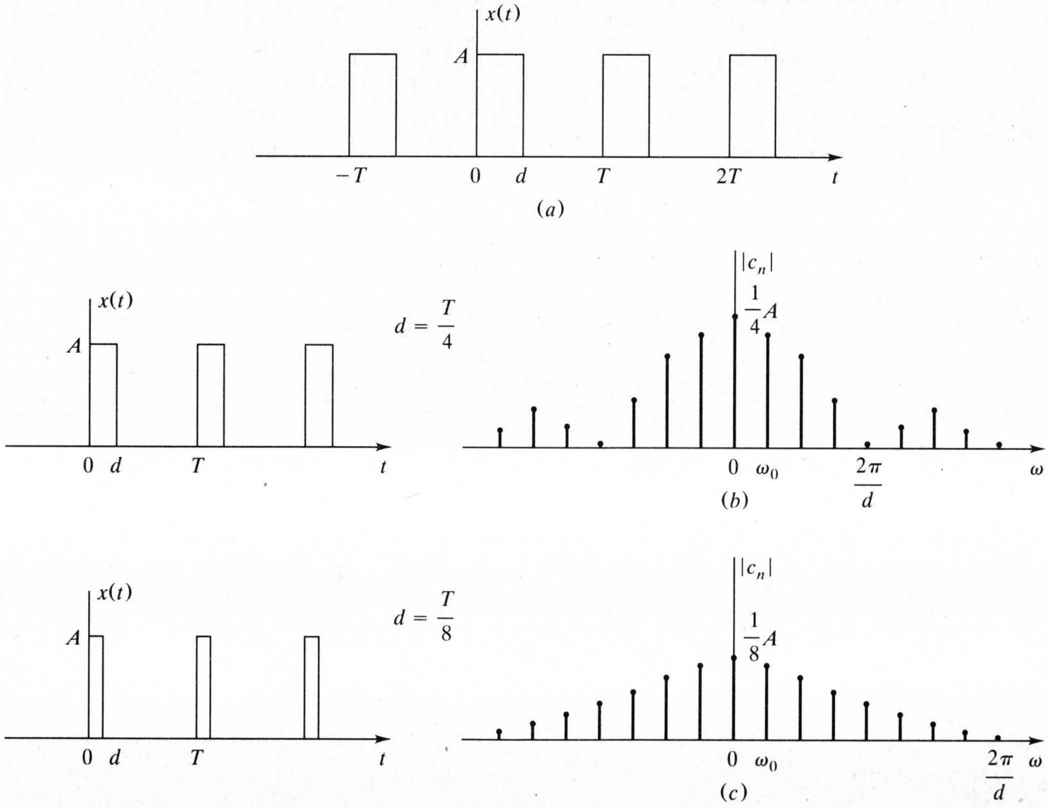

Fig. 1-17

(*b*) $d = T/8$, $n\omega_0 d/2 = n\pi\, dT = n\pi/8$

$$|c_n| = \frac{A}{8}\left|\frac{\sin(n\pi/8)}{n\pi/8}\right|$$

The magnitude spectrum for this case is shown in Fig. 1-17(*c*).

1.11. If $x_1(t)$ and $x_2(t)$ are periodic signals with period T and their complex Fourier series expressions are

$$x_1(t) = \sum_{n=-\infty}^{\infty} d_n e^{jn\omega_0 t} \qquad x_2(t) = \sum_{n=-\infty}^{\infty} e_n e^{jn\omega_0 t} \qquad \omega_0 = \frac{2\pi}{T}$$

show that the signal $x(t) = x_1(t)x_2(t)$ is periodic with the same period T and can be expressed as

$$x(t) = \sum_{n=-\infty}^{\infty} c_n e^{jn\omega_0 t}$$

where c_n is given by

$$c_n = \sum_{k=-\infty}^{\infty} d_k e_{n-k} \tag{1.109}$$

$$x(t+T) = x_1(t+T)x_2(t+T) = x_1(t)x_2(t) = x(t)$$

Thus, $x(t)$ is periodic with period T. Let

$$x(t) = \sum_{n=-\infty}^{\infty} c_n e^{jn\omega_0 t} \qquad \omega_0 = \frac{2\pi}{T}$$

Then

$$\begin{aligned} c_n &= \frac{1}{T}\int_{-T/2}^{T/2} x(t)e^{-jn\omega_0 t}\,dt \\ &= \frac{1}{T}\int_{-T/2}^{T/2} x_1(t)x_2(t)e^{-jn\omega_0 t}\,dt \\ &= \frac{1}{T}\int_{-T/2}^{T/2}\left(\sum_{k=-\infty}^{\infty} d_k e^{jk\omega_0 t}\right)x_2(t)e^{-jn\omega_0 t}\,dt \\ &= \sum_{k=-\infty}^{\infty} d_k\left[\frac{1}{T}\int_{-T/2}^{T/2} x_2(t)e^{-j(n-k)\omega_0 t}\,dt\right] \\ &= \sum_{k=-\infty}^{\infty} d_k e_{n-k} \end{aligned}$$

1.12. Let $x_1(t)$ and $x_2(t)$ be the two periodic signals of Prob. 1.11. Show that

$$\frac{1}{T}\int_{-T/2}^{T/2} x_1(t)x_2(t)\,dt = \sum_{n=-\infty}^{\infty} d_n e_{-n} \tag{1.110}$$

Equation (*1.110*) is known as *Parseval's formula*.

From Prob. 1.11 and Eq. (*1.109*), we have

$$c_n = \frac{1}{T}\int_{-T/2}^{T/2} x_1(t)x_2(t)e^{-jn\omega_0 t}\,dt = \sum_{k=-\infty}^{\infty} d_k e_{n-k}$$

Setting $n = 0$ in the above expression, we obtain

$$\frac{1}{T}\int_{-T/2}^{T/2} x_1(t)x_2(t)\,dt = \sum_{k=-\infty}^{\infty} d_k e_{-k} = \sum_{n=-\infty}^{\infty} d_n e_{-n}$$

FOURIER TRANSFORM

1.13. Find the Fourier transform of the rectangular pulse signal $x(t)$ [shown in Fig. 1-18(a)] defined by

$$x(t) = p_a(t) = \begin{cases} 1 & |t| < a \\ 0 & |t| > a \end{cases} \tag{1.111}$$

$$X(\omega) = \int_{-\infty}^{\infty} p_a(t) e^{-j\omega t}\, dt = \int_{-a}^{a} e^{-j\omega t}\, dt$$

$$= \frac{2 \sin a\omega}{\omega} = 2a \frac{\sin a\omega}{a\omega} \tag{1.112}$$

The Fourier transform of $p_a(t)$ is shown in Fig. 1-18(b).

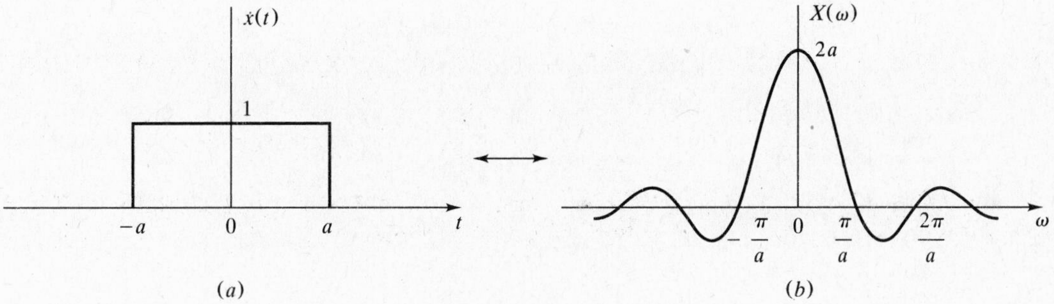

Fig. 1-18

1.14. Find the Fourier transform of the signal [Fig. 1-19(a)]

$$x(t) = \frac{\sin at}{\pi t} \tag{1.113}$$

From the result of Prob. 1.13, we have

$$\mathscr{F}[p_a(t)] = \frac{2}{\omega} \sin a\omega$$

Now from the duality property of the Fourier transform (*1.34*), we have

$$\mathscr{F}\left[\frac{2}{t} \sin at\right] = 2\pi p_a(-\omega)$$

Thus,
$$X(\omega) = \mathscr{F}\left[\frac{\sin at}{\pi t}\right] = \frac{1}{2\pi} \mathscr{F}\left[\frac{2}{t} \sin at\right] = p_a(-\omega) = p_a(\omega) \tag{1.114}$$

where $p_a(\omega)$ is defined by [see Eq. (*1.111*) and Fig. 1-19(b)]

$$p_a(\omega) = \begin{cases} 1 & |\omega| < a \\ 0 & |\omega| > a \end{cases}$$

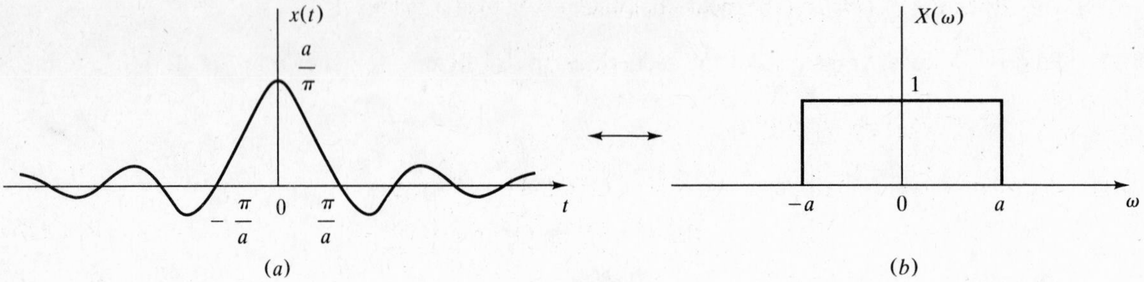

Fig. 1-19

1.15. Show that if

$$x(t) \leftrightarrow X(\omega)$$

then

$$x(t)\cos\omega_0 t \leftrightarrow \tfrac{1}{2}X(\omega-\omega_0)+\tfrac{1}{2}X(\omega+\omega_0) \tag{1.115}$$

Equation (*1.115*) is known as the *modulation theorem*.

Using *Euler's* identity

$$\cos\omega_0 t = \tfrac{1}{2}\left(e^{j\omega_0 t}+e^{-j\omega_0 t}\right)$$

and the frequency-shifting property (*1.31*), we obtain

$$\begin{aligned}\mathscr{F}[x(t)\cos\omega_0 t] &= \mathscr{F}\left[\tfrac{1}{2}x(t)e^{j\omega_0 t}+\tfrac{1}{2}x(t)e^{-j\omega_0 t}\right]\\ &= \tfrac{1}{2}X(\omega-\omega_0)+\tfrac{1}{2}X(\omega+\omega_0)\end{aligned}$$

1.16. The Fourier transform of a signal $x(t)$ is given by [Fig. 1-20(a)]

$$X(\omega) = \tfrac{1}{2}p_a(\omega-\omega_0)+\tfrac{1}{2}p_a(\omega+\omega_0)$$

Find and sketch $x(t)$.

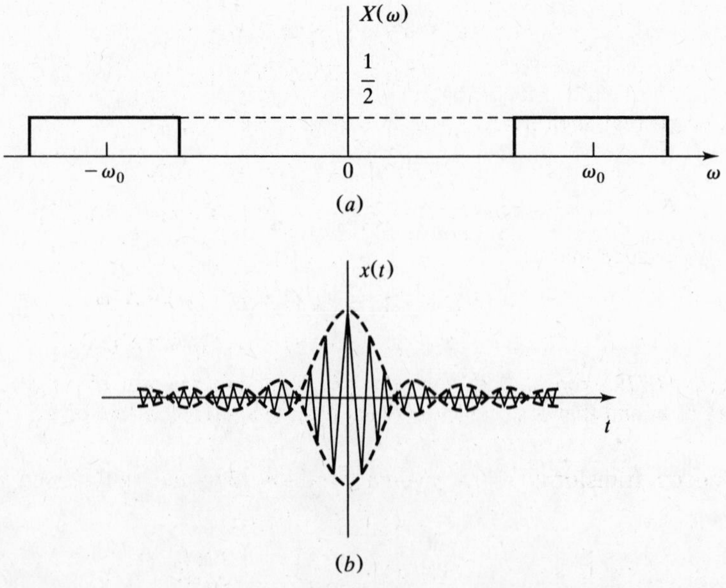

Fig. 1-20

From Eq. (*1.114*) and the modulation theorem (*1.115*) it follows that

$$x(t)=\frac{\sin at}{\pi t}\cos\omega_0 t$$

which is sketched in Fig. 1-20(*b*).

1.17. Let signal $x(t)$ be real, and $X(\omega)=\mathscr{F}[x(t)]$. Then show that

$$\mathscr{F}[x(-t)]=X(-\omega)=X^*(\omega) \tag{1.116}$$

By definition (*1.24*), we have

$$\mathscr{F}[x(-t)]=\int_{-\infty}^{\infty}x(-t)e^{-j\omega t}\,dt$$
$$=\int_{-\infty}^{\infty}x(\lambda)e^{j\omega\lambda}\,d\lambda=\int_{-\infty}^{\infty}x(\lambda)e^{-j(-\omega)\lambda}\,d\lambda=X(-\omega)$$

Thus if $x(t)$ is real, then

$$\int_{-\infty}^{\infty}x(\lambda)e^{j\omega\lambda}\,d\lambda=\left[\int_{-\infty}^{\infty}x(\lambda)e^{-j\omega\lambda}\,d\lambda\right]^*=X^*(\omega)$$

Hence,

$$X(-\omega)=X^*(\omega)$$

1.18. Consider a real signal $x(t)$, and let

$$X(\omega)=\mathscr{F}[x(t)]=A(\omega)+jB(\omega) \tag{1.117}$$

(*a*) Show that $x(t)$ can be expressed as

$$x(t)=x_e(t)+x_o(t) \tag{1.118}$$

where $x_e(t)$ and $x_o(t)$ are the even and odd components of $x(t)$, respectively.

(*b*) Show that

$$x_e(t)\leftrightarrow A(\omega) \tag{1.119}$$
$$x_o(t)\leftrightarrow jB(\omega) \tag{1.120}$$

(*a*) Let

$$x(t)=x_e(t)+x_o(t)$$

Then

$$x(-t)=x_e(-t)+x_o(-t)=x_e(t)-x_o(t)$$

and

$$x_e(t)=\tfrac{1}{2}[x(t)+x(-t)] \tag{1.121a}$$
$$x_o(t)=\tfrac{1}{2}[x(t)-x(-t)] \tag{1.121b}$$

(*b*) Now if $x(t)$ is real, then from Eq. (*1.116*) of Prob. 1.17 we have

$$\mathscr{F}[x(t)]=X(\omega)=A(\omega)+jB(\omega)$$
$$\mathscr{F}[x(-t)]=X(-\omega)=X^*(\omega)=A(\omega)-jB(\omega)$$

Thus, we conclude that

$$\mathscr{F}[x_e(t)]=\tfrac{1}{2}X(\omega)+\tfrac{1}{2}X^*(\omega)=A(\omega)$$
$$\mathscr{F}[x_o(t)]=\tfrac{1}{2}X(\omega)-\tfrac{1}{2}X^*(\omega)=jB(\omega)$$

Equations (*1.119*) and (*1.120*) show that the Fourier transform of a real, even signal is a real function of ω and that of a real, odd signal is an imaginary function of ω.

1.19. Find the Fourier transform of the *signum* function, denoted $\operatorname{sgn}(t)$ (see Fig. 1-21), which is defined as

$$\operatorname{sgn}(t)=\begin{cases}1 & t>0\\ -1 & t<0\end{cases} \tag{1.122}$$

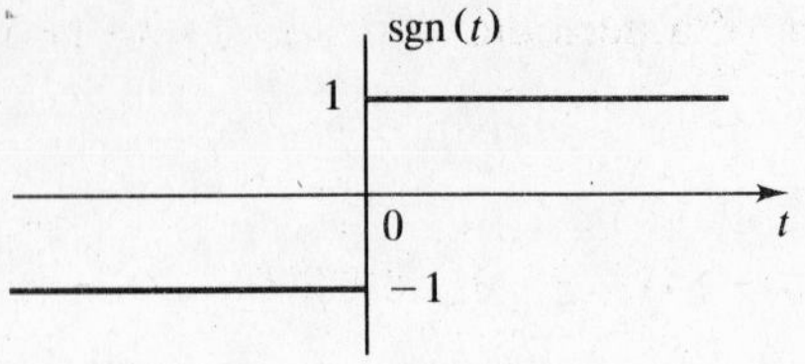

Fig. 1-21 Signum function

The signum function sgn(t) can be expressed as

$$\operatorname{sgn}(t) = 2u(t) - 1$$

Using Eq. (*1.16*), we have

$$\frac{d}{dt}\operatorname{sgn}(t) = 2\delta(t)$$

Let

$$\mathscr{F}[\operatorname{sgn}(t)] = X(\omega)$$

Then applying the differentiation theorem (*1.35*), we have

$$j\omega X(\omega) = \mathscr{F}[2\delta(t)] = 2$$

and

$$X(\omega) = \mathscr{F}[\operatorname{sgn}(t)] = \frac{2}{j\omega} \tag{1.123}$$

Note that sgn(t) is an odd function, and therefore its Fourier transform is a pure imaginary function of ω.

1.20. Verify Eq. (*1.44*), that is,

$$u(t) \leftrightarrow \pi\delta(\omega) + \frac{1}{j\omega}$$

As shown in Fig. 1-22, $u(t)$ can be expressed as

$$u(t) = \frac{1}{2} + \frac{1}{2}\operatorname{sgn}(t)$$

Note that $\frac{1}{2}$ is the even component of $u(t)$ and $\frac{1}{2}$sgn(t) is the odd component of $u(t)$. Thus,

$$\mathscr{F}[u(t)] = \frac{1}{2}\mathscr{F}[1] + \frac{1}{2}\mathscr{F}[\operatorname{sgn}(t)]$$

which, by using Eqs. (*1.40*) and (*1.123*), becomes

$$\mathscr{F}[u(t)] = \pi\delta(\omega) + \frac{1}{j\omega}$$

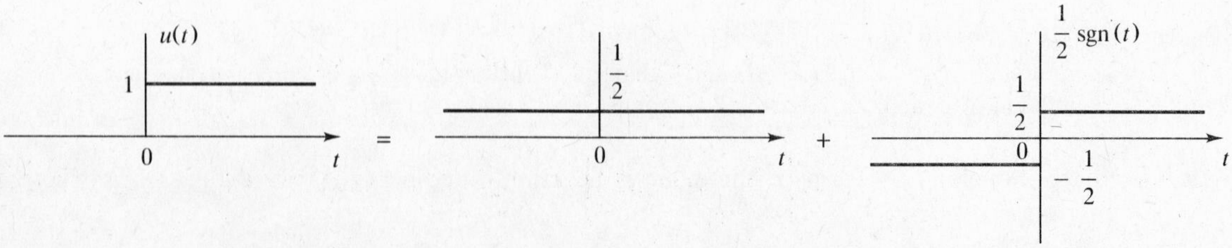

Fig. 1-22 Unit step function and its even and odd components

1.21. Find the Fourier transform of a periodic signal $x(t)$ with period T.

We express $x(t)$ as

$$x(t) = \sum_{n=-\infty}^{\infty} c_n e^{jn\omega_0 t} \qquad \omega_0 = \frac{2\pi}{T}$$

Taking the Fourier transform of both sides and using Eq. (*1.41*), we obtain

$$X(\omega) = 2\pi \sum_{n=-\infty}^{\infty} c_n \delta(\omega - n\omega_o) \tag{1.124}$$

Note that the Fourier transform of a periodic signal consists of a sequence of equidistant impulses located at the harmonic frequencies of the signal.

1.22. Find the Fourier transform of the periodic train of unit impulses $\delta_T(t)$ [Fig. 1-23(*a*) and Eq. (*1.106*)]:

$$\delta_T(t) = \sum_{n=-\infty}^{\infty} \delta(t - nT)$$

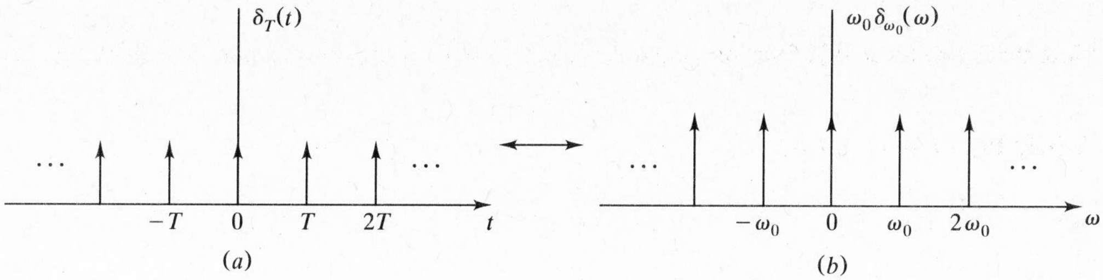

Fig. 1-23 Unit impulse train and its Fourier transform

From Eq. (*1.107*) in Prob. 1.9, the complex Fourier series of $\delta_T(t)$ is given by

$$\delta_T(t) = \sum_{n=-\infty}^{\infty} \frac{1}{T} e^{jn\omega_0 t} \qquad \omega_0 = \frac{2\pi}{T}$$

Using Eq. (*1.124*), we get

$$\mathscr{F}[\delta_T(t)] = \frac{2\pi}{T} \sum_{n=-\infty}^{\infty} \delta(\omega - n\omega_0)$$

$$= \omega_0 \sum_{n=-\infty}^{\infty} \delta(\omega - n\omega_0) = \omega_0 \delta_{\omega_0}(\omega)$$

or

$$\mathscr{F}\left[\sum_{n=-\infty}^{\infty} \delta(t - nT)\right] = \omega_0 \sum_{n=-\infty}^{\infty} \delta(\omega - n\omega_0) \tag{1.125}$$

Thus, the Fourier transform of a unit impulse train is also a similar impulse train [See Fig. 1-23(*b*)].

CONVOLUTION

1.23. Verify Eq. (*1.52*), that is,

$$x(t) * \delta(t-t_0) = x(t-t_0)$$

According to the commutative property of convolution (*1.48*) and the property (*1.8*) of the δ function, we have

$$\begin{aligned} x(t) * \delta(t-t_0) &= \delta(t-t_0) * x(t) \\ &= \int_{-\infty}^{\infty} \delta(\tau - t_0) x(t-\tau)\, d\tau \\ &= x(t-\tau)\big|_{\tau=t_0} = x(t-t_0) \end{aligned}$$

1.24. Show that

$$x(t) * \delta'(t) = x'(t) \tag{1.126}$$

First, we have to prove that

$$\delta'(-t) = -\delta'(t) \tag{1.127}$$

Now

$$\begin{aligned} \int_{-\infty}^{\infty} \delta'(-t)\phi(t)\, dt &= \int_{-\infty}^{\infty} \delta'(t)\phi(-t)\, dt \\ &= -\int_{-\infty}^{\infty} \delta(t)\phi'(-t)\, dt && \text{[by Eq. (1.15)]} \\ &= \int_{-\infty}^{\infty} \delta(t)\phi'(t)\, dt && \text{[since } \phi'(-t) = -\phi'(t)\text{]} \\ &= -\int_{-\infty}^{\infty} \delta'(t)\phi(t)\, dt && \text{[by Eq. (1.15)]} \\ &= \int_{-\infty}^{\infty} [-\delta'(t)]\phi(t)\, dt \end{aligned}$$

By the equivalence property (*1.103*),

$$\delta'(-t) = -\delta'(t)$$

and by the definition of convolution (*1.47*),

$$\begin{aligned} x(t) * \delta'(t) &= \int_{-\infty}^{\infty} x(\tau)\delta'(t-\tau)\, d\tau \\ &= -\int_{-\infty}^{\infty} x(\tau)\delta'(\tau - t)\, d\tau && \text{[by Eq. (1.127)]} \\ &= \int_{-\infty}^{\infty} x'(\tau)\delta(\tau - t)\, d\tau = x'(t) && \text{[by Eq. (1.15)]} \end{aligned}$$

1.25. Prove the time convolution theorem (*1.53*), that is,

$$x_1(t) * x_2(t) \leftrightarrow X_1(\omega) X_2(\omega)$$

$$\mathscr{F}[x_1(t) * x_2(t)] = \int_{-\infty}^{\infty} \left[\int_{-\infty}^{\infty} x_1(\tau) x_2(t-\tau)\, d\tau \right] e^{-j\omega t}\, dt$$

Changing the order of integration gives

$$\mathscr{F}[x_1(t) * x_2(t)] = \int_{-\infty}^{\infty} x_1(\tau) \left[\int_{-\infty}^{\infty} x_2(t-\tau) e^{-j\omega t}\, dt \right] d\tau$$

By the time-shifting property (*1.30*) of the Fourier transform,

$$\int_{-\infty}^{\infty} x_2(t-\tau) e^{-j\omega t}\, dt = X_2(\omega) e^{-j\omega\tau}$$

Thus, we have

$$\mathscr{F}[x_1(t)*x_2(t)] = \int_{-\infty}^{\infty} x_1(\tau)X_2(\omega)e^{-j\omega\tau}\,d\tau$$
$$= \left[\int_{-\infty}^{\infty} x_1(\tau)e^{-j\omega\tau}\,d\tau\right]X_2(\omega) = X_1(\omega)X_2(\omega)$$

1.26. Show that

$$x(t)*u(t) = \int_{-\infty}^{t} x(\tau)\,d\tau$$

and find its Fourier transform.

By definition (*1.47*) of the convolution,

$$x(t)*u(t) = \int_{-\infty}^{\infty} x(\tau)u(t-\tau)\,d\tau = \int_{-\infty}^{t} x(\tau)\,d\tau$$

since

$$u(t-\tau) = \begin{cases} 1 & \tau < t \\ 0 & \tau > t \end{cases}$$

Next, by the time convolution theorem (*1.53*) and Eq. (*1.44*), we obtain

$$\mathscr{F}\left[\int_{-\infty}^{t} x(\tau)\,d\tau\right] = X(\omega)\left[\pi\delta(\omega) + \frac{1}{j\omega}\right]$$
$$= \pi X(\omega)\delta(\omega) + \frac{1}{j\omega}X(\omega)$$
$$= \pi X(0)\delta(\omega) + \frac{1}{j\omega}X(\omega)$$

since $X(\omega)\delta(\omega) = X(0)\delta(\omega)$ by Eq. (*1.11*).

1.27. Show that if $x(t)$ is band-limited, that is,

$$X(\omega) = 0 \qquad \text{for } |\omega| > \omega_c$$

then

$$x(t)*\frac{\sin at}{\pi t} = x(t) \qquad \text{if } a > \omega_c$$

From Prob. 1.14, we have

$$\frac{\sin at}{\pi t} \longleftrightarrow p_a(\omega) = \begin{cases} 1 & |\omega| < a \\ 0 & |\omega| > a \end{cases}$$

By the time convolution theorem (*1.53*),

$$x(t)*\frac{\sin at}{\pi t} \longleftrightarrow X(\omega)p_a(\omega) = X(\omega) \qquad \text{if } a > \omega_c$$

Hence,

$$x(t)*\frac{\sin at}{\pi t} = x(t) \qquad \text{if } a > \omega_c$$

1.28. Use the frequency convolution theorem (*1.54*) to derive the modulation theorem (*1.115*). (See Prob. 1.15.)

From Eq. (*1.42*), we have

$$\cos\omega_0 t \longleftrightarrow \pi\delta(\omega-\omega_0) + \pi\delta(\omega+\omega_0)$$

By the frequency convolution theorem (*1.54*),

$$x(t)\cos\omega_0 t \longleftrightarrow \frac{1}{2\pi}X(\omega)*[\pi\delta(\omega-\omega_0)+\pi\delta(\omega+\omega_0)] = \frac{1}{2}X(\omega-\omega_0)+\frac{1}{2}X(\omega+\omega_0)$$

The last equality follows from Eq. (*1.52*).

1.29. If

$$x_1(t) \longleftrightarrow X_1(\omega) \qquad \text{and} \qquad x_2(t) \longleftrightarrow X_2(\omega)$$

show that

$$\int_{-\infty}^{\infty} x_1(t)x_2(t)\,dt = \frac{1}{2\pi}\int_{-\infty}^{\infty} X_1(\omega)X_2(-\omega)\,d\omega \tag{1.128}$$

From the frequency convolution theorem (*1.54*), we have

$$\mathscr{F}[x_1(t)x_2(t)] = \frac{1}{2\pi}\int_{-\infty}^{\infty} X_1(\lambda)X_2(\omega-\lambda)\,d\lambda$$

that is,

$$\int_{-\infty}^{\infty}[x_1(t)x_2(t)]e^{-j\omega t}\,dt = \frac{1}{2\pi}\int_{-\infty}^{\infty} X_1(\lambda)X_2(\omega-\lambda)\,d\lambda$$

Setting $\omega = 0$, we get

$$\int_{-\infty}^{\infty} x_1(t)x_2(t)\,dt = \frac{1}{2\pi}\int_{-\infty}^{\infty} X_1(\lambda)X_2(-\lambda)\,d\lambda$$

By changing the dummy variable of integration,

$$\int_{-\infty}^{\infty} x_1(t)x_2(t)\,dt = \frac{1}{2\pi}\int_{-\infty}^{\infty} X_1(\omega)X_2(-\omega)\,d\omega$$

1.30. Prove *Parseval's* theorem [Eq. (*1.65*)] for the Fourier transform.

If $x(t)$ is real, then from Eq. (*1.116*) of Prob. 1.17 we have

$$X(-\omega) = X^*(\omega)$$

Then, from Eq. (*1.128*) of Prob. 1.29 we have

$$\begin{aligned}\int_{-\infty}^{\infty}[x_1(t)]^2\,dt &= \frac{1}{2\pi}\int_{-\infty}^{\infty} X_1(\omega)X_1(-\omega)\,d\omega \\ &= \frac{1}{2\pi}\int_{-\infty}^{\infty} X_1(\omega)X_1^*(\omega)\,d\omega = \frac{1}{2\pi}\int_{-\infty}^{\infty}|X_1(\omega)|^2\,d\omega\end{aligned}$$

CORRELATION AND SPECTRAL DENSITY

1.31. Find and sketch the autocorrelation function $R_{11}(\tau)$ for

$$x_1(t) = e^{-at}u(t) \qquad a > 0$$

From definition (*1.55*) of autocorrelation,

$$\begin{aligned}R_{11}(\tau) &= \int_{-\infty}^{\infty} x_1(t)x_1(t-\tau)\,dt \\ &= \int_{-\infty}^{\infty} e^{-at}u(t)e^{-a(t-\tau)}u(t-\tau)\,dt \\ &= e^{a\tau}\int_{-\infty}^{\infty} e^{-2at}u(t)u(t-\tau)\,dt\end{aligned}$$

For $\tau > 0$,
$$u(t)u(t-\tau) = \begin{cases} 1 & t > \tau \\ 0 & t < \tau \end{cases}$$

Thus,

$$R_{11}(\tau) = e^{a\tau}\int_{\tau}^{\infty} e^{-2at}\,dt = \frac{1}{2a}e^{-a\tau}$$

Since $R_{11}(\tau)$ is an even function of τ [see Eq. (*1.58*)], we conclude that

$$R_{11}(\tau) = \frac{1}{2a}e^{-a|\tau|} \qquad a > 0$$

which is sketched in Fig. 1-24.

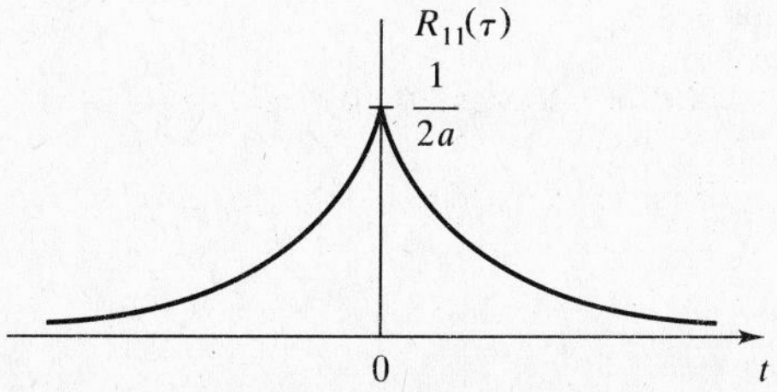

Fig. 1-24

1.32. Show that the cross-correlation function of $x_1(t)$ and $x_2(t)$ can be written in terms of convolution as

$$R_{12}(\tau) = x_1(\tau) * x_2(-\tau) \tag{1.129}$$

By definition (*1.47*) of convolution,

$$x_1(\tau) * x_2(-\tau) = \int_{-\infty}^{\infty} x_1(\lambda)x_2[-(\tau-\lambda)]\,d\lambda$$
$$= \int_{-\infty}^{\infty} x_1(\lambda)x_2(\lambda-\tau)\,d\lambda = R_{12}(\tau)$$

1.33. Show that if

$$x_1(t) \longleftrightarrow X_1(\omega) \qquad \text{and} \qquad x_2(t) \longleftrightarrow X_2(\omega)$$

then
$$\mathscr{F}[R_{12}(\tau)] = X_1(\omega)X_2(-\omega) \tag{1.130}$$
$$\mathscr{F}[R_{11}(\tau)] = X_1(\omega)X_1(-\omega) \tag{1.131}$$

By the time-reversal property (*1.33*) of the Fourier transform,

$$x(-\tau) \longleftrightarrow X(-\omega)$$

Applying the time convolution theorem (*1.53*) to Eq. (*1.129*), we obtain

$$\mathscr{F}[R_{12}(\tau)] = X_1(\omega)X_2(-\omega)$$

Setting $x_2(t) = x_1(t)$, we get

$$\mathscr{F}[R_{11}(\tau)] = X_1(\omega)X_1(-\omega)$$

1.34. Verify Eq. (*1.62*); that is, if $x_1(t)$ is real, then

$$S_{11}(\omega) = \mathscr{F}[R_{11}(\tau)] = |X_1(\omega)|^2$$

If $x_1(t)$ is real, then from Eq. (*1.115*) of Prob. 1.17

$$x_1(-t) \longleftrightarrow X_1(-\omega) = X_1^*(\omega)$$

Thus, by Eq. (*1.131*) we obtain

$$S_{11}(\omega) = \mathscr{F}[R_{11}(\tau)] = X_1(\omega)X_1(-\omega) = X_1(\omega)X_1^*(\omega) = |X_1(\omega)|^2$$

1.35. Verify Eq. (*1.62*) for $x_1(t) = e^{-at}u(t)$.

By Eq. (*1.45*),

$$X_1(\omega) = \frac{1}{j\omega + a}$$

Then

$$|X_1(\omega)|^2 = \frac{1}{\omega^2 + a^2}$$

Next, from Prob. 1.31, we have

$$R_{11}(\tau) = \frac{1}{2a}e^{-a|\tau|} \qquad a > 0$$

Then by Eq. (*1.46*),

$$S_{11}(\omega) = \mathscr{F}[R_{11}(\tau)] = \frac{1}{\omega^2 + a^2} = |X_1(\omega)|^2$$

1.36. Show that the average autocorrelation function of a periodic signal with period T_1 is periodic with the same period.

Let $x_1(t)$ be a periodic signal with period T_1. Then

$$x_1(t) = x_1(t + T_1)$$
$$x_1(t - \tau) = x_1(t - \tau + T_1)$$

Now, from Eq. (*1.68*)

$$\begin{aligned}\bar{R}_{11}(\tau - T_1) &= \frac{1}{T_1}\int_{-T_1/2}^{T_1/2} x_1(t)x_1[t - (\tau - T_1)]\,dt \\ &= \frac{1}{T_1}\int_{-T_1/2}^{T_1/2} x_1(t)x_1(t - \tau + T_1)\,dt \\ &= \frac{1}{T_1}\int_{-T_1/2}^{T_1/2} x_1(t)x_1(t - \tau)\,dt \\ &= \bar{R}_{11}(\tau)\end{aligned}$$

which indicates that $\bar{R}_{11}(\tau)$ is periodic with period T_1.

1.37. Find the average autocorrelation function of the sine wave signal

$$x_1(t) = A\sin(\omega_1 t + \phi) \qquad \omega_1 = \frac{2\pi}{T_1}$$

From Eq. (*1.68*),

$$\begin{aligned}\bar{R}_{11}(\tau) &= \frac{1}{T_1}\int_{-T_1/2}^{T_1/2} x_1(t)x_1(t - \tau)\,dt \\ &= \frac{A^2}{T_1}\int_{-T_1/2}^{T_1/2} \sin(\omega_1 t + \phi)\sin[\omega_1(t - \tau) + \phi]\,dt \\ &= \frac{A^2}{2T_1}\int_{-T_1/2}^{T_1/2} [\cos\omega_1\tau - \cos(2\omega_1 t + 2\phi - \omega_1\tau)]\,dt \\ &= \frac{A^2}{2T_1}\cos\omega_1\tau\int_{-T_1/2}^{T_1/2} dt = \frac{A^2}{2}\cos\omega_1\tau\end{aligned} \qquad (1.132)$$

1.38. Verify Eq. (*1.72*) for the sine wave signal $x_1(t)$ of Prob. 1.37.

From the result of Prob. 1.37, Eq. (*1.132*), and using Eq. (*1.42*), the power spectral density $\bar{S}_{11}(\omega)$ of $x_1(t)$ is given by

$$\bar{S}_{11}(\omega) = \mathscr{F}\left[\bar{R}_{11}(\tau)\right] = \frac{\pi A^2}{2}\delta(\omega - \omega_1) + \frac{\pi A^2}{2}\delta(\omega + \omega_1)$$

Then

$$\frac{1}{2\pi}\int_{-\infty}^{\infty} \bar{S}_{11}(\omega)\,d\omega = \frac{1}{2\pi}\int_{-\infty}^{\infty}\left[\frac{\pi A^2}{2}\delta(\omega - \omega_1) + \frac{\pi A^2}{2}\delta(\omega + \omega_1)\right]d\omega$$

$$= \frac{A^2}{4}\int_{-\infty}^{\infty}\left[\delta(\omega - \omega_1) + \delta(\omega + \omega_1)\right]d\omega$$

$$= \frac{A^2}{4}(1+1) = \frac{A^2}{2} = P_1$$

where P_1 is the normalized average power of $x_1(t)$ over a period.

SYSTEM REPRESENTATION AND CLASSIFICATION

1.39. Consider the system whose input-output relation is given by the linear equation

$$y(t) = ax(t) + b$$

where $x(t)$ and $y(t)$ are input and output of the system, respectively, and a and b are constants. Is this system linear?

We represent the input-output relation of the system by an operator $\mathscr{T}$ such that

$$y(t) = \mathscr{T}[x(t)] = ax(t) + b$$

Consider two inputs $x_1(t)$ and $x_2(t)$. The corresponding outputs are

$$y_1(t) = \mathscr{T}[x_1(t)] = ax_1(t) + b$$

$$y_2(t) = \mathscr{T}[x_2(t)] = ax_2(t) + b$$

Now apply an input $x(t) = x_1(t) + x_2(t)$. Then the output is given by

$$y(t) = \mathscr{T}[x_1(t) + x_2(t)] = a[x_1(t) + x_2(t)] + b$$

and

$$y(t) \neq y_1(t) + y_2(t)$$

which indicates that the additivity condition (*1.74*) is not satisfied. Thus, the system is not linear.

The system also does not satisfy the homogeneity condition (*1.75*) since

$$\mathscr{T}[2x(t)] = 2ax(t) + b \neq 2y(t)$$

1.40. For each of the following systems, determine whether the system is linear.

(*a*) $\mathscr{T}[x(t)] = x(t)\cos\omega_c t$

(*b*) $\mathscr{T}[x(t)] = [A + x(t)]\cos\omega_c t$

(*a*)

$$\mathscr{T}[x_1(t) + x_2(t)] = [x_1(t) + x_2(t)]\cos\omega_c t$$

$$= x_1(t)\cos\omega_c t + x_2(t)\cos\omega_c t$$

$$= \mathscr{T}[x_1(t)] + \mathscr{T}[x_2(t)]$$

$$\mathscr{T}[\alpha x(t)] = [\alpha x(t)]\cos\omega_c t = \alpha\mathscr{T}[x(t)]$$

Hence the system represented by (*a*) is linear.

(*b*)
$$\begin{aligned}\mathscr{T}[x_1(t)+x_2(t)] &= [A+x_1(t)+x_2(t)]\cos\omega_c t\\ &\neq \mathscr{T}[x_1(t)]+\mathscr{T}[x_2(t)]\\ &= [A+x_1(t)]\cos\omega_c t+[A+x_2(t)]\cos\omega_c t\\ &= [2A+x_1(t)+x_2(t)]\cos\omega_c t\end{aligned}$$

Thus, the system represented by (*b*) is not linear. The system also does not satisfy the homogeneity condition (*1.75*). Note that the system represented by (*a*) is called the *balanced modulator* for DSB (double-sideband) signals in amplitude modulation (see Sec. 2.3), and the system represented by (*b*) is the generator for *ordinary* AM (amplitude modulation) signals (see Sec. 2.4).

1.41. Consider the system which is represented by
$$\mathscr{T}[x(t)] = x^*(t)$$
where $x^*(t)$ is the complex conjugate of $x(t)$. Is this system linear?

$$\begin{aligned}\mathscr{T}[x_1(t)+x_2(t)] &= [x_1(t)+x_2(t)]^* = x_1^*(t)+x_2^*(t)\\ &= \mathscr{T}[x_1(t)]+\mathscr{T}[x_2(t)]\end{aligned}$$

which indicates that the additivity condition (*1.74*) is satisfied. We must check whether the homogeneity condition (*1.75*) is also satisfied. Let α be any arbitrary complex-valued constant.

$$\mathscr{T}[\alpha x(t)] = [\alpha x(t)]^* = \alpha^* x^*(t) = \alpha^*\mathscr{T}[x(t)] \neq \alpha\mathscr{T}[x(t)]$$

which indicates that the homogeneity condition is not satisfied. Thus, the system is not linear.

1.42. Consider a system with input $x(t)$ and output $y(t)$ given by

$$y(t) = x(t)\delta_T(t) = x(t)\sum_{n=-\infty}^{\infty}\delta(t-nT)$$

(*a*) Is this system linear?

(*b*) Is this system time-invariant?

(*a*) Let $x(t) = x_1(t)+x_2(t)$. Then

$$\begin{aligned}y(t) &= [x_1(t)+x_2(t)]\delta_T(t) = x_1(t)\delta_T(t)+x_2(t)\delta_T(t)\\ &= y_1(t)+y_2(t)\end{aligned}$$

Similarly, let $x(t) = \alpha x_1(t)$. Then

$$y(t) = [\alpha x_1(t)]\delta_T(t) = \alpha[x_1(t)\delta_T(t)] = \alpha y_1(t)$$

Thus the system is linear.

(*b*) Let
$$x_1(t) = \cos\left(\frac{2\pi}{T}t\right)$$

Then
$$\begin{aligned}y_1(t) &= \sum_{n=-\infty}^{\infty} x_1(nT)\delta(t-nT)\\ &= \sum_{n=-\infty}^{\infty}\cos\left(\frac{2\pi}{T}nT\right)\delta(t-nT) = \sum_{n=-\infty}^{\infty}\delta(t-nT)\end{aligned}$$

Next, let the input be

$$x_2(t) = x_1\left(t-\frac{T}{4}\right) = \sin\left(\frac{2\pi}{T}t\right)$$

Then $$y_2(t) = \sum_{n=-\infty}^{\infty} \sin\left(\frac{2\pi}{T} nT\right) \delta(t - nT) = 0 \neq y_1\left(t - \frac{T}{4}\right)$$

Thus, the system is not time-invariant.

Note that this system is known as the ideal sampler (see Sec. 4.4).

IMPULSE RESPONSE AND FREQUENCY RESPONSE

1.43. Derive Eq. (*1.79*), that is,

$$y(t) = x(t) * h(t) = \int_{-\infty}^{\infty} x(\tau) h(t - \tau)\, d\tau$$

where $y(t)$ and $x(t)$ are the output and input, respectively, of an LTI system whose impulse response is $h(t)$.

If the system is time-invariant, then from Eq. (*1.76*), we have

$$\mathscr{T}[\delta(t - \tau)] = h(t - \tau)$$

Now, from the property (*1.8*) of $\delta(t)$, we can express $x(t)$ as

$$x(t) = \int_{-\infty}^{\infty} x(\tau) \delta(t - \tau)\, d\tau$$

Then from the linearity of the $\mathscr{T}$ operator, we obtain

$$y(t) = \mathscr{T}[x(t)] = \int_{-\infty}^{\infty} x(\tau) \mathscr{T}[\delta(t - \tau)]\, d\tau = \int_{-\infty}^{\infty} x(\tau) h(t - \tau)\, d\tau$$

1.44. The response of an LTI system to a unit step function $u(t)$ is called the *unit step response* of the system and is denoted by $a(t)$. Show that $a(t)$ can be obtained as

$$a(t) = \int_{-\infty}^{t} h(\tau)\, d\tau \tag{1.133}$$

and if the system is causal, then

$$a(t) = \int_{0}^{t} h(\tau)\, d\tau \tag{1.134}$$

From Eq. (*1.80*),

$$a(t) = h(t) * u(t) = \int_{-\infty}^{\infty} h(\tau) u(t - \tau)\, d\tau$$

Since

$$u(t - \tau) = \begin{cases} 1 & \tau < t \\ 0 & \tau > t \end{cases}$$

we have

$$a(t) = \int_{-\infty}^{t} h(\tau)\, d\tau$$

For a causal system, since $h(\tau) = 0$ for $\tau < 0$,

$$a(t) = \int_{0}^{t} h(\tau)\, d\tau$$

1.45. Let an LTI system with impulse response $h(t)$ be represented by an operator $\mathscr{T}$. If

$$\mathscr{T}[x(t)] = \lambda x(t) \tag{1.135}$$

then λ is called the *eigenvalue* of $\mathscr{T}$ and $x(t)$ is called the associated *eigenfunction* of $\mathscr{T}$. Show that the frequency response $H(\omega) = \mathscr{F}[h(t)]$ is the eigenvalue of the LTI system and $e^{j\omega t}$ is the associated eigenfunction.

Using Eq. (*1.80*), we have

$$\mathscr{T}[e^{j\omega t}] = \int_{-\infty}^{\infty} h(\tau) e^{j\omega(t-\tau)}\, d\tau$$

$$= \left[\int_{-\infty}^{\infty} h(\tau) e^{-j\omega\tau}\, d\tau\right] e^{j\omega t} = H(\omega) e^{j\omega t} \tag{1.136}$$

Thus, we see that $H(\omega)$ is the eigenvalue of the LTI system and $e^{j\omega t}$ is the associated eigenfunction.

1.46. Consider the RC network shown in Fig. 1-25(a).

(a) Find the differential equation relating the output voltage $y(t)$ and the input voltage $x(t)$.

(b) Find the frequency response $H(\omega)$ and the impulse response $h(t)$ of the RC network.

(a) From *Kirchhoff's* voltage law, we have

$$Ri(t) + y(t) = x(t)$$

or

$$RC\frac{dy(t)}{dt} + y(t) = x(t)$$

which is the required differential equation relating the input voltage signal and the output voltage signal.

(b) Taking the Fourier transforms of both sides of the above differential equation, we have

$$j\omega RCY(\omega) + Y(\omega) = X(\omega)$$

or

$$(j\omega RC + 1)Y(\omega) = X(\omega)$$

Thus, by Eq. (*1.84*), we have

$$H(\omega) = \frac{Y(\omega)}{X(\omega)} = \frac{1}{j\omega RC + 1} = \frac{1}{RC}\frac{1}{j\omega + 1/(RC)} \tag{1.137}$$

Next, by Eq. (*1.45*), we obtain

$$h(t) = \mathscr{F}^{-1}[H(\omega)] = \frac{1}{RC} e^{-[1/(RC)]t} u(t) \tag{1.138}$$

which is sketched in Fig. 1-25(b).

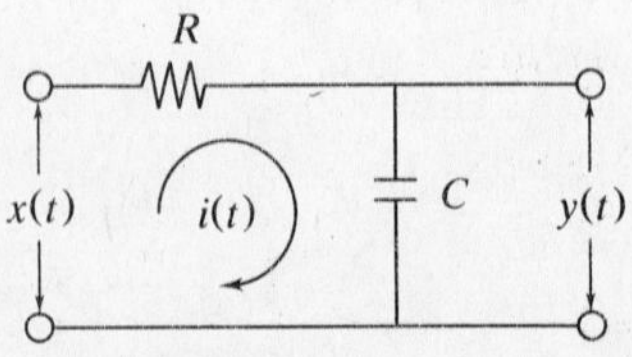

(a) RC network

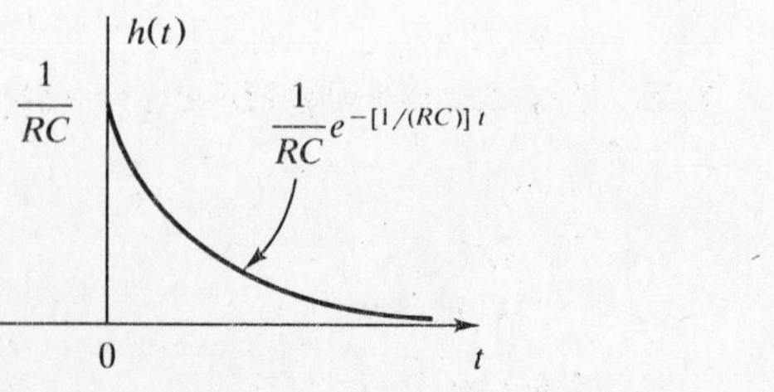

(b) Unit impulse response of the RC network

Fig. 1-25

1.47. Show that the output $\hat{m}(t)$ of a $-\pi/2$ radian (rad) phase shifter of Fig. 1-26 due to the input $m(t)$ can be expressed as

$$\hat{m}(t) = m(t) * \frac{1}{\pi t} = \frac{1}{\pi} \int_{-\infty}^{\infty} \frac{m(\tau)}{t-\tau}\, d\tau \tag{1.139}$$

The signal $\hat{m}(t)$ is called the *Hilbert transform* of $m(t)$.

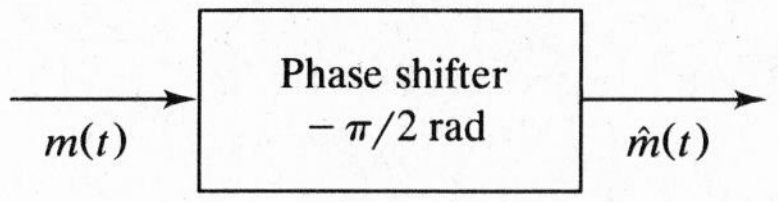

Fig. 1-26 $-\pi/2$ rad phase shifter

By definition of $-\pi/2$ rad phase shift and Eqs. (*1.88a*) and (*1.88b*), the frequency response $H(\omega)$ of the $-\pi/2$ rad phase shifter is given by

$$H(\omega) = \begin{cases} e^{-j\pi/2} & \omega > 0 \\ e^{j\pi/2} & \omega < 0 \end{cases}$$

or

$$H(\omega) = -j\,\mathrm{sgn}\,(\omega) \tag{1.140}$$

since $e^{-j\pi/2} = -j$, $e^{j\pi/2} = j$, and

$$\mathrm{sgn}\,(\omega) = \begin{cases} 1 & \omega > 0 \\ -1 & \omega < 0 \end{cases}$$

If $\mathscr{F}[m(t)] = M(\omega)$, we have

$$\mathscr{F}[\hat{m}(t)] = \hat{M}(\omega) = -j\,\mathrm{sgn}\,(\omega) M(\omega) \tag{1.141}$$

Now, from Eq. (*1.123*) of Prob. 1.19,

$$\mathrm{sgn}\,(t) \longleftrightarrow \frac{2}{j\omega}$$

and by the duality property (*1.34*) of the Fourier transform, we have

$$\frac{2}{jt} \longleftrightarrow 2\pi\,\mathrm{sgn}\,(-\omega) = -2\pi\,\mathrm{sgn}\,(\omega)$$

or

$$\frac{1}{\pi t} \longleftrightarrow -j\,\mathrm{sgn}\,(\omega) \tag{1.142}$$

Then, by the time convolution theorem (*1.53*) and convolution definition (*1.47*), we obtain

$$\hat{m}(t) = \mathscr{F}^{-1}\{M(\omega)[-j\,\mathrm{sgn}\,(\omega)]\}$$

$$= m(t) * \frac{1}{\pi t} = \frac{1}{\pi} \int_{-\infty}^{\infty} \frac{m(\tau)}{t-\tau}\, d\tau$$

1.48. Let $\hat{m}(t)$ be the Hilbert transform of $m(t)$. Then show that

$$\hat{\hat{m}}(t) = -m(t) \tag{1.143}$$

Let

$$m(t) \longleftrightarrow M(\omega)$$

Then by Eq. (*1.141*),

$$\hat{m}(t) \longleftrightarrow -j[\mathrm{sgn}\,(\omega)]M(\omega)$$

and

$$\hat{\hat{m}}(t) \longleftrightarrow -j[\operatorname{sgn}(\omega)]\{-j[\operatorname{sgn}(\omega)]M(\omega)\} = j^2[\operatorname{sgn}(\omega)]^2 M(\omega) = -M(\omega)$$

since $j^2 = -1$ and $[\operatorname{sgn}(\omega)]^2 = 1$. Hence, we conclude

$$\hat{\hat{m}}(t) = \mathscr{F}^{-1}[-M(\omega)] = -\mathscr{F}^{-1}[M(\omega)] = -m(t)$$

FILTER CHARACTERISTICS OF LINEAR SYSTEMS

1.49. Show that the RC network of Prob. 1.46 [Fig. 1-25(a)] is a low-pass filter. Also find its 3-dB bandwidth $W_{3\,\mathrm{dB}}$.

From Prob. 1.46, the frequency response $H(\omega)$ is given by

$$H(\omega) = \frac{1}{1 + j\omega RC} = \frac{1}{1 + j\omega/\omega_0}$$

where $\omega_0 = 1/(RC)$. Writing

$$H(\omega) = |H(\omega)|e^{j\theta_h(\omega)}$$

we have

$$|H(\omega)| = \frac{1}{\sqrt{1 + (\omega/\omega_0)^2}} \qquad \text{and} \qquad \theta_h(\omega) = -\tan^{-1}\frac{\omega}{\omega_0}$$

The amplitude spectrum $|H(\omega)|$ and phase spectrum $\theta_h(\omega)$ are plotted in Fig. 1-27. From Fig. 1-27 we see that the RC network of Fig. 1-25(a) is a low-pass filter.

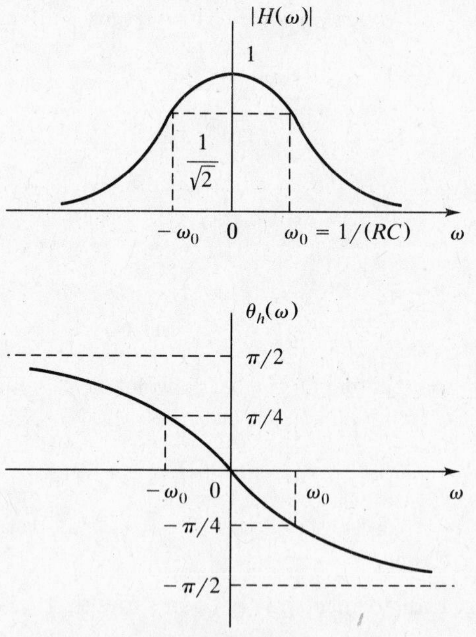

Fig. 1-27

When $\omega = \omega_0 = 1/(RC)$, $|H(\omega)| = 1/\sqrt{2}$. Thus,

$$W_{3\,\mathrm{dB}} = \omega_0 = \frac{1}{RC}$$

1.50. The rise time t_r of the low-pass RC filter of Fig. 1-25(a) is defined as the time required for a unit step response to go from 10 to 90% of its final value. Show that

$$t_r = \frac{0.35}{f_{3\text{ dB}}}$$

where $f_{3\text{ dB}} = W_{3\text{ dB}}/(2\pi) = 1/(2\pi RC)$ is the 3-dB bandwidth (in hertz) of the filter.

From Eq. (*1.138*) of Prob. 1.46 and Eq. (*1.134*) of Prob. 1.44, the unit step response of the low-pass RC filter is found to be

$$a(t) = \int_0^t h(\tau)\, d\tau = \int_0^t \frac{1}{RC} e^{-\tau/(RC)}\, d\tau = (1 - e^{-t/(RC)})u(t)$$

which is sketched in Fig. 1-28. By definition of the rise time,

$$t_r = t_2 - t_1$$

where

$$a(t_1) = 1 - e^{-t_1/(RC)} = 0.1 \longrightarrow e^{-t_1/(RC)} = 0.9$$

$$a(t_2) = 1 - e^{-t_2/(RC)} = 0.9 \longrightarrow e^{-t_2/(RC)} = 0.1$$

Dividing the first equation by the second equation on the right-hand side, we obtain

$$e^{(t_2 - t_1)/(RC)} = 9$$

and

$$t_r = t_2 - t_1 = RC \ln 9 = 2.197RC = \frac{2.197}{2\pi f_{3\text{ dB}}} = \frac{0.35}{f_{3\text{ dB}}}$$

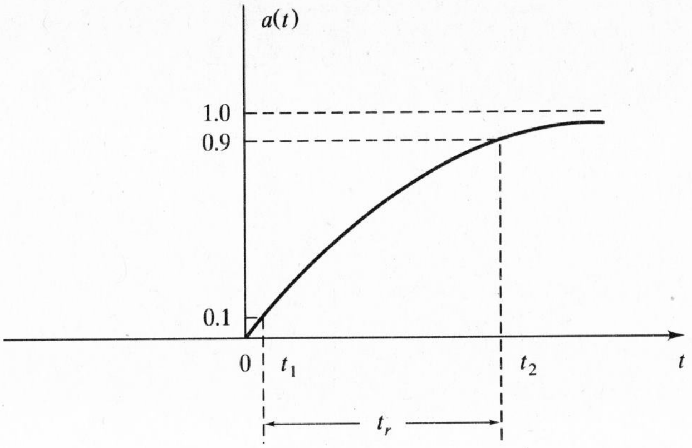

Fig. 1-28

1.51. A multipath transmission occurs when a transmitted signal arrives at the receiver by two or more paths of different delays. A simple model for a multipath communication channel is illustrated in Fig. 1-29(a).

(a) Find the frequency system function $H(\omega)$ for this channel and plot $|H(\omega)|$ for $\alpha = 1$ and 0.5.

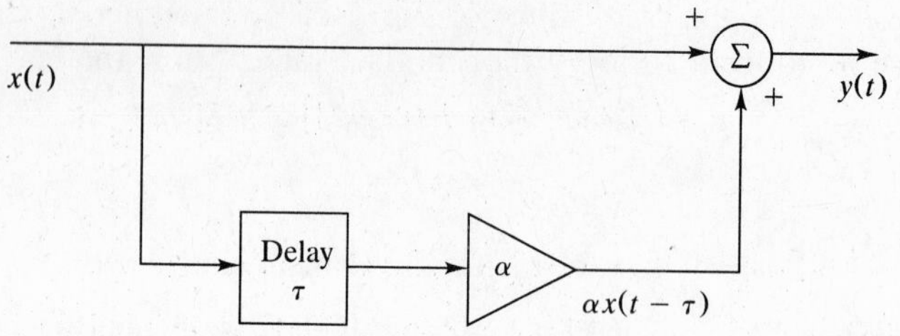

(a) Model for multipath transmission

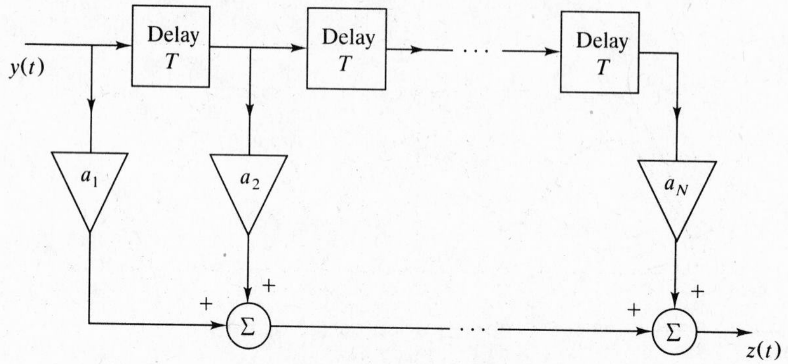

(b) Tapped delay-line filter

Fig. 1-29

(b) To compensate for the channel-induced distortion, an *equalization* filter is often utilized. Ideally, the frequency system function of the equalization filter should be

$$H_{\text{eq}}(\omega) = \frac{1}{H(\omega)}$$

A *tapped delay-line* or *transversal filter*, as shown in Fig. 1-29(b), is commonly used to approximate this equalization filter. Find the values for $a_1, a_2, \ldots, a_N$, assuming $\tau = T$ and $\alpha \ll 1$.

(a)
$$y(t) = x(t) + \alpha x(t - \tau)$$

Taking the Fourier transform of both sides, we have

$$Y(\omega) = X(\omega) + \alpha e^{-j\omega\tau} X(\omega) = (1 + \alpha e^{-j\omega\tau}) X(\omega)$$

By Eq. (*1.84*),

$$H(\omega) = \frac{Y(\omega)}{X(\omega)} = 1 + \alpha e^{-j\omega\tau}$$

Using *Euler's* identity for $e^{-j\omega\tau}$ gives

$$H(\omega) = 1 + \alpha \cos \omega\tau - j\alpha \sin \omega\tau$$

Thus,

$$|H(\omega)| = \left[(1 + \alpha \cos \omega\tau)^2 + (\alpha \sin \omega\tau)^2\right]^{1/2}$$

$$= \left[(1 + \alpha^2 + 2\alpha \cos \omega\tau)\right]^{1/2}$$

When $\alpha = 1$,

$$|H(\omega)| = [2(1+\cos\omega\tau)]^{1/2} = 2\left|\cos\frac{\omega\tau}{2}\right|$$

When $\alpha = \frac{1}{2}$,

$$|H(\omega)| = (1.25 + \cos\omega\tau)^{1/2}$$

Amplitude spectra $|H(\omega)|$ for $\alpha = 1$ and $\alpha = \frac{1}{2}$ are plotted in Fig. 1-30.

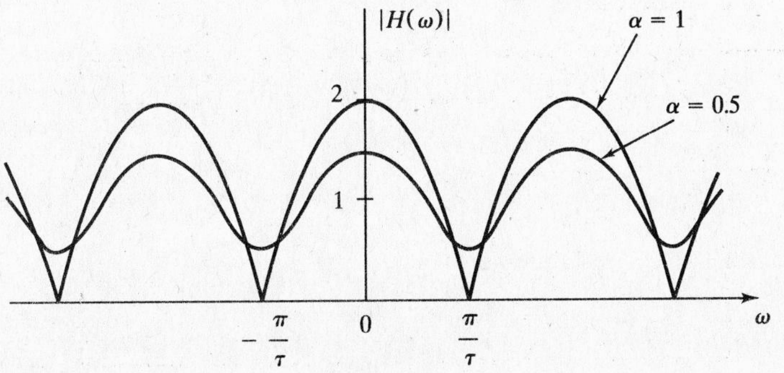

Fig. 1-30

(*b*) From Fig. 1-29(*b*), we have

$$z(t) = \sum_{k=1}^{N} a_k y[t-(k-1)T]$$

Taking the Fourier transform of both sides gives

$$Z(\omega) = \sum_{k=1}^{N} a_k e^{-j\omega(k-1)T} Y(\omega)$$

Thus the frequency response $H_{tr}(\omega)$ of the transversal filter is given by

$$H_{tr}(\omega) = \frac{Z(\omega)}{Y(\omega)} = \sum_{k=1}^{N} a_k e^{-j\omega(k-1)T}$$

$$= a_1 + a_2 e^{-j\omega T} + a_3 e^{-j\omega 2T} + \cdots + a_N e^{-j\omega(N-1)T}$$

Now

$$H_{eq}(\omega) = \frac{1}{H(\omega)} = \frac{1}{1+\alpha e^{-j\omega\tau}}$$

Using

$$\frac{1}{1+x} = 1 - x + x^2 - x^3 + \cdots \qquad |x| < 1$$

we can express $H_{eq}(\omega)$ as

$$H_{eq}(\omega) = 1 - \alpha e^{-j\omega\tau} + \alpha^2 e^{-j\omega 2\tau} + \cdots$$

Thus, if $\tau = T$ and $|\alpha| < 1$, we have

$$a_1 = 1,\ a_2 = -\alpha,\ a_3 = \alpha^2, \ldots, a_N = (-\alpha)^{N-1}$$

1.52. Consider a filter with $H(\omega) = 1/(1 + j\omega)$ and input $x(t) = e^{-2t}u(t)$.

(*a*) Find the energy spectral density of the output.

(*b*) Show that the total energy output is one-third of the input energy.

(*a*)

$$x(t) = e^{-2t}u(t) \longleftrightarrow X(\omega) = \frac{1}{j\omega + 2}$$

$$S_{xx}(\omega) = |X(\omega)|^2 = \frac{1}{\omega^2 + 4}$$

The energy spectral density of the output is, by Eq. (*1.99*),

$$S_{yy}(\omega) = |H(\omega)|^2 |X(\omega)|^2 = \frac{1}{(\omega^2 + 1)(\omega^2 + 4)}$$

$$= \frac{1}{3}\frac{1}{\omega^2 + 1} - \frac{1}{3}\frac{1}{\omega^2 + 4}$$

(*b*) Using *Parseval's* theorem (*1.65*) and

$$x(t) = e^{-at}u(t) \qquad a > 0 \longleftrightarrow X(\omega) = \frac{1}{j\omega + a} \qquad |X(\omega)|^2 = \frac{1}{\omega^2 + a^2}$$

we have

$$E_{\text{input}} = \frac{1}{2\pi}\int_{-\infty}^{\infty} \frac{d\omega}{\omega^2 + 4} = \int_0^{\infty} e^{-4t}\,dt = \frac{1}{4}$$

$$E_{\text{output}} = \frac{1}{3}\int_0^{\infty} e^{-2t}\,dt - \frac{1}{3}\int_0^{\infty} e^{-4t}\,dt$$

$$= \frac{1}{3}\left(\frac{1}{2} - \frac{1}{4}\right) = \frac{1}{3}\cdot\frac{1}{4} = \frac{1}{3}E_{\text{input}}$$

1.53. Consider the ideal differentiator shown in Fig. 1-31(*a*). A power signal $x(t)$ whose power spectral density is shown in Fig. 1-31(*b*) is applied to the differentiator. Determine and sketch the output power spectral density. Also find the mean square value of the output of the differentiator.

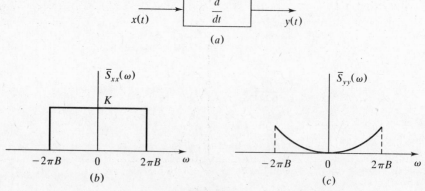

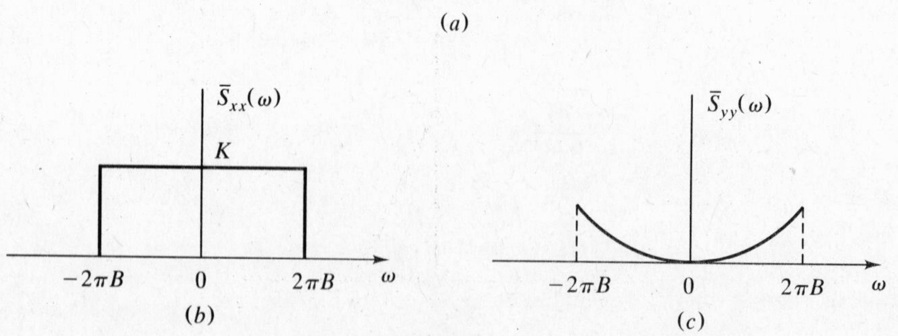

Fig. 1-31

From Eq. (*1.35*) we have

$$H(\omega) = j\omega$$

Then by Eq. (*1.100*),

$$\bar{S}_{yy}(\omega) = |j\omega|^2 \bar{S}_{xx}(\omega) = \omega^2 \bar{S}_{xx}(\omega)$$

which is sketched in Fig. 1-31(*c*).

From Eq. (*1.72*), the mean square value of $y(t)$ is given by

$$\lim_{T \to \infty} \frac{1}{T} \int_{-T/2}^{T/2} [y(t)]^2 \, dt = \frac{1}{2\pi} \int_{-\infty}^{\infty} \bar{S}_{yy}(\omega) \, d\omega$$

$$= \frac{1}{2\pi} \int_{-2\pi B}^{2\pi B} K\omega^2 \, d\omega$$

$$= \frac{K}{\pi} \int_0^{2\pi B} \omega^2 \, d\omega = \frac{8\pi^2 K B^3}{3}$$

Supplementary Problems

1.54. Show that if

$$x(t) \longleftrightarrow X(\omega)$$

then

$$x^{(n)}(t) = \frac{d^n x(t)}{dt^n} \longleftrightarrow (j\omega)^n X(\omega)$$

Hint: Repeat time-differentiation theorem (*1.35*).

1.55. Use the differentiation technique and the result of Prob. 1.54 to obtain the Fourier transform of the triangular pulse signal shown in Fig. 1-32.

Ans. $Ad\left[\dfrac{\sin(\omega d/2)}{\omega d/2}\right]^2$

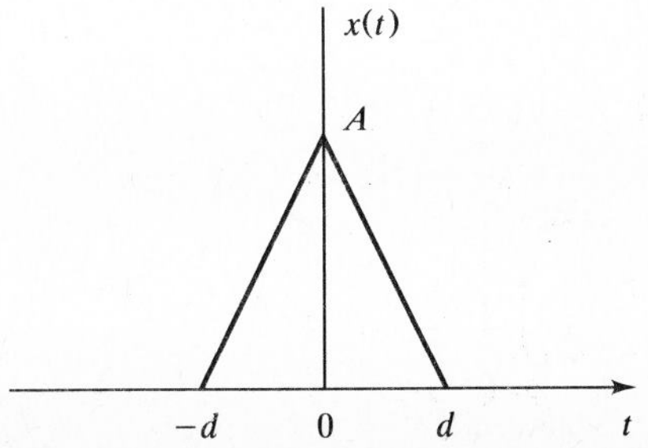

Fig. 1-32

1.56. If $R_{11}(\tau)$ and $R_{22}(\tau)$ are the autocorrelation functions of $x_1(t)$ and $x_2(t)$, respectively, and $R_{12}(\tau)$ is their cross-correlation function, then show that the following relations are true for all τ.

(*a*) $R_{11}(0) \geq |R_{11}(\tau)|$

(*b*) $R_{22}(0) \geq |R_{22}(\tau)|$

(*c*) $R_{11}(0) + R_{22}(0) \geq 2|R_{12}(\tau)|$

Hint: Use the fact that for all τ,

$$\int_{-\infty}^{\infty} [x_1(t) \pm x_2(t-\tau)]^2 \, dt \geq 0$$

1.57. Find the average autocorrelation function and the power spectral density of

$$x(t) = \cos 2\pi t + 2\cos 4\pi t$$

Ans. $\overline{R}_{xx}(\tau) = \frac{1}{2}\cos 2\pi\tau + 2\cos 4\pi\tau$

$\overline{S}_{xx}(\omega) = \frac{1}{2}\pi[\delta(\omega - 2\pi) + \delta(\omega + 2\pi)] + 2\pi[\delta(\omega - 4\pi) + \delta(\omega + 4\pi)]$

1.58. A system is called *BIBO stable* if every *b*ounded *i*nput produces a *b*ounded *o*utput. Show that the system is stable if its impulse response is absolutely integrable, that is,

$$\int_{-\infty}^{\infty} |h(\tau)| \, d\tau < \infty$$

Hint: Take the absolute value of both sides of Eq. (*1.82*) and use the fact that $|x(t-\tau)| < K$.

1.59. If the unit impulse response of a causal LTI system contains no impulse at the origin, then show that with

$$H(\omega) = A(\omega) + jB(\omega)$$

$A(\omega)$ and $B(\omega)$ satisfy the following equations:

$$A(\omega) = \frac{1}{\pi}\int_{-\infty}^{\infty} \frac{B(\lambda)}{\omega - \lambda} \, d\lambda$$

$$B(\omega) = -\frac{1}{\pi}\int_{-\infty}^{\infty} \frac{A(\lambda)}{\omega - \lambda} \, d\lambda$$

These equations are known as the *Hilbert transform pair* (see Prob. 1.47).

Hint: Let $h(t) = h_e(t) + h_o(t)$ and use the causality of $h(t)$ to show that $h_e(t) = h_o(t)[\operatorname{sgn}(t)]$, $h_o(t) = h_e(t)[\operatorname{sgn}(t)]$.

1.60. Show that

(*a*) $\displaystyle\int_{-\infty}^{\infty} [m(t)]^2 \, dt = \int_{-\infty}^{\infty} [\hat{m}(t)]^2 \, dt$

(*b*) $\displaystyle\int_{-\infty}^{\infty} m(t)\hat{m}(t) \, dt = 0$

where $\hat{m}(t)$ is the Hilbert transform of $m(t)$.

Hint: (*a*) Use Eq. (*1.141*) and apply *Parseval's* theorem (*1.65*) for the Fourier transform.
(*b*) Use Eq. (*1.141*) and apply relation (*1.128*) of Prob. 1.29.

1.61. Determine the impulse response and the 3-dB bandwidth of the filter whose frequency response is $H(\omega) = 10/(\omega^2 + 100)$.

Ans. $h(t) = \frac{1}{2}e^{-10|t|}$, $W_{3\text{ dB}} = 6.44$ radians per second (rad/s)

1.62. A *gaussian filter* is a linear system whose frequency response is given by

$$H(\omega) = e^{-a\omega^2} e^{-j\omega t_0}$$

Calculate (a) the 3-dB bandwidth $W_{3\text{ dB}}$ and (b) the equivalent bandwidth W_{eq} defined by

$$W_{\text{eq}} = \frac{1}{2}\frac{1}{H(0)}\int_{-\infty}^{\infty} |H(\omega)|\,d\omega$$

Ans. (a) $W_{3\text{ dB}} = \dfrac{0.59}{\sqrt{a}}$ (b) $W_{\text{eq}} = \dfrac{0.886}{\sqrt{a}}$

1.63. A *Butterworth* low-pass filter has

$$|H(\omega)| = \frac{1}{\sqrt{1 + (\omega/\omega_0)^{2n}}}$$

where n is the number of reactive components (i.e., inductors or capacitors).

(a) Show that as $n \to \infty$, $|H(\omega)|$ approaches the characteristics of the ideal low-pass filter, shown in Fig. 1-6(a) with $\omega_0 = \omega_c$.

(b) Find n so that $|H(\omega)|^2$ is constant to within 1 dB over the frequency range of $|\omega| = 0.8\omega_0$.

Ans. (a) Note that

$$\lim_{n\to\infty}\left(\frac{\omega}{\omega_0}\right)^{2n} = \begin{cases} \infty & \text{for } \omega > \omega_0 \\ 0 & \text{for } \omega < \omega_0 \end{cases}$$

(b) $n = 3$

1.64. Let $x(t)$ and $y(t)$ be the input and output power signals, respectively, for an LTI system characterized by $H(\omega)$. Show that the average autocorrelation of the input and the average autocorrelation of the output are related by

$$\bar{R}_{yy}(\tau) = \int_{-\infty}^{\infty} h(\beta)\int_{-\infty}^{\infty} h(\sigma)\bar{R}_{xx}(\tau + \sigma - \beta)\,d\sigma\,d\beta$$

Hint: Use Eqs. (*1.80*) and (*1.66*).

1.65. Derive Eq. (*1.101*).

Hint: Use the result of Prob. 1.64 and Eq. (*1.69*).

1.66. A power signal $x(t)$ whose power spectral density is a constant K is applied to a low-pass RC filter of Prob. 1.46 [Fig. 1-25(a)]. Find the mean square value of the output.

Ans. $\dfrac{K}{2RC}$

Chapter 2

Amplitude Modulation

2.1 INTRODUCTION

The transmission of an information-bearing signal (or the message signal) over a bandpass communication channel, such as a telephone line or a satellite channel, usually requires a shift of the range of frequencies contained in the signal to another frequency range suitable for transmission. A shift in the signal frequency range is accomplished by modulation. *Modulation* is defined as the process by which some characteristic of a carrier signal is varied in accordance with a modulating signal. Here the message signal is referred to as the *modulating signal*, and the result of modulation is referred to as the *modulated signal*.

The basic types of analog modulation are continuous-wave (CW) modulation and pulse modulation. In continuous-wave modulation, a sinusoidal signal $A_c \cos(\omega_c t + \phi)$ is used as a *carrier signal*. Then a general modulated carrier signal can be represented mathematically as

$$x_c(t) = A(t)\cos[\omega_c t + \phi(t)] \qquad \omega_c = 2\pi f_c \tag{2.1}$$

In Eq. (*2.1*), ω_c [or $f_c = \omega_c/(2\pi)$] is known as the *carrier frequency*. And $A(t)$ and $\phi(t)$ are called the *instantaneous amplitude* and *phase angle* of the carrier, respectively. When $A(t)$ is linearly related to the message signal $m(t)$, the result is *amplitude modulation*. If $\phi(t)$ or its derivative is linearly related to $m(t)$, then we have *phase* or *frequency modulation*. Collectively, phase and frequency modulation are referred to as *angle modulation*, which is discussed in Chap. 3.

In *pulse modulation*, a periodic train of short pulses act as the carrier signal.

2.2 AMPLITUDE MODULATION

In amplitude modulation, the modulated carrier is represented by [setting $\phi(t) = 0$ in Eq. (*2.1*) without loss of generality],

$$x_c(t) = A(t)\cos\omega_c t \tag{2.2}$$

in which the carrier amplitude $A(t)$ is linearly related to the message signal $m(t)$. Amplitude modulation is sometimes referred to as *linear modulation*. Depending on the nature of the spectral relationship between $m(t)$ and $A(t)$, we have the following types of amplitude modulation schemes: *double-sideband* (DSB) modulation, *ordinary* amplitude modulation (AM), *single-sideband* (SSB) modulation, and *vestigial-sideband* (VSB) modulation.

2.3 DOUBLE-SIDEBAND MODULATION

DSB modulation results when $A(t)$ is proportional to the message signal $m(t)$, that is,

$$x_{\text{DSB}}(t) = m(t)\cos\omega_c t \tag{2.3}$$

where we assumed that the constant of proportionality is 1. Equation (*2.3*) indicates that DSB modulation is simply the multiplication of a carrier, $\cos\omega_c t$, by the message signal $m(t)$. By application of the modulation theorem, Eq. (*1.115*), the spectrum of a DSB signal is given by

$$X_{\text{DSB}}(\omega) = \tfrac{1}{2}M(\omega - \omega_c) + \tfrac{1}{2}M(\omega + \omega_c) \tag{2.4}$$

A. Generation of DSB Signals:

The process of DSB modulation is illustrated in Fig. 2-1(*a*). The time-domain waveforms are shown in Fig. 2-1(*b*) and (*c*) for an assumed message signal. The frequency-domain representations

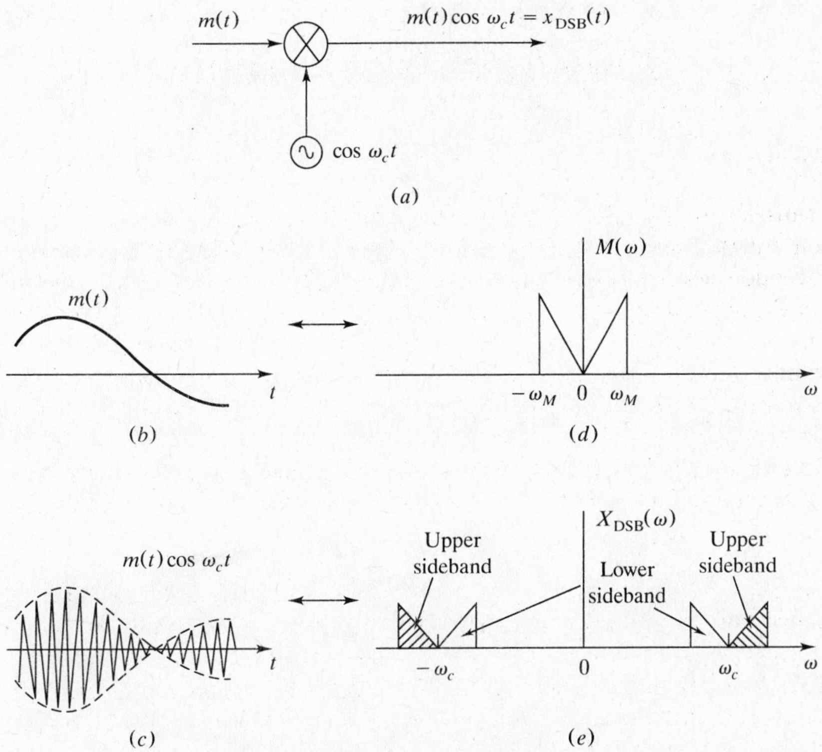

Fig. 2-1 Double-sideband modulation

of $m(t)$ and $x_{\mathrm{DSB}}(t)$ are shown in Fig. 2-1(d) and (e) for an assumed $M(\omega)$ having bandwidth ω_M. The spectra $M(\omega - \omega_c)$ and $M(\omega + \omega_c)$ are the message spectrum translated to $\omega = \omega_c$ and $\omega = -\omega_c$, respectively. The part of the spectrum that lies above ω_c is called the *upper sideband*, and the part below ω_c is called the *lower sideband*. The spectral range occupied by the message signal is called the *baseband*, and thus the message signal is often referred to as the *baseband signal*. As seen in Fig. 2-1(e), the spectrum of $x_{\mathrm{DSB}}(t)$ has no identifiable carrier in it. Thus, this type of modulation is also known as *double-sideband suppressed-carrier* (DSB-SC) modulation. The carrier frequency ω_c is normally much higher than the bandwidth ω_M of the message signal $m(t)$; that is, $\omega_c \gg \omega_M$.

B. Demodulation of DSB Signals:

Recovery of the message signal from the modulated signal is called *demodulation*, or *detection*. The message signal $m(t)$ can be recovered from the modulated signal $x_{\mathrm{DSB}}(t)$ by multiplying $x_{\mathrm{DSB}}(t)$ by a local carrier and using a low-pass filter (LPF) on the product signal, as shown in Fig. 2-2 (see Prob. 2.1).

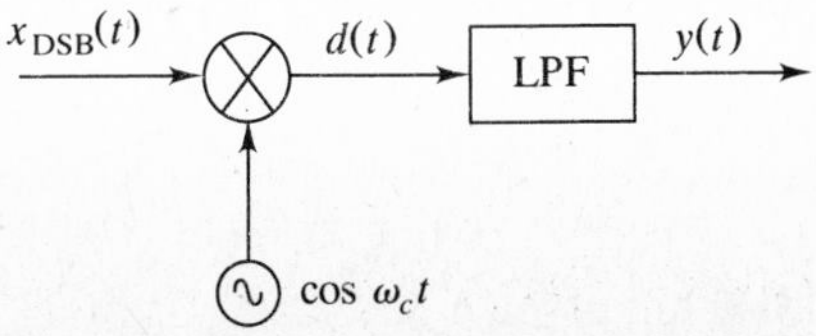

Fig. 2-2 Synchronous demodulator

The basic difficulty associated with the DSB modulation is that for demodulation, the receiver must generate a local carrier which is in phase and frequency synchronism with the incoming carrier (Probs. 2.2 and 2.3). This type of demodulation is known as *synchronous demodulation* or *coherent detection*.

2.4 ORDINARY AMPLITUDE MODULATION

An ordinary amplitude-modulated signal is generated by adding a large carrier signal to the DSB signal. The ordinary AM signal (or simply AM signal) has the form

$$x_{AM}(t) = m(t)\cos\omega_c t + A\cos\omega_c t = [A + m(t)]\cos\omega_c t \tag{2.5}$$

The spectrum of $x_{AM}(t)$ is given by

$$X_{AM}(\omega) = \tfrac{1}{2}M(\omega-\omega_c) + \tfrac{1}{2}M(\omega+\omega_c) + \pi A[\delta(\omega-\omega_c) + \delta(\omega+\omega_c)] \tag{2.6}$$

An example of an AM signal, in both time domain and frequency domain, is shown in Fig. 2-3.

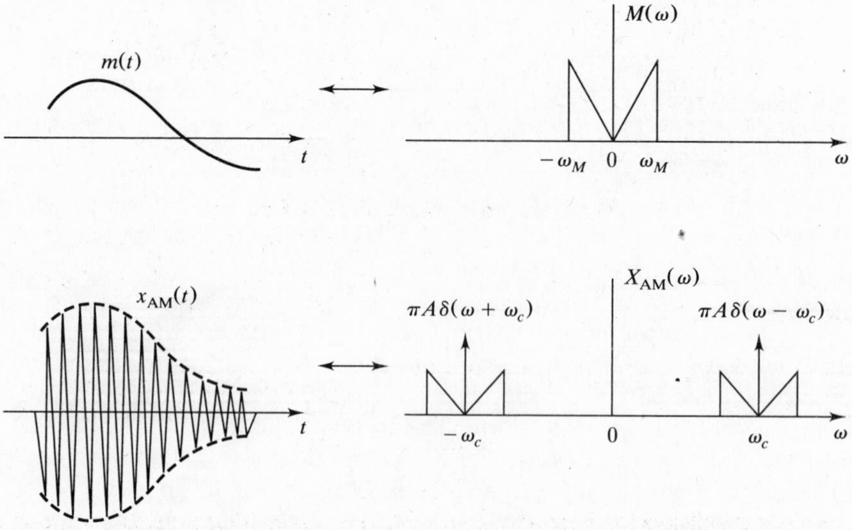

Fig. 2-3 Amplitude modulation

A. Demodulation of AM Signals:

The advantage of AM over DSB modulation is that a very simple scheme, known as *envelope detection*, can be used for demodulation if sufficient carrier power is transmitted. In Eq. (2.5), if A is large enough, the envelope (amplitude) of the modulated waveform given by $A + m(t)$ will be proportional to $m(t)$. Demodulation in this case simply reduces to the detection of the envelope of a modulated carrier with no dependence on the exact phase or frequency of the carrier. If A is not large enough, then the envelope of $x_{AM}(t)$ is not always proportional to $m(t)$, as illustrated in Fig. 2-4. Thus, the condition for demodulation of AM by an envelope detector is

$$A + m(t) > 0 \qquad \text{for all } t \tag{2.7}$$

or

$$A \geq |\min\{m(t)\}| \tag{2.8}$$

where $\min\{m(t)\}$ is the minimum value of $m(t)$.

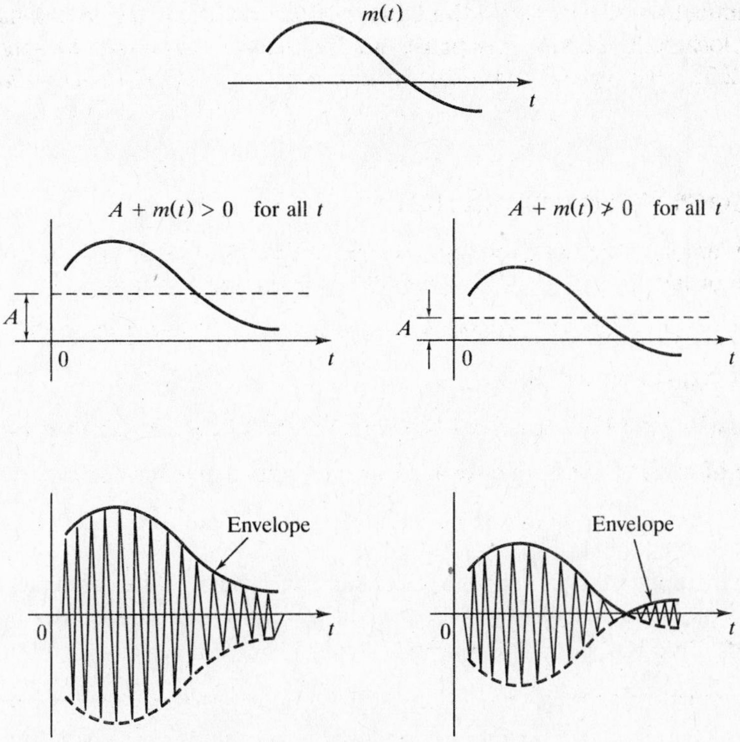

Fig. 2-4 AM signal and its envelope

B. Modulation Index:

The *modulation index* μ for AM is defined as

$$\mu = \frac{|\min\{m(t)\}|}{A} \tag{2.9}$$

From Eq. (*2.8*), the condition for demodulation of AM by an envelope detector can be expressed as

$$\mu \leq 1 \tag{2.10}$$

When $\mu > 1$, the carrier is said to be *overmodulated*, resulting in envelope distortion.

C. Envelope Detector:

Figure 2-5(*a*) shows the simplest form of an envelope detector consisting of a diode and a resistor-capacitor combination. The operation of the envelope detector is as follows. During the positive half-cycle of the input signal, the diode is forward-biased, and the capacitor C charges up rapidly to the peak value of the input signal. As the input signal falls below its maximum, the diode turns off. This is followed by a slow discharge of the capacitor through resistor R until the next positive half-cycle, when the input signal becomes greater than the capacitor voltage and the diode turns on again. The capacitor charges to the new peak value, and the process is repeated.

For proper operation of the envelope detector, the discharge time constant RC must be chosen properly (see Prob. 2.8). In practice, satisfactory operation requires that $1/f_c \ll 1/f_M$, where f_M is the message signal bandwidth.

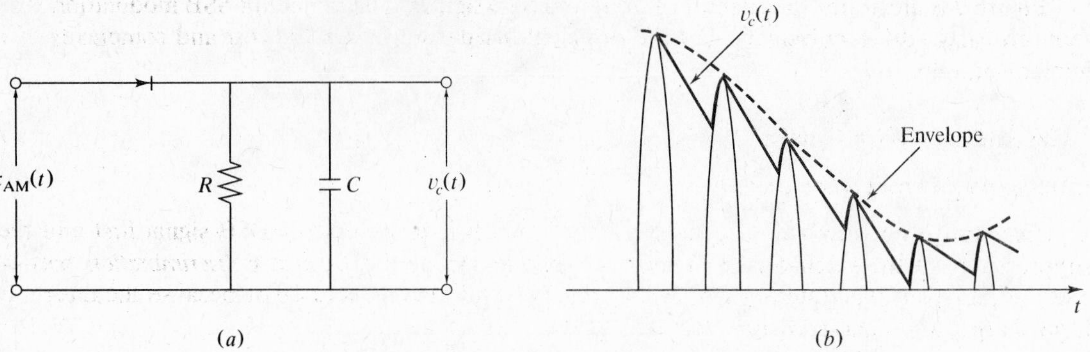

Fig. 2-5 Envelope detector for AM

2.5 SINGLE-SIDEBAND MODULATION

Ordinary AM modulation and DSB modulation waste bandwidth because they both require a transmission bandwidth equal to twice the message bandwidth. (See Figs. 2-1 and 2-3.)

Since either the upper sideband or the lower sideband contains the complete information of the message signal, only one sideband is necessary for information transmission. When only one sideband is transmitted, the modulation is referred to as *single-sideband* (SSB) modulation.

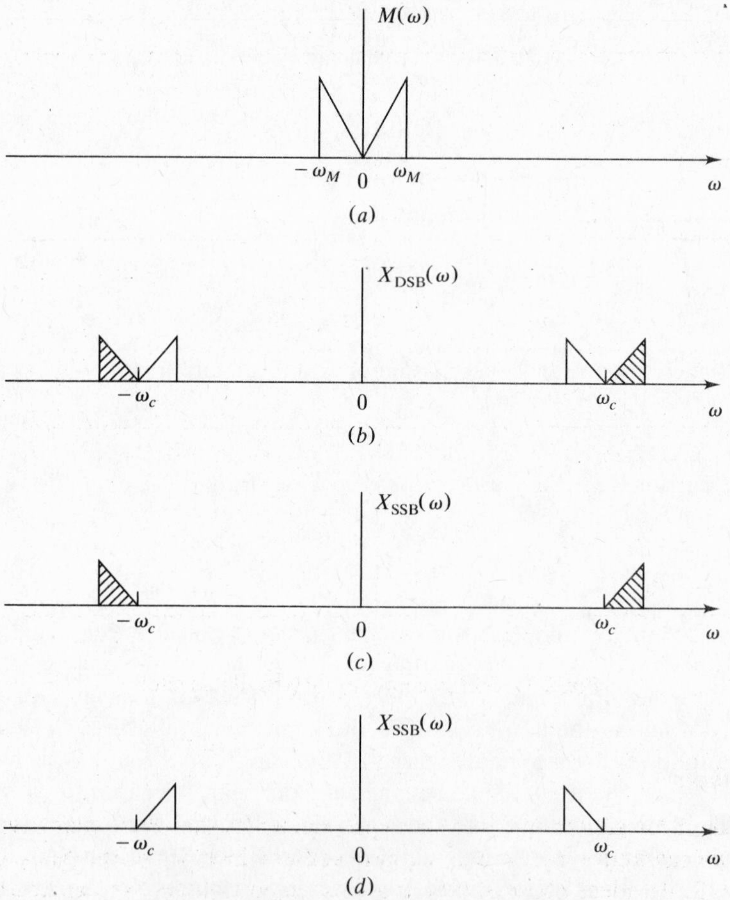

Fig. 2-6 Spectra of DSB and SSB signals

Figure 2-6 illustrates the spectra of DSB and SSB signals. The benefit of SSB modulation is the reduced bandwidth requirement, but the principal disadvantages are the cost and complexity of its implementation.

A. Generation of SSB Signals:

1. *Frequency Discrimination Method:*

The straightforward way to generate an SSB signal is to generate a DSB signal first and then suppress one of the sidebands by filtering. This is known as the *frequency discrimination* method, and the process is illustrated in Fig. 2-7. In practice, this method is not easy because the filter must have sharp cutoff characteristics.

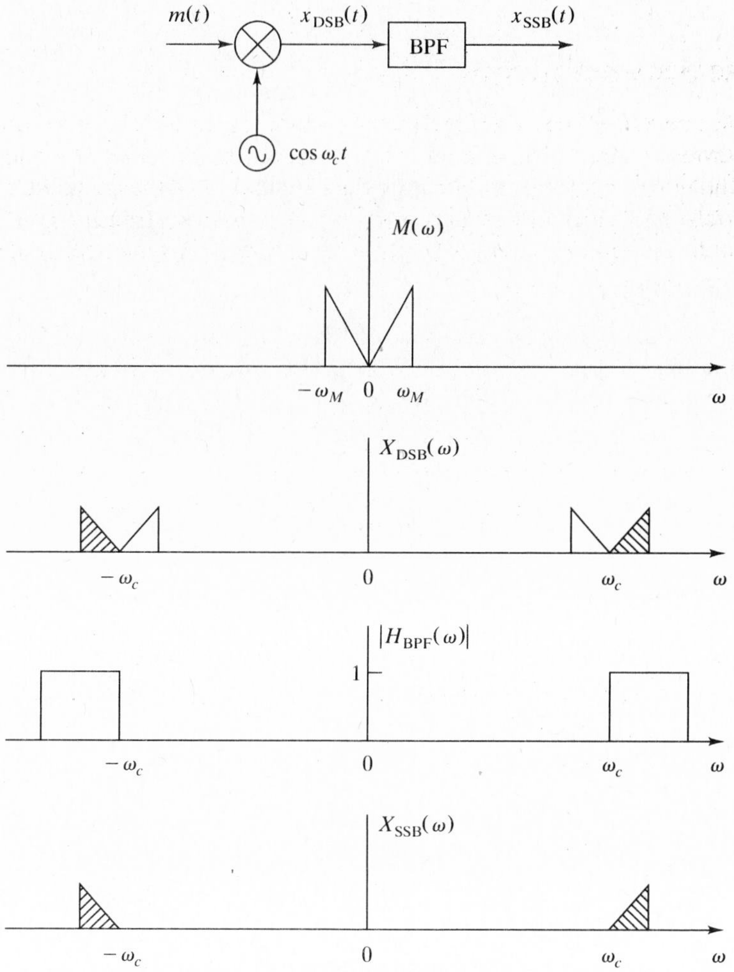

Fig. 2-7 SSB generation using bandpass filter

2. *Phase-Shift Method:*

Another method for generating an SSB signal, known as the *phase-shift* method, is illustrated in Fig. 2-8. The box marked $-\pi/2$ is a $\pi/2$ phase shifter which delays the phase of every frequency component by $\pi/2$. An ideal phase shifter is almost impossible to implement exactly. But we can approximate it over a finite frequency band.

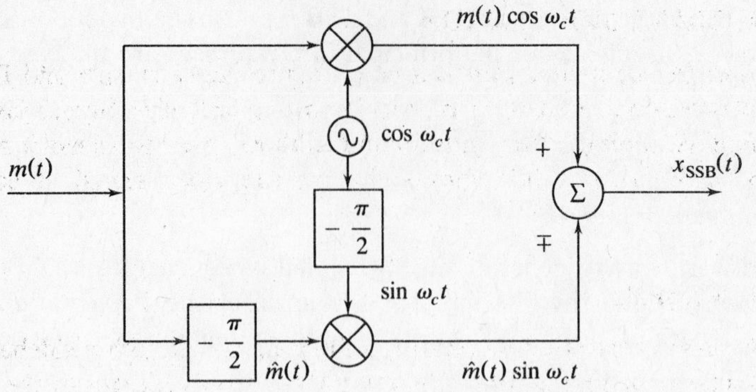

Fig. 2-8 SSB generation using phase shifters

If we let $\hat{m}(t)$ be the output of the $-\pi/2$ phase shifter due to the input $m(t)$ (see Prob. 1.47), then from Fig. 2-8 the SSB signal $x_{\text{SSB}}(t)$ can be represented by

$$x_{\text{SSB}}(t) = m(t)\cos\omega_c t \mp \hat{m}(t)\sin\omega_c t \tag{2.11}$$

The difference represents the upper-sideband SSB signal, and the sum represents the lower-sideband SSB signal. (See Prob. 2.10.)

B. Demodulation of SSB Signals:

Demodulation of SSB signals can be achieved easily by using the coherent detector as used in the DSB demodulation, that is, by multiplying $x_{\text{SSB}}(t)$ by a local carrier and passing the resulting signal through a low-pass filter. This is illustrated in Fig. 2-9.

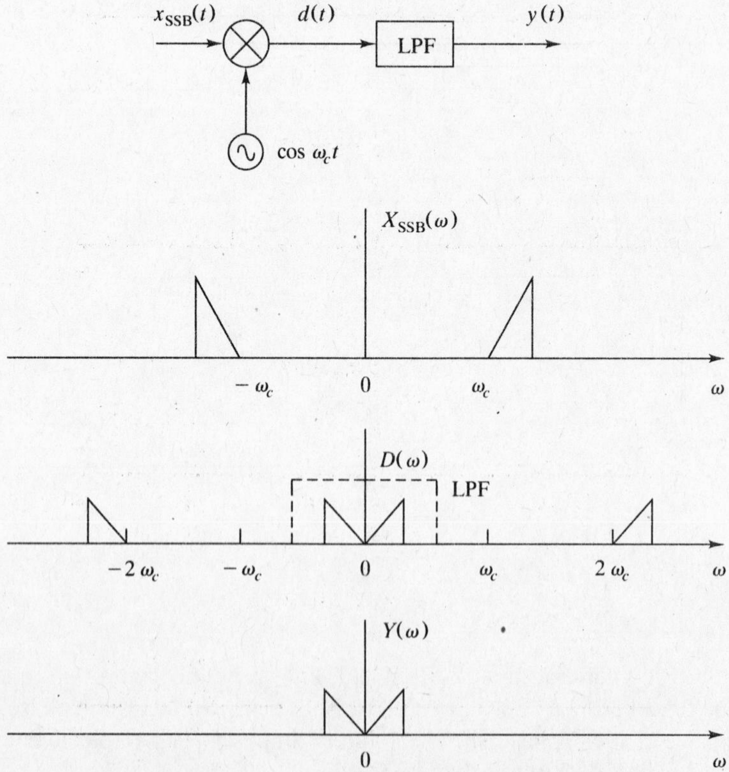

Fig. 2-9 Synchronous demodulation of SSB signals

2.6 VESTIGIAL-SIDEBAND MODULATION

Vestigial-sideband (VSB) modulation is a compromise between SSB and DSB modulations. In this modulation scheme, one sideband is passed almost completely whereas just a trace, or vestige, of the other sideband is retained. The typical bandwidth required to transmit a VSB signal is about 1.25 that of SSB. VSB is used for transmission of the video signal in commercial television broadcasting.

A. Generation of VSB Signals:

A VSB signal can be generated by passing a DSB signal through a sideband shaping filter (or vestigial filter), as shown in Fig. 2-10(*a*). Figure 2-10(*b*) to (*e*) illustrates the spectrum of a VSB signal [$x_{VSB}(t)$] in relation to that of the message signal $m(t)$, assuming that the lower sideband is transformed to vestigial sideband.

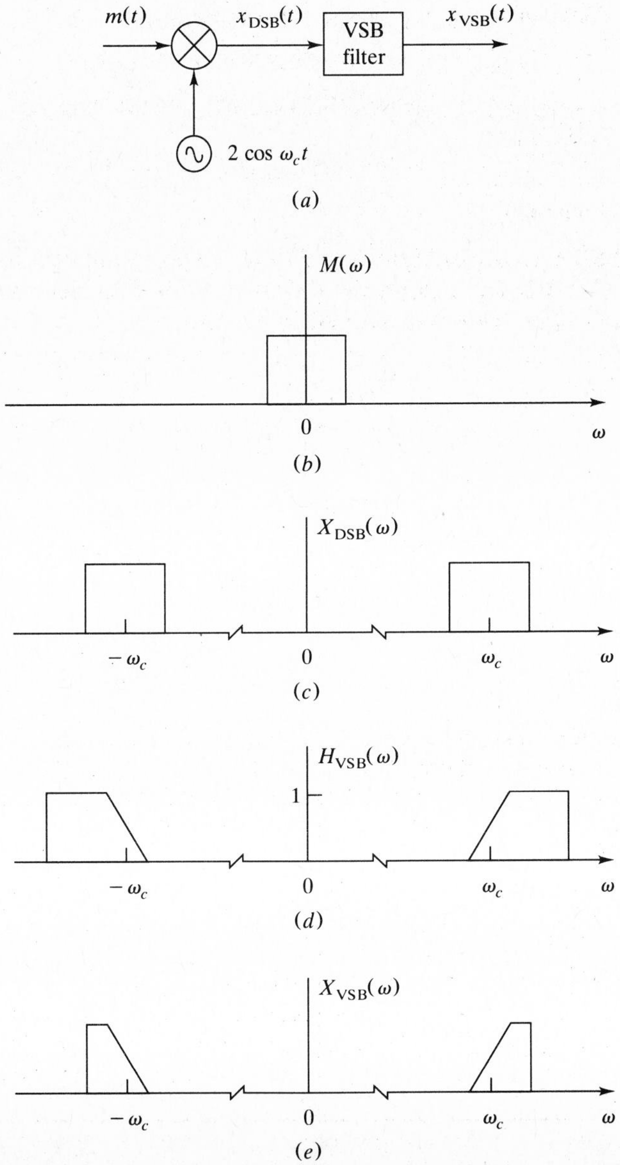

Fig. 2-10 VSB modulation

B. Demodulation of VSB Signals:

For VSB signals, $m(t)$ can be recovered by synchronous or coherent demodulation (see Fig. 2-11); this determines the requirements of the frequency response $H(\omega)$. It can be shown that for distortionless recovery of $m(t)$, it is required that

$$H(\omega+\omega_c)+H(\omega-\omega_c)=\text{constant} \qquad \text{for } |\omega| \le \omega_M \tag{2.12}$$

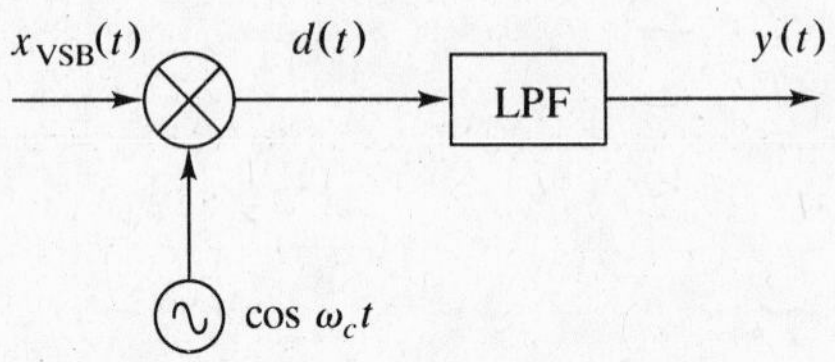

Fig. 2-11 VSB demodulator

where ω_M is the maximum frequency of $m(t)$ (Prob. 2.13). By letting the constant in Eq. (2.12) be $2H(\omega_c)$, Eq. (2.12) becomes

$$H(\omega-\omega_c)-H(\omega_c)=-[H(\omega+\omega_c)-H(\omega_c)] \tag{2.13}$$

which shows that $H(\omega)$ will have antisymmetry about the carrier frequency (Fig. 2-12). Figure 2-12(*a*) and (*b*) shows two possible forms of $|H(\omega)|$ that satisfy Eq. (2.13). The filters in Fig. 2-12(*a*) and (*b*) correspond to VSB filters that retain the lower sideband (LSB) and the upper sideband (USB), respectively.

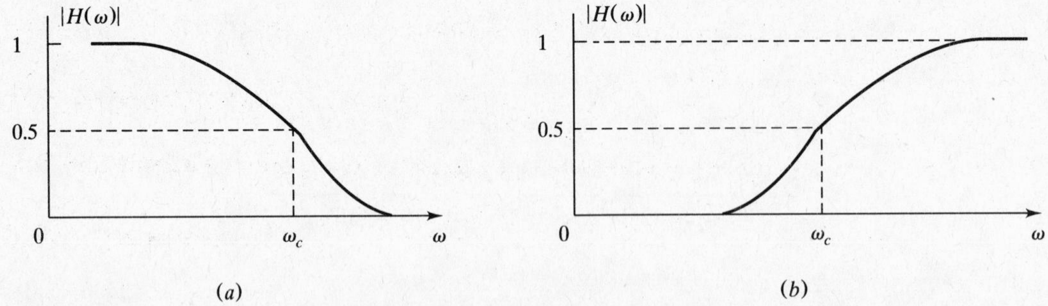

Fig. 2-12 VSB filter characteristics

Figure 2-13 illustrates the demodulation of the VSB signal shown in Fig. 2-10 by the synchronous detector of Fig. 2-11.

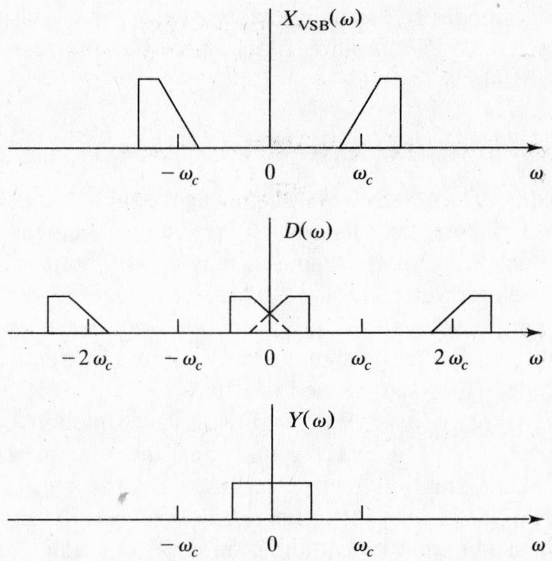

Fig. 2-13 Synchronous demodulation of VSB signals

2.7 FREQUENCY TRANSLATION AND MIXING

In the processing of signals in communication systems, it is often desirable to translate or shift the modulated signal to a new frequency band. For example, in most commercial AM radio receivers, the received radio-frequency (RF) signal [540 to 1600 kilohertz (kHz)] is shifted to the intermediate-frequency (IF) (455-kHz) band for processing. The received signal, now translated to a fixed IF, can easily be amplified, filtered, and demodulated.

A device that performs the frequency translation of a modulated signal is called a *frequency mixer* (Fig. 2-14). The operation is often called *frequency mixing*, *frequency conversion*, or *heterodyning*.

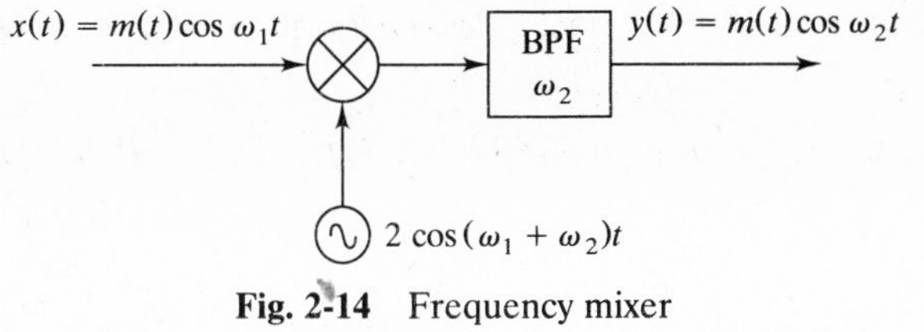

Fig. 2-14 Frequency mixer

A common problem associated with frequency mixing is the presence of the *image frequency*. For example, in an AM superheterodyne receiver (see Prob. 2.15), the locally generated frequency is chosen to be 455 kHz higher than the incoming signal. Suppose that the reception of an AM station at 600 kHz is desired. Then the locally generated signal is at 1055 kHz. Now if there is another station at 1510 kHz, it also will be received (note that 1510 kHz − 1055 kHz = 455 kHz). This second frequency, 1510 kHz = 600 kHz + 2(455 kHz), is called the image frequency of the first, and after the heterodyning operation it is impossible to distinguish the two. Note that the image frequency is separated from the desired signal by exactly twice the IF. Usually, the image frequency signal is attenuated by a selective RF amplifier placed before the mixer.

2.8 FREQUENCY-DIVISION MULTIPLEXING

Multiplexing is a technique whereby several message signals are combined into a composite signal for transmission over a common channel. To transmit a number of these signals over the same channel, the signals must be kept apart so that they do not interfere with each other, and thus they can be separated at the receiver end.

There are two basic multiplexing techniques: *frequency-division multiplexing* (FDM) and *time-division multiplexing* (TDM). In FDM the signals are separated in frequency, whereas in TDM the signals are separated in time. TDM is discussed in Chap. 4.

The FDM scheme is illustrated in Fig. 2-15 with the simultaneous transmission of three message signals. The spectra of the message signals and the sum of the modulated carriers are indicated in the figure. DSB modulation is used in illustrating the spectra of Fig. 2-15. Any type of modulation can be used in FDM as long as the carrier spacing is sufficient to avoid spectral overlap. However, the most widely used method of modulation is SSB modulation. At the receiving end of the channel the three modulated signals are separated by bandpass filters (BPFs) and then demodulated.

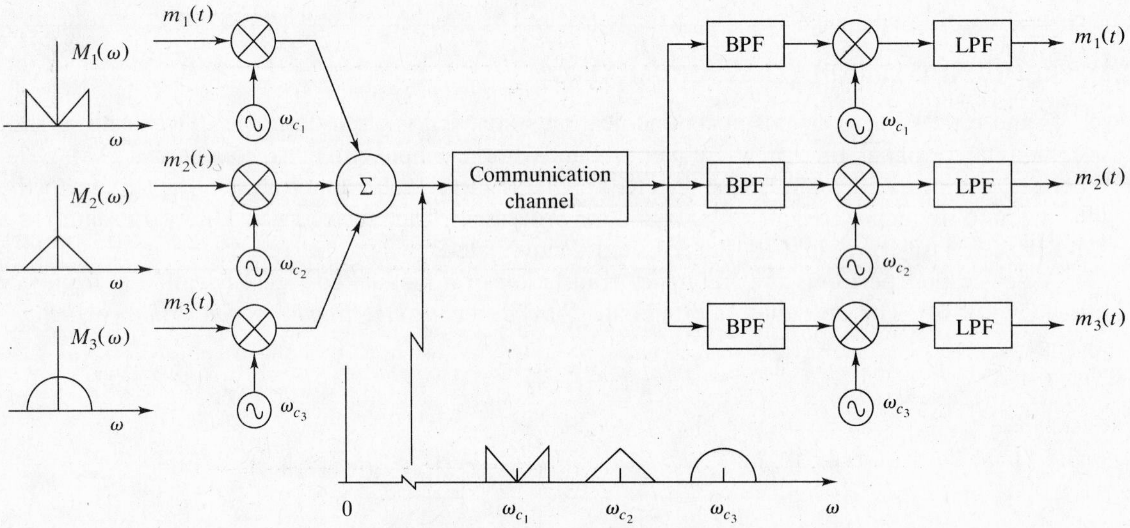

Fig. 2-15 Frequency-division multiplexing

FDM is used in telephone system, telemetry, commercial broadcast, television, and communication networks. Commercial AM broadcast stations use carrier frequency spaced 10 kHz apart in the frequency range from 540 to 1600 kHz. This separation is not sufficient to avoid spectral overlap for AM with a reasonably high-fidelity (50 Hz to 15 kHz) audio signal. Therefore, AM stations on adjacent carrier frequencies are placed geographically far apart to minimize interference. Commercial FM (frequency-modulation) broadcast (Chap. 3) uses carrier frequencies spaced 200 kHz apart. In a long-distance telephone system, up to 600 or more voice signals (200 Hz to 3.2 kHz) are transmitted over a coaxial cable or microwave links by using SSB modulation with carrier frequencies spaced 4 kHz apart. In practice, the composite signal formed by spacing several signals in frequency may, in turn, be modulated by using another carrier frequency. In this case, the first carrier frequencies are often called *subcarriers*.

Solved Problems

DOUBLE-SIDEBAND MODULATION

2.1. Verify that the message signal $m(t)$ is recovered from a modulated DSB signal by first multiplying it by a local sinusoidal carrier and then passing the resultant signal through a low-pass filter, as shown in Fig. 2-2, (*a*) in the time domain and (*b*) in the frequency domain.

(*a*) Referring to Fig. 2-2, the output of the multiplier is

$$\begin{aligned} d(t) &= x_{\mathrm{DSB}}(t)\cos\omega_c t = [m(t)\cos\omega_c t]\cos\omega_c t \\ &= m(t)\cos^2\omega_c t \\ &= \tfrac{1}{2}m(t) + \tfrac{1}{2}m(t)\cos 2\omega_c t \end{aligned}$$

After low-pass filtering of $d(t)$, we obtain

$$y(t) = \tfrac{1}{2}m(t) \tag{2.14}$$

Thus by proper amplification (multiplying by 2) we can recover the message signal $m(t)$.

(*b*) Demodulation of $x_{\mathrm{DSB}}(t)$ by the process shown in Fig. 2-2 in frequency domain is illustrated in Fig. 2-16.

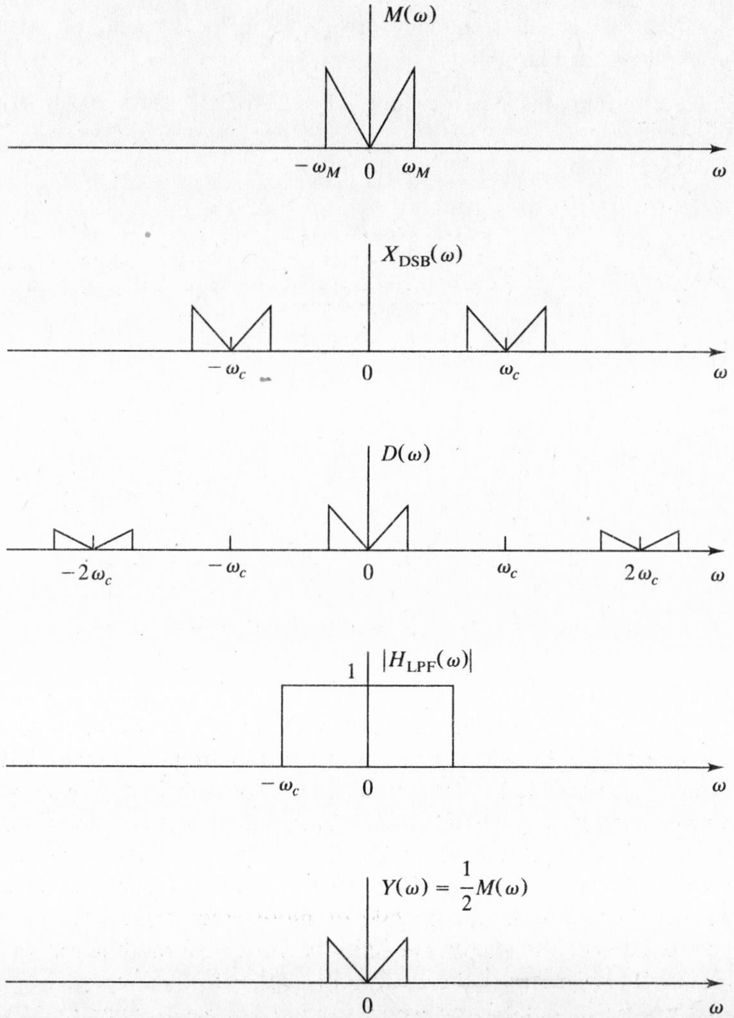

Fig. 2-16 Demodulation of a DSB signal

2.2. Evaluate the effect of a phase error in the local oscillator on synchronous DSB demodulation as shown in Fig. 2-2.

Let the phase error of the local oscillator in Fig. 2-2 be ϕ. Then the local carrier is expressed as $\cos(\omega_c t + \phi)$. Now

$$x_{DSB}(t) = m(t)\cos\omega_c t$$

and

$$\begin{aligned} d(t) &= [m(t)\cos\omega_c t]\cos(\omega_c t + \phi) \\ &= \tfrac{1}{2}m(t)[\cos\phi + \cos(2\omega_c t + \phi)] \\ &= \tfrac{1}{2}m(t)\cos\phi + \tfrac{1}{2}m(t)\cos(2\omega_c t + \phi) \end{aligned}$$

The second term on the right-hand side is filtered out by the low-pass filter, and we obtain

$$y(t) = \tfrac{1}{2}m(t)\cos\phi \tag{2.15}$$

This output is proportional to $m(t)$ when ϕ is constant. The output is completely lost when $\phi = \pm\pi/2$. Thus, the phase error in the local carrier causes attenuation of the output signal without any distortion as long as ϕ is constant and not equal to $\pm\pi/2$. If the phase error ϕ varies randomly with time, then the output also will vary randomly and is undesirable.

2.3. Evaluate the effect of a small frequency error in the local oscillator on synchronous DSB demodulation as shown in Fig. 2-2.

Let the frequency error of the local oscillator in Fig. 2-2 be $\Delta\omega$. The local carrier is then expressed as $\cos(\omega_c + \Delta\omega)t$. Then

$$\begin{aligned} d(t) &= m(t)\cos\omega_c t\cos(\omega_c + \Delta\omega)t \\ &= \tfrac{1}{2}m(t)\cos(\Delta\omega)t + \tfrac{1}{2}m(t)\cos 2\omega_c t \end{aligned}$$

Thus

$$y(t) = \tfrac{1}{2}m(t)\cos(\Delta\omega)t \tag{2.16}$$

The output is the signal $m(t)$ multiplied by a low-frequency sinusoid. This is a "beating" effect and is a very undesirable distortion.

ORDINARY AMPLITUDE MODULATION

2.4. Sketch the ordinary AM signal for a single-tone modulation with modulation indices of $\mu = 0.5$ and $\mu = 1$.

For a single-tone modulation

$$m(t) = a_m\cos\omega_m t$$

Then, by Eq. (*2.9*), the modulation index is

$$\mu = \frac{|\min\{m(t)\}|}{A} = \frac{a_m}{A} \tag{2.17}$$

Hence,

$$m(t) = a_m\cos\omega_m t = \mu A\cos\omega_m t \tag{2.18}$$

and by Eq. (*2.5*),

$$\begin{aligned} x_{AM}(t) &= [A + m(t)]\cos\omega_c t \\ &= A[1 + \mu\cos\omega_m t]\cos\omega_c t \end{aligned} \tag{2.19}$$

Figure 2-17(*a*) and (*b*) shows the ordinary AM signals corresponding to $\mu = 0.5$ and $\mu = 1$, respectively.

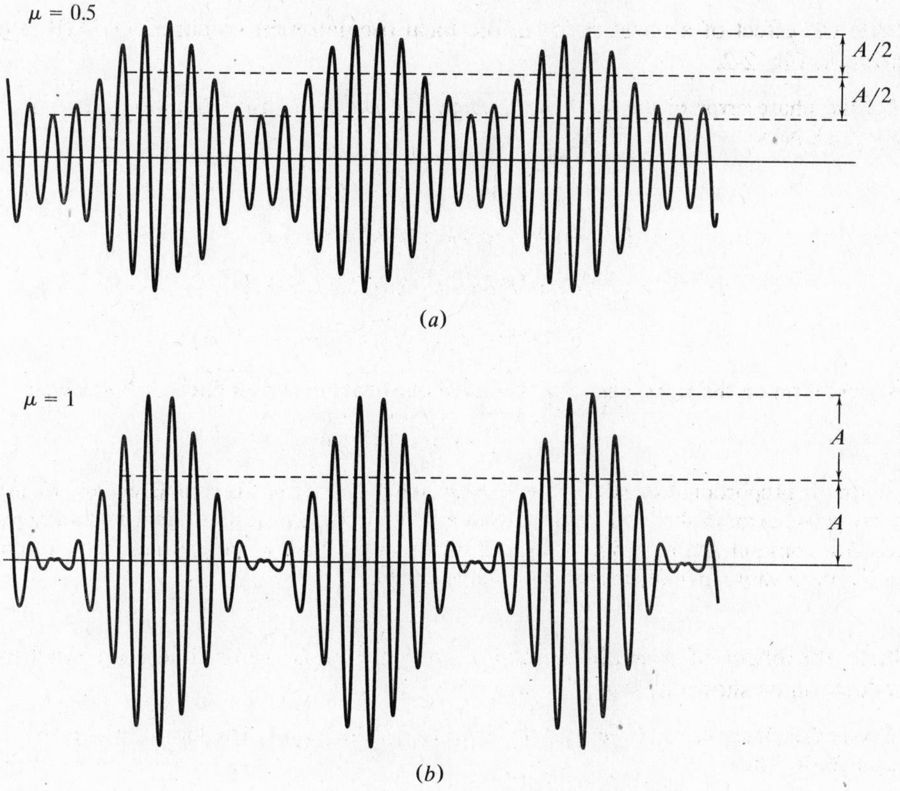

Fig. 2-17

2.5. The *efficiency* η of ordinary AM is defined as the percentage of the total power carried by the sidebands, that is,

$$\eta = \frac{P_s}{P_t} \times 100\% \tag{2.20}$$

where P_s is the power carried by the sidebands and P_t is the total power of the AM signal.

(a) Find η for $\mu = 0.5$ (50 percent modulation).

(b) Show that for a single-tone AM, $\eta_{\max}$ is 33.3 percent at $\mu = 1$.

From Eq. (2.19), a single-tone AM signal can be expressed as

$$\begin{aligned} x_{\text{AM}}(t) &= A\cos\omega_c t + \mu A \cos\omega_m t \cos\omega_c t \\ &= A\cos\omega_c t + \tfrac{1}{2}\mu A\cos(\omega_c - \omega_m)t + \tfrac{1}{2}\mu A\cos(\omega_c + \omega_m)t \end{aligned}$$

$$P_c = \text{carrier power} = \tfrac{1}{2}A^2 \tag{2.21}$$

$$P_s = \text{sideband power} = \tfrac{1}{2}\left[\left(\tfrac{1}{2}\mu A\right)^2 + \left(\tfrac{1}{2}\mu A\right)^2\right] = \tfrac{1}{4}\mu^2 A^2 \tag{2.22}$$

The total power P_t is

$$P_t = P_c + P_s = \tfrac{1}{2}A^2 + \tfrac{1}{4}\mu^2 A^2 = \tfrac{1}{2}\left(1 + \tfrac{1}{2}\mu^2\right)A^2 \tag{2.23}$$

Thus

$$\eta = \frac{P_s}{P_t} \times 100\% = \frac{\tfrac{1}{4}\mu^2 A^2}{\left(\tfrac{1}{2} + \tfrac{1}{4}\mu^2\right)A^2} \times 100\% = \frac{\mu^2}{2 + \mu^2} \times 100\% \tag{2.24}$$

with the condition that $\mu \le 1$.

(*a*) For $\mu = 0.5$,

$$\eta = \frac{(0.5)^2}{2 + (0.5)^2} \times 100\% = 11.1\%$$

(*b*) Since $\mu \le 1$, it can be seen that $\eta_{\max}$ occurs at $\mu = 1$ and is given by

$$\eta = \tfrac{1}{3} \times 100\% = 33.3\%$$

2.6. Show that a synchronous demodulator (Fig. 2-18) can demodulate an AM signal $x_{AM}(t) = [A + m(t)] \cos \omega_c t$ regardless of the value of A.

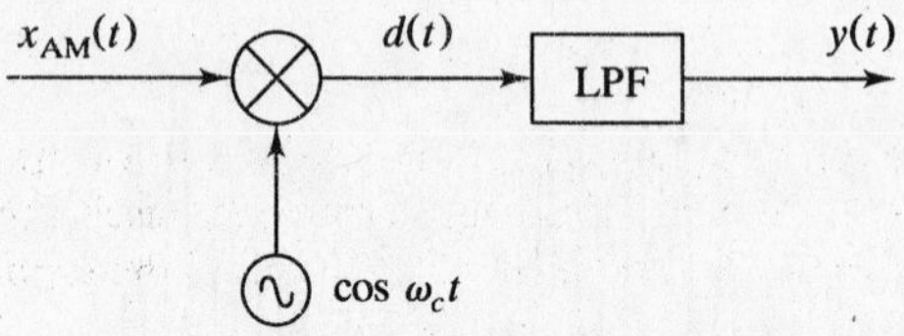

Fig. 2-18 Synchronous detector

From Fig. 2-18,

$$d(t) = x_{AM}(t) \cos \omega_c t = [A + m(t)] \cos^2 \omega_c t$$
$$= \tfrac{1}{2}[A + m(t)] + \tfrac{1}{2}[A + m(t)] \cos 2\omega_c t$$

Hence, after low-pass filtering, we obtain

$$y(t) = \tfrac{1}{2}[A + m(t)] = \tfrac{1}{2}m(t) + \tfrac{1}{2}A \tag{2.25}$$

A blocking capacitor will suppress the direct-current (dc) term $\tfrac{1}{2}A$, yielding the output $\tfrac{1}{2}m(t)$.

2.7. Show that an AM signal with large carrier can be demodulated by squaring it and then passing the resulting signal through a low-pass filter, as shown in Fig. 2-19. This type of detector is known as a *square-law detector*.

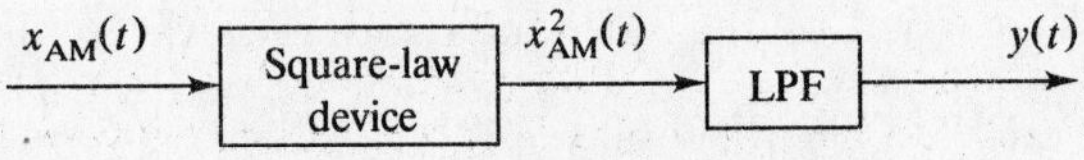

Fig. 2-19 Square-law detector

From Eq. (*2.5*),

$$x_{AM}(t) = [A + m(t)] \cos \omega_c t$$

then

$$x_{AM}^2(t) = [A + m(t)]^2 \cos^2 \omega_c t$$
$$= \frac{1}{2}[A^2 + 2Am(t) + m^2(t)](1 + \cos 2\omega_c t)$$

The low-pass filter output $y(t)$ is

$$y(t) = \frac{A^2}{2}\left\{1 + 2\frac{m(t)}{A} + \left[\frac{m(t)}{A}\right]^2\right\} \tag{2.26}$$

With a large carrier, the $[m(t)/A]^2$ term can be neglected and

$$y(t) \approx \frac{A^2}{2} + Am(t) \tag{2.27}$$

A blocking capacitor will suppress the dc term, $A^2/2$, yielding the output $Am(t)$.

2.8. The input to an envelope detector (Fig. 2-5) is a single-tone AM signal $x_{AM}(t) = A(1 + \mu \cos \omega_m t) \cos \omega_c t$, where μ is a constant, $0 < \mu < 1$, and $\omega_c \gg \omega_m$.

(*a*) Show that if the detector output is to follow the envelope of $x_{AM}(t)$, it is required that at any time t_0

$$\frac{1}{RC} \geq \omega_m \left(\frac{\mu \sin \omega_m t_0}{1 + \mu \cos \omega_m t_0} \right) \tag{2.28}$$

(*b*) Show that if the detector output is to follow the envelope at all times, it is required that

$$RC \leq \frac{1}{\omega_m} \frac{\sqrt{1 - \mu^2}}{\mu} \tag{2.29}$$

(*a*) Figure 2-20 shows the envelope of $x_{AM}(t)$ and the output of the detector (the voltage across the capacitor of Fig. 2-5). Assume that the capacitor discharge from the peak value $E_0 = A(1 + \mu \cos \omega_m t_0)$ at $t_0 = 0$. Then the voltage $v_c(t)$ across the capacitor of Fig. 2-5 is given by

$$v_c(t) = E_0 e^{-t/(RC)} \tag{2.30}$$

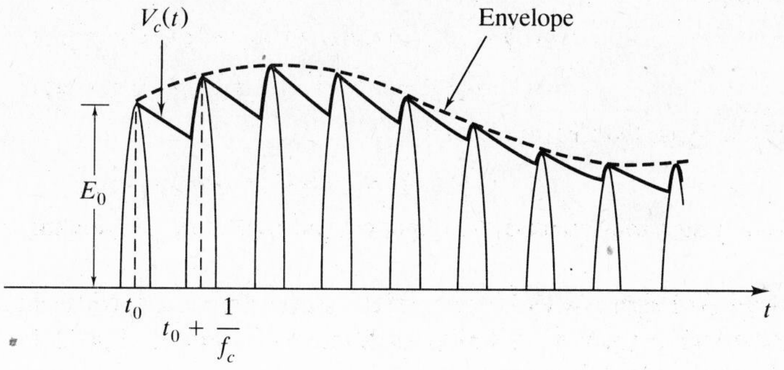

Fig. 2-20

The interval between two successive carrier peaks is $1/f_c = 2\pi/\omega_c$ and $RC \gg 1/\omega_c$. This means that the time constant RC is much larger than the interval between two successive carrier peaks. Therefore, $v_c(t)$ can be approximated by

$$v_c(t) \approx E_0 \left(1 - \frac{t}{RC} \right) \tag{2.31}$$

Thus if the $v_c(t)$ is to follow the envelope of $x_{AM}(t)$, it is required that at any time t_0

$$(1 + \mu \cos \omega_m t_0) \left(1 - \frac{1}{RCf_c} \right) \leq 1 + \mu \cos \omega_m \left(t_0 + \frac{1}{f_c} \right) \tag{2.32}$$

Now if $\omega_m \ll \omega_c$, then

$$\begin{aligned} 1 + \mu \cos \omega_m \left(t_0 + \frac{1}{f_c} \right) &= 1 + \mu \cos \left(\omega_m t_0 + \frac{\omega_m}{f_c} \right) \\ &= 1 + \mu \cos \omega_m t_0 \cos \frac{\omega_m}{f_c} - \mu \sin \omega_m t_0 \sin \frac{\omega_m}{f_c} \\ &\approx 1 + \mu \cos \omega_m t_0 - \mu \frac{\omega_m}{f_c} \sin \omega_m t_0 \end{aligned} \tag{2.33}$$

Hence
$$(1+\mu\cos\omega_m t_0)\left(\frac{1}{RCf_c}\right) \geq \frac{\mu\omega_m}{f_c}\sin\omega_m t_0 \tag{2.34}$$

or
$$\frac{1}{RC} \geq \omega_m\left(\frac{\mu\sin\omega_m t_0}{1+\mu\cos\omega_m t_0}\right)$$

(*b*) Rewriting Eq. (*2.28*), we have

$$\frac{1}{RC} + \frac{\mu}{RC}\cos\omega_m t_0 \geq \mu\omega_m \sin\omega_m t_0 \tag{2.35}$$

or
$$\mu\left(\omega_m\sin\omega_m t_0 - \frac{1}{RC}\cos\omega_m t_0\right) \leq \frac{1}{RC} \tag{2.36}$$

or
$$\mu\sqrt{\omega_m^2+\left(\frac{1}{RC}\right)^2}\sin\left(\omega_m t_0 - \tan^{-1}\frac{1}{\omega_m RC}\right) \leq \frac{1}{RC} \tag{2.37}$$

Since this inequality must hold for every t_0, we must have

$$\mu\sqrt{\omega_m^2+\left(\frac{1}{RC}\right)^2} \leq \frac{1}{RC} \tag{2.38}$$

or
$$\mu^2\left[\omega_m^2+\left(\frac{1}{RC}\right)^2\right] \leq \left(\frac{1}{RC}\right)^2 \tag{2.39}$$

From which we obtain

$$RC \leq \frac{1}{\omega_m}\frac{\sqrt{1-\mu^2}}{\mu}$$

SINGLE-SIDEBAND MODULATION

2.9. Using the single-tone modulating signal $\cos\omega_m t$, verify that the output of the SSB generator of Fig. 2-8 is indeed an SSB signal, and show that an upper-sideband (USB) or a lower-sideband (LSB) signal results from subtraction or addition at the summer.

Referring to Fig. 2-21, which is the redrawing of Fig. 2-8 with a single-tone modulating signal, we have

$$m(t) = \cos\omega_m t$$
$$\cos\left(\omega_c t - \frac{\pi}{2}\right) = \sin\omega_c t$$
$$\hat{m}(t) = \cos\left(\omega_m t - \frac{\pi}{2}\right) = \sin\omega_m t$$

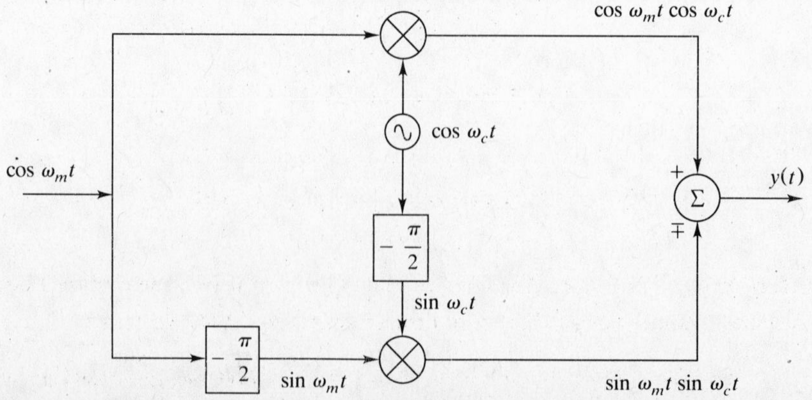

Fig. 2-21

Hence
$$y(t) = \cos \omega_m t \cos \omega_c t \mp \sin \omega_m t \sin \omega_c t$$
$$= \cos(\omega_c \pm \omega_m)t \tag{2.40}$$

Thus, with subtraction we have
$$y(t) = x_{\text{USB}}(t) = \cos(\omega_c + \omega_m)t \tag{2.41}$$

and with addition we have
$$y(t) = x_{\text{LSB}}(t) = \cos(\omega_c - \omega_m)t \tag{2.42}$$

2.10. Show that if the output of the phase-shift modulator (Fig. 2-8) is an SSB signal, (*a*) the difference of the signals at the summing junction produces the upper-sideband SSB signal and (*b*) the sum produces the lower-sideband SSB signal. That is,
$$x_c(t) = x_{\text{USB}}(t) = m(t) \cos \omega_c t - \hat{m}(t) \sin \omega_c t \tag{2.43}$$
is an upper-sideband SSB signal, and
$$x_c(t) = x_{\text{LSB}}(t) = m(t) \cos \omega_c t + \hat{m}(t) \sin \omega_c t \tag{2.44}$$
is a lower-sideband SSB signal.

(*a*) Let
$$m(t) \longleftrightarrow M(\omega) \quad \text{and} \quad \hat{m}(t) \longleftrightarrow \hat{M}(\omega)$$
Then applying the modulation theorem (*1.115*) of Prob. 1.15, or the frequency-shifting property (*1.31*) of the Fourier transform, we have
$$m(t) \cos \omega_c t \longleftrightarrow \frac{1}{2} M(\omega - \omega_c) + \frac{1}{2} M(\omega + \omega_c)$$
and
$$\hat{m}(t) \sin \omega_c t \longleftrightarrow \frac{1}{2j} \hat{M}(\omega - \omega_c) - \frac{1}{2j} \hat{M}(\omega + \omega_c)$$
Taking the Fourier transform of Eq. (*2.43*), we have
$$X_c(\omega) = \frac{1}{2} M(\omega - \omega_c) + \frac{1}{2} M(\omega + \omega_c) - \left[\frac{1}{2j} \hat{M}(\omega - \omega_c) - \frac{1}{2j} \hat{M}(\omega + \omega_c)\right] \tag{2.45}$$
From Eq. (*1.141*) of Prob. 1.47, we have
$$\hat{M}(\omega - \omega_c) = -j \operatorname{sgn}(\omega - \omega_c) M(\omega - \omega_c) \tag{2.46a}$$
and
$$\hat{M}(\omega + \omega_c) = -j \operatorname{sgn}(\omega + \omega_c) M(\omega + \omega_c) \tag{2.46b}$$
Thus
$$X_c(\omega) = \frac{1}{2} M(\omega - \omega_c) + \frac{1}{2} M(\omega + \omega_c) - \left[-\frac{1}{2} \operatorname{sgn}(\omega - \omega_c) M(\omega - \omega_c) + \frac{1}{2} \operatorname{sgn}(\omega + \omega_c) M(\omega + \omega_c)\right]$$
$$= \frac{1}{2} M(\omega - \omega_c)[1 + \operatorname{sgn}(\omega - \omega_c)] + \frac{1}{2} M(\omega + \omega_c)[1 - \operatorname{sgn}(\omega + \omega_c)] \tag{2.47}$$
Since
$$1 + \operatorname{sgn}(\omega - \omega_c) = \begin{cases} 2 & \omega > \omega_c \\ 0 & \omega < \omega_c \end{cases}$$
and
$$1 - \operatorname{sgn}(\omega - \omega_c) = \begin{cases} 2 & \omega < -\omega_c \\ 0 & \omega > -\omega_c \end{cases}$$
we have
$$X_c(\omega) = \begin{cases} 0 & |\omega| < \omega_c \\ M(\omega + \omega_c) & \omega < -\omega_c \\ M(\omega - \omega_c) & \omega > \omega_c \end{cases} \tag{2.48}$$
which is sketched in Fig. 2-22(*b*). We see that $x_c(t)$ is an upper sideband SSB signal.

(*b*) In a similar manner, taking the Fourier transform of Eq. (*2.44*), we have
$$X_c(\omega) = \frac{1}{2} M(\omega - \omega_c)[1 - \operatorname{sgn}(\omega - \omega_c)] + \frac{1}{2} M(\omega + \omega_c)[1 + \operatorname{sgn}(\omega + \omega_c)] \tag{2.49}$$

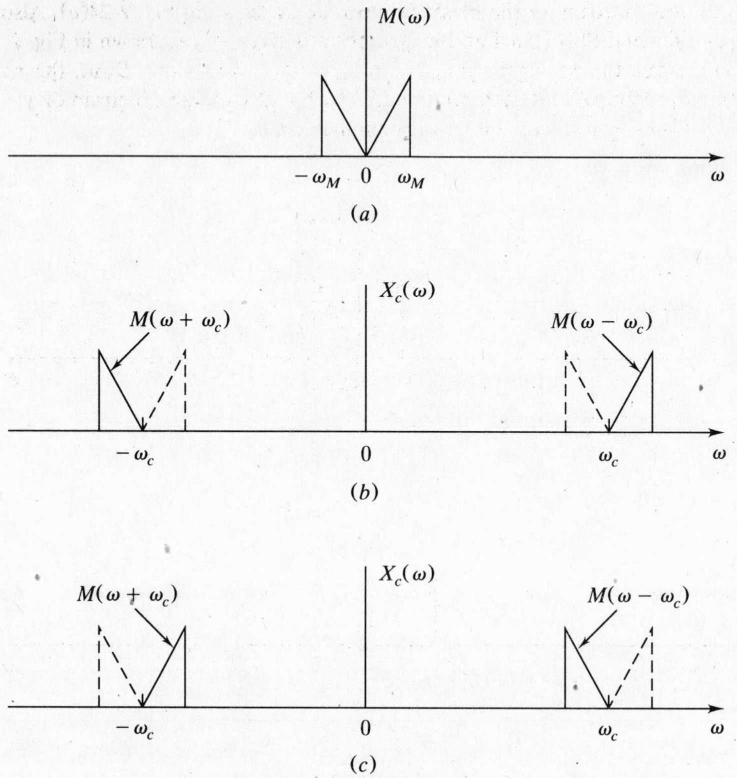

Fig. 2-22

Since
$$1 - \operatorname{sgn}(\omega - \omega_c) = \begin{cases} 2 & \omega < \omega_c \\ 0 & \omega > \omega_c \end{cases}$$

and
$$1 + \operatorname{sgn}(\omega + \omega_c) = \begin{cases} 2 & \omega > -\omega_c \\ 0 & \omega < -\omega_c \end{cases}$$

We have
$$X_c(\omega) = \begin{cases} 0 & |\omega| > \omega_c \\ M(\omega - \omega_c) & \omega < \omega_c \\ M(\omega + \omega_c) & \omega > -\omega_c \end{cases} \tag{2.50}$$

which is sketched in Fig. 2-22(c). We see that $x_c(t)$ is a lower-sideband SSB signal.

2.11. Show that an SSB signal can be demodulated by the synchronous detector of Fig. 2-23 (a) by sketching the spectrum of the signal at each point and (b) by the time-domain expression of the signals at each point.

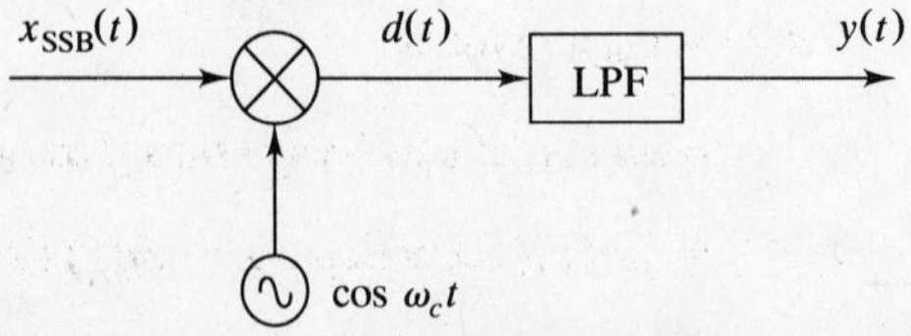

Fig. 2-23 Synchronous detector

(*a*) Let $M(\omega)$, the spectrum of the message $m(t)$, be as shown in Fig. 2-24(*a*). Also assume that $x_{SSB}(t)$ is a lower-sideband SSB signal and its spectrum is $X_{SSB}(\omega)$, as shown in Fig. 2-24(*b*). Multiplication by $\cos \omega_c t$ shifts the spectrum $X_{SSB}(\omega)$ to $\pm\omega_c$ and we obtain $D(\omega)$, the spectrum of $d(t)$ [Fig. 2-24(*c*)]. After low-pass filtering, we obtain $Y(\omega) = \frac{1}{2}M(\omega)$, the spectrum of $y(t)$ [Fig. 2-24(*d*)]. Thus we obtain $y(t) = \frac{1}{2}m(t)$ which is proportional to $m(t)$.

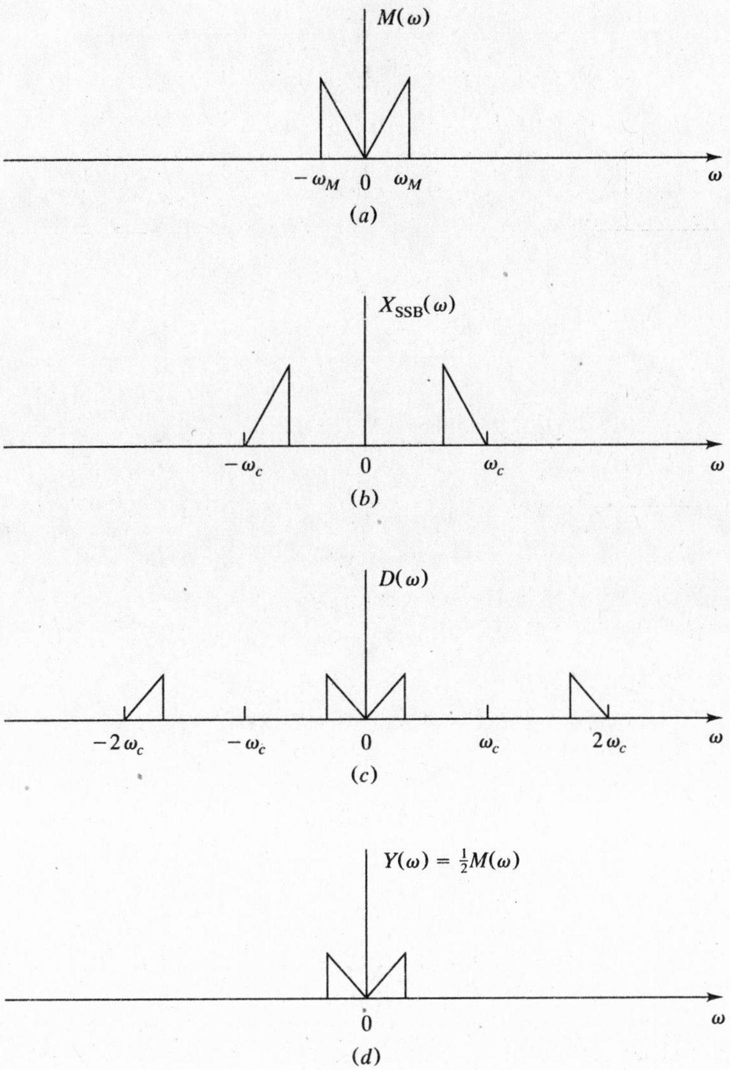

Fig. 2-24

(*b*) From Eq. (*2.11*), $x_{SSB}(t)$ can be expressed as

$$x_{SSB}(t) = m(t)\cos\omega_c t \mp \hat{m}(t)\sin\omega_c t$$

Thus,

$$\begin{aligned} d(t) &= x_{SSB}(t)\cos\omega_c t \\ &= m(t)\cos^2\omega_c t \mp \hat{m}(t)\sin\omega_c t\cos\omega_c t \\ &= \tfrac{1}{2}m(t)(1+\cos 2\omega_c t) \mp \tfrac{1}{2}\hat{m}(t)\sin 2\omega_c t \\ &= \tfrac{1}{2}m(t) + \tfrac{1}{2}m(t)\cos 2\omega_c t \mp \tfrac{1}{2}\hat{m}(t)\sin 2\omega_c t \end{aligned}$$

Hence, after low-pass filtering, we obtain

$$y(t) = \tfrac{1}{2}m(t) \tag{2.51}$$

2.12. Show that the system depicted in Fig. 2-25 can be used to demodulate an SSB signal.

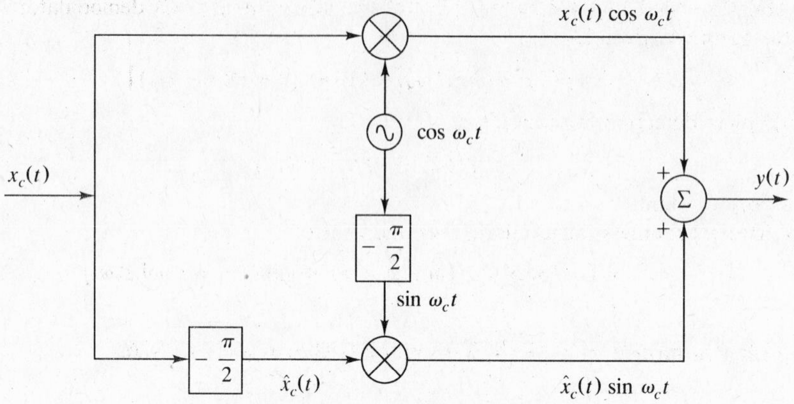

Fig. 2-25 Phase-shift SSB demodulator

From Eq. (*2.43*) of Prob. 2.10, the upper-sideband SSB signal $x_c(t)$ is

$$x_c(t) = m(t)\cos\omega_c t - \hat{m}(t)\sin\omega_c t$$

Then using Eq. (*1.141*) of Prob. 1.47, we have

$$\hat{x}_c(t) = m(t)\sin\omega_c t + \hat{m}(t)\cos\omega_c t \tag{2.52}$$

From Fig. 2-25, we have

$$y(t) = x_c(t)\cos\omega_c t + \hat{x}_c(t)\sin\omega_c t$$
$$= m(t)\left(\cos^2\omega_c t + \sin^2\omega_c t\right) = m(t) \tag{2.53}$$

In a similar manner, the lower-sideband SSB signal $x_c(t)$ is

$$x_c(t) = m(t)\cos\omega_c t + \hat{m}(t)\sin\omega_c t$$

and

$$\hat{x}_c(t) = m(t)\sin\omega_c t - \hat{m}(t)\cos\omega_c t \tag{2.54}$$

Again, from Fig. 2-25 we obtain

$$y(t) = x_c(t)\cos\omega_c t + \hat{x}_c(t)\sin\omega_c t$$
$$= m(t)\left(\cos^2\omega_c t + \sin^2\omega_c t\right) = m(t) \tag{2.55}$$

Note that Fig. 2-25 is exactly the same as Fig. 2-8 except in the addition at the summing junction for both cases.

VESTIGIAL-SIDEBAND MODULATION

2.13. Show that for distortionless demodulation of a VSB signal using the synchronous detector of Fig. 2-11 the frequency response $H(\omega)$ of the VSB filter of Fig. 2-10(*a*) must satisfy Eq. (*2.12*), that is,

$$H(\omega + \omega_c) + H(\omega - \omega_c) = \text{constant} \qquad \text{for } |\omega| \le \omega_M$$

The spectrum of $x_{\text{DSB}}(t)$, $X_{\text{DSB}}(\omega)$, is given by [Fig. 2-10(*c*)]

$$X_{\text{DSB}}(\omega) = M(\omega - \omega_c) + M(\omega + \omega_c) \tag{2.56}$$

From Fig. 2-10(e), the spectrum of $x_{\text{VSB}}(t)$, $X_{\text{VSB}}(\omega)$, is

$$X_{\text{VSB}}(\omega) = [M(\omega - \omega_c) + M(\omega + \omega_c)]H(\omega) \qquad (2.57)$$

Next, $d(t)$ of the VSB demodulator (Fig. 2-11) and its Fourier transform are given by

$$d(t) = x_{\text{VSB}}(t)\cos\omega_c t \longleftrightarrow \tfrac{1}{2}[X_{\text{VSB}}(\omega - \omega_c) + X_{\text{VSB}}(\omega + \omega_c)] \qquad (2.58)$$

Substituting Eq. (2.57) in the above equation and eliminating the spectra at $\pm 2\omega_c$ (suppressed by a low-pass filter), we find that the output $y(t)$ of the synchronous VSB demodulator (Fig. 2-11) and its Fourier transform are given by

$$y(t) \longleftrightarrow \tfrac{1}{2}M(\omega)[H(\omega + \omega_c) + H(\omega - \omega_c)] \qquad (2.59)$$

For distortionless detection, we must have

$$y(t) = km(t) \longleftrightarrow kM(\omega) \qquad (2.60)$$

where k is any constant.

Thus, for distortionless demodulation, we must have

$$H(\omega + \omega_c) + H(\omega - \omega_c) = \text{constant} \qquad \text{for } |\omega| \le \omega_M$$

2.14. The frequency response $H(\omega)$ of a VSB filter is shown in Fig. 2-26.

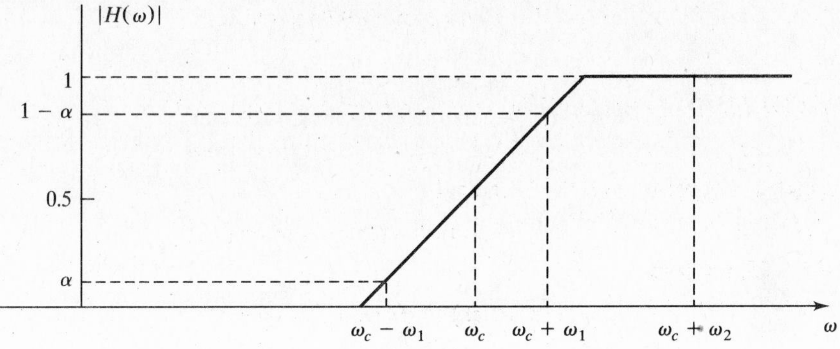

Fig. 2-26

(a) Find the VSB signal $x_{\text{VSB}}(t)$ when

$$m(t) = a_1\cos\omega_1 t + a_2\cos\omega_2 t$$

(b) Show that $x_{\text{VSB}}(t)$ can be demodulated by the synchronous demodulator of Fig. 2-11.

(a) Referring to Fig. 2-10, we have

$$\begin{aligned} x_{\text{DSB}}(t) &= m(t)\cos\omega_c t \\ &= (a_1\cos\omega_1 t + a_2\cos\omega_2 t)\cos\omega_c t \\ &= \tfrac{1}{2}a_1\cos(\omega_c - \omega_1)t + \tfrac{1}{2}a_1\cos(\omega_c + \omega_1)t + \tfrac{1}{2}a_2\cos(\omega_c - \omega_2)t + \tfrac{1}{2}a_2\cos(\omega_c + \omega_2)t \end{aligned}$$

These sinusoids are transmitted through $H(\omega)$, shown in Fig. 2-26, which has gains of 0, α, $1 - \alpha$, and 1 at $\omega_c - \omega_2$, $\omega_c - \omega_1$, $\omega_c + \omega_1$, and $\omega_c + \omega_2$, respectively. Thus the VSB filter output $x_{\text{VSB}}(t)$ is

$$x_{\text{VSB}}(t) = \tfrac{1}{2}a_1\alpha\cos(\omega_c - \omega_1)t + \tfrac{1}{2}a_1(1-\alpha)\cos(\omega_c + \omega_1)t + \tfrac{1}{2}a_2\cos(\omega_c + \omega_2)t \qquad (2.61)$$

(b) Referring to Fig. 2-11, we get

$$\begin{aligned} d(t) &= x_{\text{VSB}}(t)\cos\omega_c t \\ &= \tfrac{1}{4}(a_1\cos\omega_1 t + a_2\cos\omega_2 t) + \tfrac{1}{4}[a_1\alpha\cos(2\omega_c - \omega_1)t \\ &\qquad + a_1(1-\alpha)\cos(2\omega_c + \omega_1)t + a_2\cos(2\omega_c + \omega_2)t] \end{aligned}$$

Using low-pass filtering to eliminate the double-frequency terms, we obtain

$$y(t) = \tfrac{1}{4}(a_1 \cos \omega_1 t + a_2 \cos \omega_2 t) = \tfrac{1}{4}m(t) \tag{2.62}$$

FREQUENCY MIXING AND FDM

2.15. A radio receiver used in the AM system is shown in Fig. 2-27. The mixer translates the carrier frequency f_c to a fixed IF of 455 kHz by using a local oscillator of frequency f_{LO}. The broadcast-band frequencies range from 540 to 1600 kHz.

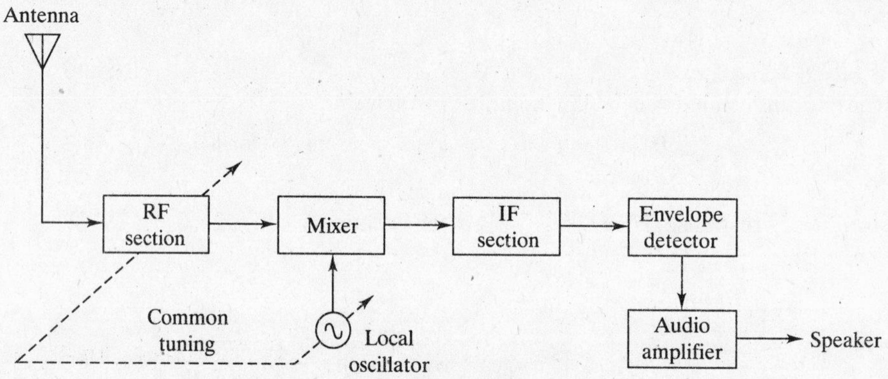

Fig. 2-27 A superheterodyne AM receiver

(*a*) Determine the range of tuning that must be provided in the local oscillator (i) when f_{LO} is higher than f_c (superheterodyne receiver) and (ii) when f_{LO} is lower than f_c.

(*b*) Explain why the usual AM radio receiver uses a superheterodyne system.

(*a*) (i) When $f_{LO} > f_c$,

$$540 < f_c < 1600$$

$$f_{LO} - f_c = 455$$

where both f_c and f_{LO} are expressed in kilohertz. Thus

$$f_{LO} = f_c + 455$$

When $f_c = 540$ kHz, we get $f_{LO} = 995$ kHz; and when $f_c = 1600$ kHz, we get $f_{LO} = 2055$ kHz. Thus the required tuning range of the local oscillator is 995 – 2055 kHz.

(ii) When $f_{LO} < f_c$,

$$f_{LO} = f_c - 455$$

When $f_c = 540$ kHz, we get $f_{LO} = 85$ kHz; and when $f_c = 1600$ kHz, we get $f_{LO} = 1145$ kHz. Thus the required tuning range of the local oscillator for this case is 85 – 1145 kHz.

(*b*) The frequency ratio, that is, the ratio of the highest f_{LO} to the lowest f_{LO}, is 2.07 for case (i) and 13.47 for case (ii). It is much easier to design an oscillator that is tunable over a smaller frequency ratio; that is the reason why the usual AM radio receiver uses the superheterodyne system.

2.16. The spectrum of a message signal $m(t)$ is shown in Fig. 2-28(*a*). To ensure communication privacy, this signal is applied to a system (known as a *scrambler*) shown in Fig. 2-28(*b*). Analyze the system and sketch the spectrum of the output $x(t)$.

The spectrum of the signal at each point is shown in Fig. 2-29. We see that the spectrum of the output $x(t)$, $X(\omega)$, consists of the two reversed lobes of $M(\omega)$.

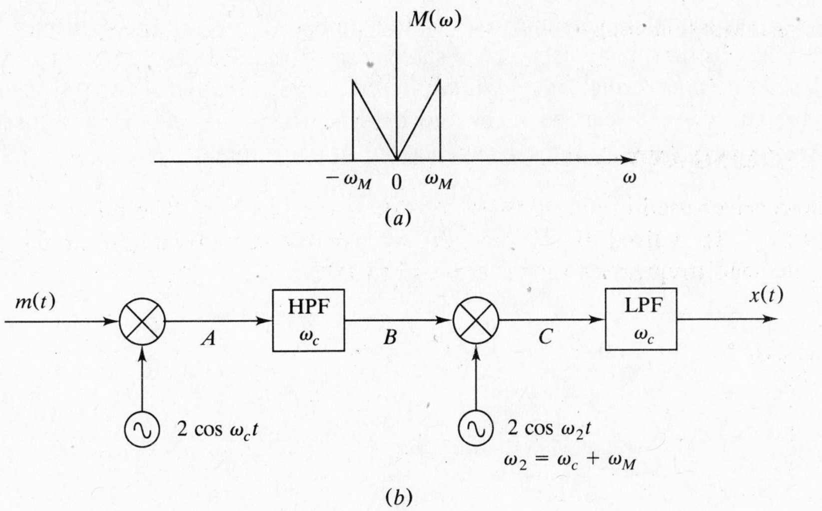

Fig. 2-28

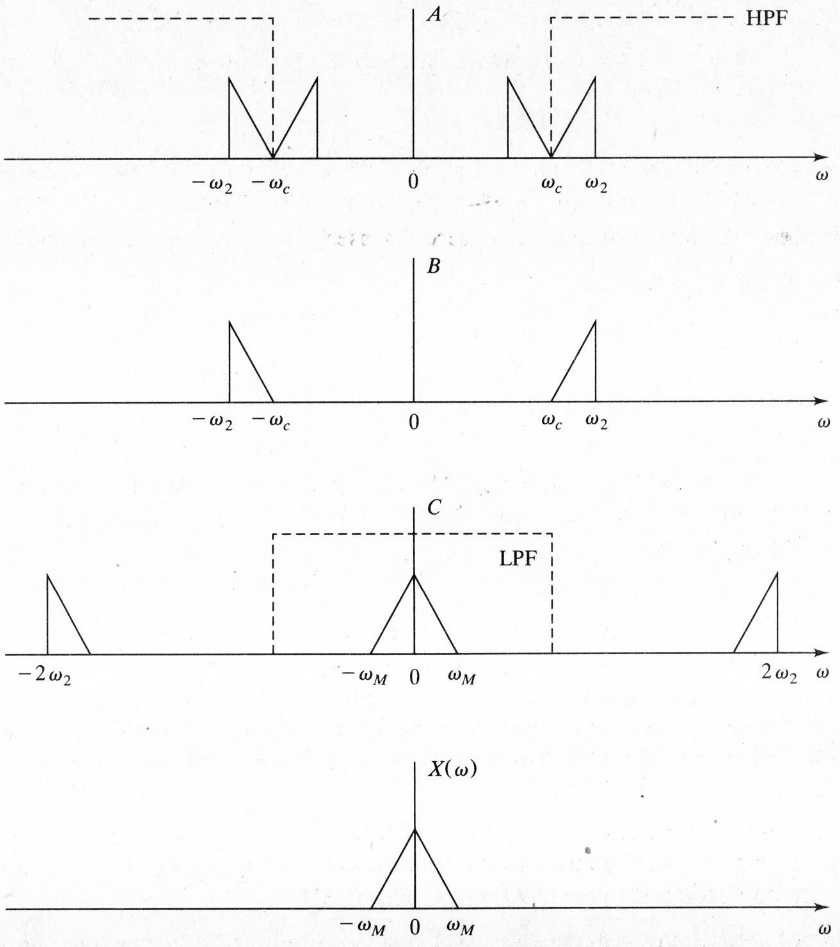

Fig. 2-29

2.17. Using the orthogonality of sine and cosine makes it possible to transmit and receive two different signals simultaneously on the same carrier frequency. A scheme for doing this, known as *quadrature multiplexing*, or *quadrature amplitude modulation* (QAM), is shown in Fig. 2-30. Show that each signal can be recovered by synchronous detection of the received signal by using two local carriers of same frequency but in phase quadrature.

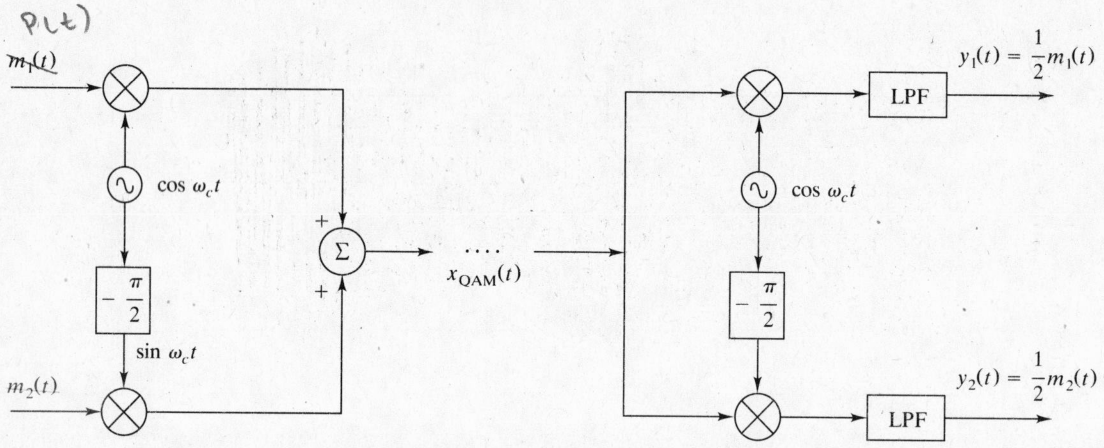

Fig. 2-30 Quadrature multiplexing system

$$x_{QAM}(t) = m_1(t)\cos \omega_c t + m_2(t)\sin \omega_c t$$

$$x_{QAM}(t)\cos \omega_c t = m_1(t)\cos^2 \omega_c t + m_2(t)\sin \omega_c t \cos \omega_c t$$

$$= \tfrac{1}{2}m_1(t) + \tfrac{1}{2}m_1(t)\cos 2\omega_c t + \tfrac{1}{2}m_2(t)\sin 2\omega_c t$$

$$x_{QAM}(t)\sin \omega_c t = m_1(t)\cos \omega_c t \sin \omega_c t + m_2(t)\sin^2 \omega_c t$$

$$= \tfrac{1}{2}m_1(t)\sin 2\omega_c t + \tfrac{1}{2}m_2(t) - \tfrac{1}{2}m_2(t)\cos 2\omega_c t$$

All terms at $2\omega_c$ are filtered out by the low-pass filter, yielding

$$y_1(t) = \tfrac{1}{2}m_1(t) \quad \text{and} \quad y_2(t) = \tfrac{1}{2}m_2(t)$$

Note that quadrature multiplexing is an efficient method of transmitting two message signals within the same bandwidth. It is used in the transmission of color information signals in commercial television broadcasts.

Supplementary Problems

2.18. A signal $m(t)$ is band-limited to ω_M. It is frequency-translated by multiplying it by the signal $\cos \omega_c t$. Find ω_c so that the bandwidth of the transmitted signal is 1 percent of the carrier frequency ω_c.

Ans. $\omega_c = 200\omega_M$

2.19. A sinusoidally modulated ordinary AM waveform is shown in Fig. 2-31. (*a*) Find the modulation index. (*b*) Calculate the efficiency. (*c*) Determine the amplitude of the carrier which must be added to attain a modulation index of 0.1.

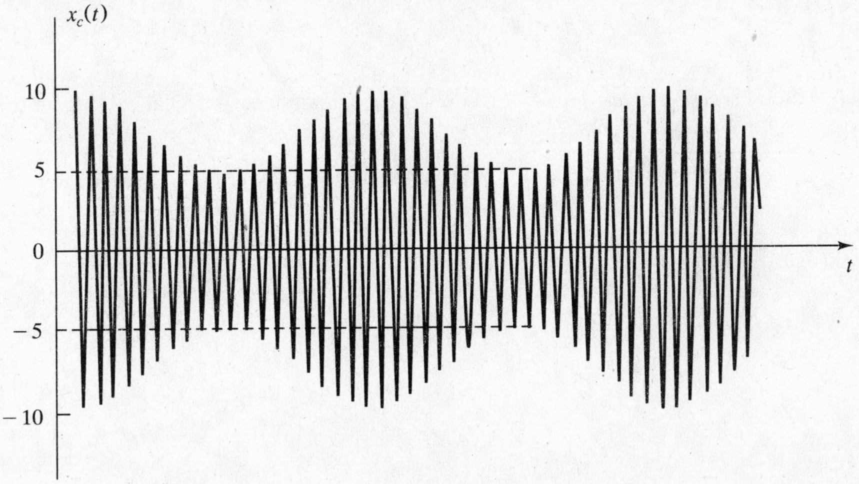

Fig. 2-31

Ans. (*a*) $\mu = \frac{1}{2}$, (*b*) $\eta = 5.26$ percent, (*c*) 17.5 V

2.20. Given a real signal $m(t)$, define a signal

$$M_+(t) = m(t) + j\hat{m}(t)$$

where $\hat{m}(t)$ is the Hilbert transform of $m(t)$ (see Prob. 1.47). And $m_+(t)$ is called an *analytic signal.*

(*a*) Show that

$$\mathscr{F}[m_+(t)] = M_+(\omega) = \begin{cases} 2M(\omega) & \omega > 0 \\ 0 & \omega < 0 \end{cases}$$

(*b*) Show that

$$\text{Re}\left[m_+(t)e^{j\omega_c t}\right]$$

is an upper-sideband SSB signal and

$$\text{Re}\left[m_+(t)e^{-j\omega_c t}\right]$$

is a lower-sideband SSB signal.

Hint: (*a*) Use Eq. (*1.141*) of Prob. 1.47.

(*b*) Use Euler's identity; take the real part of the expression, then compare it with Eqs. (*2.43*) and (*2.44*).

2.21. The frequency response $H(\omega)$ of the VSB filter of Fig. 2-26 (Prob. 2.14) is characterized by

$$H(\omega_c - \omega_1) = \alpha e^{j\phi}$$

$$H(\omega_c + \omega_1) = (1 - \alpha)e^{j\theta_1}$$

$$H(\omega_c + \omega_2) = 1e^{j\theta_2}$$

The message signal is

$$m(t) = a_1 \cos \omega_1 t + a_2 \cos \omega_2 t$$

and is to be demodulated by a synchronous detector. Derive the expressions for θ_1 and θ_2 as functions of ϕ for distortionless demodulation.

Ans. $\theta_1 = -\phi$ and $\theta_2 = -\dfrac{\omega_2}{\omega_1}\phi$

2.22. Design a system, a *descrambler*, that will recover the original message signal $m(t)$ from the output of the scrambler $x(t)$ of Fig. 2-28, Prob. 2.16.

Hint: Consider a system identical to the one shown in Fig. 2-28.

Chapter 3

Angle Modulation

3.1 INTRODUCTION

As we mentioned in Sec. 2.1, *angle modulation* encompasses *phase modulation* (PM) and *frequency modulation* (FM) and refers to the process by which the phase angle of a sinusoidal carrier wave is varied according to the message signal. As we studied in Chap. 2, in amplitude modulation the spectrum of the modulated signal is essentially the translated message spectrum, and the transmission bandwidth never exceeds twice the message bandwidth. In angle modulation, the spectral components of the modulated waveform are not related in any simple fashion to the message spectrum. Furthermore, superposition does not apply (see Prob. 3.4), and the bandwidth of the angle-modulated signal is usually much greater than twice the message bandwidth. The increase in bandwidth and system complexity is compensated for by the improved performance in the face of noise and interference (see Chap. 7).

3.2 ANGLE MODULATION AND INSTANTANEOUS FREQUENCY

For angle modulation, the modulated carrier is represented by [see Eq. (*2.1*)]

$$x_c(t) = A\cos[\omega_c t + \phi(t)] \tag{3.1}$$

where A and ω_c are constants and the phase angle $\phi(t)$ is a function of the message signal $m(t)$.

If we rewrite Eq. (*3.1*) as

$$x_c(t) = A\cos\theta(t) \tag{3.2}$$

where

$$\theta(t) = \omega_c t + \phi(t) \tag{3.3}$$

then we can define the instantaneous radian frequency of $x_c(t)$, denoted by ω_i, as

$$\omega_i = \frac{d\theta(t)}{dt} = \omega_c + \frac{d\phi(t)}{dt} \tag{3.4}$$

Note that when $\phi(t) =$ constant, then $\omega_i = \omega_c$.

The function $\phi(t)$ and $d\phi(t)/dt$ are known as the *instantaneous phase deviation* and *instantaneous frequency deviation* of $x_c(t)$. The quantity $\Delta\omega$ defined by

$$\Delta\omega = |\omega_i - \omega_c|_{\max} \tag{3.5}$$

is called the *maximum* (or *peak*) *radian frequency deviation* of the angle-modulated signal.

3.3 PHASE AND FREQUENCY MODULATION

The two basic types of angle modulation are phase modulation and frequency modulation. In PM, the instantaneous phase deviation of the carrier is proportional to the message signal; that is,

$$\phi(t) = k_p m(t) \tag{3.6}$$

where k_p is the phase deviation constant, expressed in radians per unit of $m(t)$.

In FM, the instantaneous frequency deviation of the carrier is proportional to the message signal; that is,

$$\frac{d\phi(t)}{dt} = k_f m(t) \tag{3.7}$$

or

$$\phi(t) = k_f \int_{t_0}^{t} m(\lambda)\, d\lambda + \phi(t_0) \tag{3.8}$$

where k_f is the frequency deviation constant, expressed in radians per second per unit of $m(t)$, and $\phi(t_0)$ is the initial phase angle at $t = t_0$. It is usually assumed that $t_0 = -\infty$ and $\phi(-\infty) = 0$.

Thus, we can express the angle-modulated signal as

$$x_{\mathrm{PM}}(t) = A \cos\left[\omega_c t + k_p m(t)\right] \tag{3.9}$$

$$x_{\mathrm{FM}}(t) = A \cos\left[\omega_c t + k_f \int_{-\infty}^{t} m(\lambda)\, d\lambda\right] \tag{3.10}$$

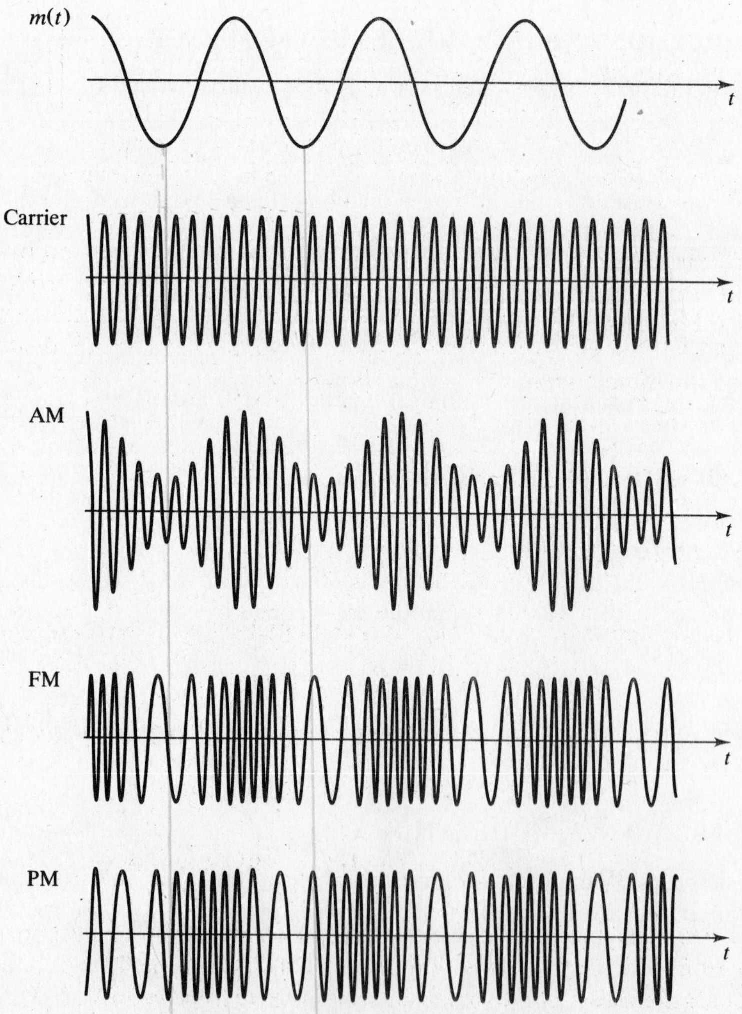

Fig. 3-1 AM, FM, and PM waveforms

From definition (3.4), we have

$$\omega_i = \omega_c + k_p \frac{dm(t)}{dt} \qquad \text{for PM} \tag{3.11}$$

$$\omega_i = \omega_c + k_f m(t) \qquad \text{for FM} \tag{3.12}$$

Thus, in PM, the instantaneous frequency ω_i varies linearly with the derivative of the modulating signal, and in FM, ω_i varies linearly with the modulating signal.

Figure 3-1 illustrates AM, FM, and PM waveforms produced by a sinusoidal message waveform.

3.4 FOURIER SPECTRA OF ANGLE-MODULATED SIGNALS

An angle-modulated carrier can be represented in exponential form by writing Eq. (3.1) as

$$x_c(t) = \text{Re}\left(Ae^{j(\omega_c t + \phi(t))}\right) = \text{Re}\left(Ae^{j\omega_c t}e^{j\phi(t)}\right) \tag{3.13}$$

where Re means the "real part of." Because of this representation the angle modulation is also referred to as the *exponential modulation*.

Expanding $e^{j\phi(t)}$ in a power series yields

$$x_c(t) = \text{Re}\left\{Ae^{j\omega_c t}\left[1 + j\phi(t) - \frac{\phi^2(t)}{2!} - \cdots + j^n\frac{\phi^n(t)}{n!} + \cdots\right]\right\}$$

$$= A\left[\cos\omega_c t - \phi(t)\sin\omega_c t - \frac{\phi^2(t)}{2!}\cos\omega_c t + \frac{\phi^3(t)}{3!}\sin\omega_c t + \cdots\right] \tag{3.14}$$

Thus the angle-modulated signal consists of an unmodulated carrier plus various amplitude-modulated terms, such as $\phi(t)\sin\omega_c t$, $\phi^2(t)\cos\omega_c t$, $\phi^3(t)\sin\omega_c t, \ldots,$ etc. Hence its Fourier spectrum consists of an unmodulated carrier plus spectra of $\phi(t), \phi^2(t), \phi^3(t), \ldots,$ etc., centered at ω_c.

It is clear that the Fourier spectrum of an angle-modulated signal is not related to the message signal spectrum in any simple way, as was the case in AM.

3.5 NARROWBAND ANGLE MODULATION

If

$$|\phi(t)|_{\max} \ll 1 \tag{3.15}$$

then Eq. (3.14) can be approximated by [neglecting all higher-power terms of $\phi(t)$]

$$x_c(t) \approx A\cos\omega_c t - A\phi(t)\sin\omega_c t \tag{3.16}$$

The signal represented by Eq. (3.16) is called the *narrowband* (NB) angle-modulated signal. Thus,

$$x_{\text{NBPM}}(t) \approx A\cos\omega_c t - Ak_p m(t)\sin\omega_c t \tag{3.17}$$

$$x_{\text{NBFM}}(t) \approx A\cos\omega_c t - A\left[k_f\int_{-\infty}^{t} m(\lambda)\,d\lambda\right]\sin\omega_c t \tag{3.18}$$

Equation (3.16) indicates that a narrowband angle-modulated signal contains an unmodulated carrier plus a term in which $\phi(t)$ [a function of $m(t)$] multiplies a $\pi/2$ (rad) phase-shifted carrier. This multiplication generates a pair of sidebands, and if $\phi(t)$ has a bandwidth W_B, the bandwidth of an NB angle-modulated signal is $2W_B$. This is reminiscent of AM.

3.6 SINUSOIDAL (OR TONE) MODULATION

A. Modulation Index:

If the message signal $m(t)$ is a pure sinusoid, that is,

$$m(t) = \begin{cases} a_m \sin \omega_m t & \text{for PM} \\ a_m \cos \omega_m t & \text{for FM} \end{cases} \tag{3.19}$$

then Eqs. (3.6) and (3.8) both give

$$\phi(t) = \beta \sin \omega_m t \tag{3.20}$$

where

$$\beta = \begin{cases} k_p a_m & \text{for PM} \\ \dfrac{k_f a_m}{\omega_m} & \text{for FM} \end{cases} \tag{3.21}$$

The parameter β is known as the *modulation index* for angle modulation and is the maximum value of phase deviation for both PM and FM. Note that β is defined only for sinusoidal modulation and it can be expressed as

$$\beta = \frac{\Delta\omega}{\omega_m} \tag{3.22}$$

where $\Delta\omega$ is the maximum frequency deviation, defined in Eq. (3.5).

B. Fourier Spectrum:

Substituting Eq. (3.20) into Eq. (3.1), we obtain

$$x_c(t) = A \cos(\omega_c t + \beta \sin \omega_m t) \tag{3.23}$$

which is the angle-modulated signal with sinusoidal modulation. It can be shown by the use of the Fourier series that this signal can also be written as (see Prob. 3.5)

$$x_c(t) = A \sum_{n=-\infty}^{\infty} J_n(\beta) \cos(\omega_c + n\omega_m)t \tag{3.24}$$

where $J_n(\beta)$ is the Bessel function of the first kind of order n and argument β. Table B-1 (in App. B) lists some selected values of $J_n(\beta)$. From Eq. (3.24) and Table B-1, we observe that

1. The spectrum consists of a carrier-frequency component plus an infinite number of sideband components at frequencies $\omega_c \pm n\omega_m$ $(n = 1, 2, 3, \ldots)$.
2. The relative amplitudes of the spectral lines depend on the value of $J_n(\beta)$, and the value of $J_n(\beta)$ becomes very small for large values of n.
3. The number of significant spectral lines (that is, having appreciable relative amplitude) is a function of the modulation index β. With $\beta \ll 1$, only J_0 and J_1 are significant, so the spectrum will consist of carrier and two sideband lines. But if $\beta \gg 1$, there will be many sideband lines.

Figure 3-2 shows the amplitude spectra of angle-modulated signals for several values of β.

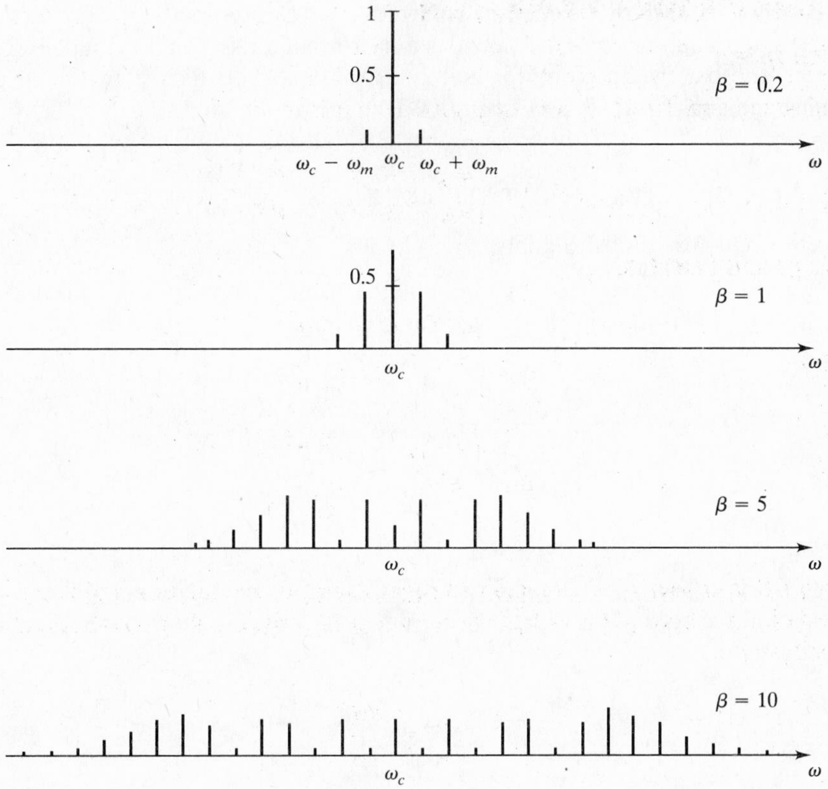

Fig. 3-2 Amplitude spectra of sinusoidally modulated FM signals (ω_m fixed)

3.7 BANDWIDTH OF ANGLE-MODULATED SIGNALS

A. Sinusoidal Modulation:

From Fig. 3-2 and Table B-1 we see that the bandwidth of the angle-modulated signal with sinusoidal modulation depends on β and ω_m. In fact, it can be shown that 98 percent of the normalized total signal power is contained in the bandwidth

$$W_B \approx 2(\beta + 1)\omega_m \tag{3.25}$$

When $\beta \ll 1$, the signal is an NB angle-modulated signal and its bandwidth is approximately equal to $2\omega_m$. Usually a value of $\beta < 0.2$ is taken to be sufficient to satisfy this condition. All the bandwidths can be expressed in hertz (Hz) simply by replacing $\Delta\omega$ with Δf and ω_m with f_m.

B. Arbitrary Modulation:

For an angle-modulated signal with an arbitrary modulating signal $m(t)$ band-limited to ω_M radians per second (rad/s), we define the *deviation ratio D* as

$$D = \frac{\text{maximum frequency deviation}}{\text{bandwidth of } m(t)} = \frac{\Delta\omega}{\omega_M} \tag{3.26}$$

The deviation ratio D plays the same role for arbitrary modulation as the modulation index β plays for sinusoidal modulation. Replacing β by D and ω_m by ω_M in Eq. (*3.25*), we have

$$W_B \approx 2(D + 1)\omega_M \tag{3.27}$$

This expression for bandwidth is generally referred to as *Carson's rule*. If $D \ll 1$, the bandwidth is approximately $2\omega_M$, and the signal is known as a narrowband (NB) angle-modulated signal (see Sec. 3.4). If $D \gg 1$, the bandwidth is approximately $2D\omega_M = 2\Delta\omega$, which is twice the peak frequency deviation. Such a signal is called a *wideband* (WB) angle-modulated signal.

3.8 GENERATION OF ANGLE-MODULATED SIGNALS

A. Narrowband Angle-Modulated Signals:

The generation of narrowband angle-modulated signals is easily accomplished in view of Eq. (*3.16*) or Eqs. (*3.17*) and (*3.18*). This is illustrated in Fig. 3-3.

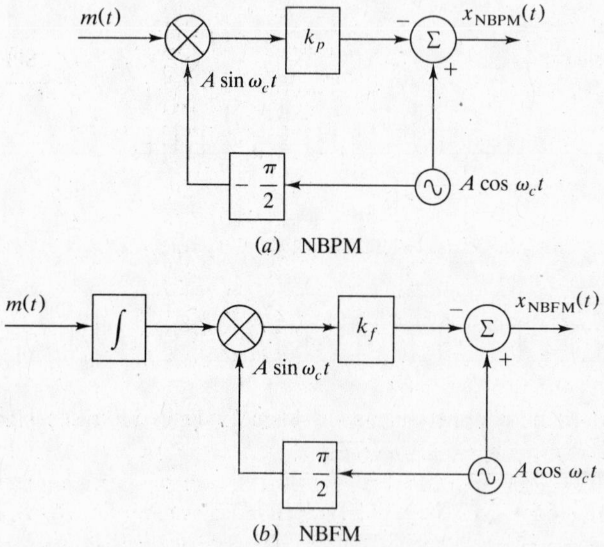

Fig. 3-3 Generation of narrowband angle-modulated signals

B. Wideband Angle-Modulated Signals:

There are two methods of generating wideband (WB) angle-modulated signals; the indirect method and the direct method.

1. Indirect Method:

In this method, an NB angle-modulated signal is produced first (see Fig. 3-3) and then converted to a WB angle-modulated signal by using frequency multipliers (Fig. 3-4). The frequency multiplier multiplies the argument of the input sinusoid by n. Thus, if the input of a frequency multiplier is

$$x(t) = A\cos\left[\omega_c t + \phi(t)\right] \tag{3.28}$$

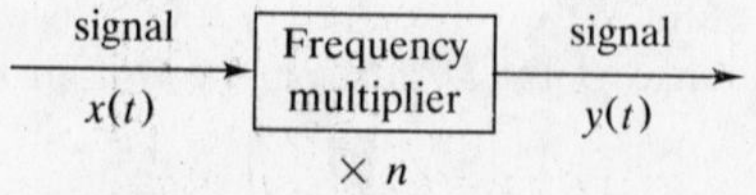

Fig. 3-4 Frequency multiplier

then the output of the frequency multiplier is

$$y(t) = A \cos [n\omega_c t + n\phi(t)] \tag{3.29}$$

Use of frequency multiplication normally increases the carrier frequency to an impractically high value. To avoid this, a frequency conversion (using a mixer or DSB modulator) is necessary (Fig. 3-5) to shift the spectrum.

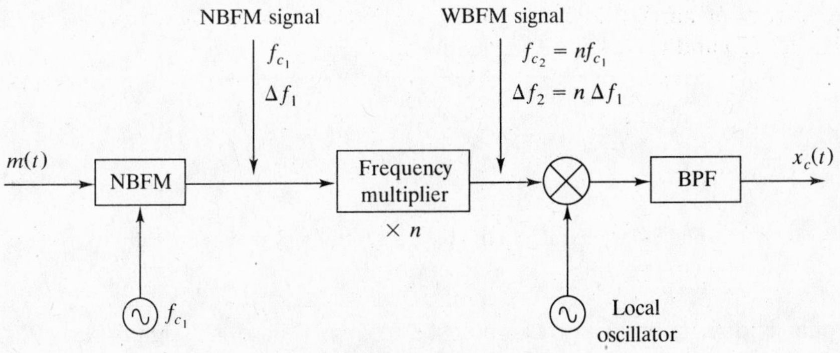

Fig. 3-5 NB-to-WB conversion

2. *Direct Method:*

In the direct method of generating an FM signal, the modulating signal directly controls the carrier frequency. A common method used for generating FM directly is to vary the inductance or capacitance of a tuned electric oscillator. Any oscillator whose frequency is controlled by the modulating signal voltage is called a *voltage controlled oscillator* (VCO). The main advantage of direct FM is that large frequency deviations are possible and thus less frequency multiplication is required. The major disadvantage is that the carrier frequency tends to drift and so additional circuitry is required for frequency stabilization.

3.9 DEMODULATION OF ANGLE-MODULATED SIGNALS

Demodulation of an FM signal requires a system that produces an output proportional to the instantaneous frequency deviation of the input signal. Such a system is called a *frequency discriminator*. If the input to an ideal discriminator is an angle-modulated signal

$$x_c(t) = A \cos [\omega_c t + \phi(t)]$$

then the output of the discriminator is

$$y_d(t) = k_d \frac{d\phi(t)}{dt} \tag{3.30}$$

where k_d is the discriminator sensitivity.

For FM, $\phi(t)$ is given by [Eq. (3.8)]

$$\phi(t) = k_f \int_{-\infty}^{t} m(\lambda)\, d\lambda$$

so that Eq. (*3.30*) becomes

$$y_d(t) = k_d k_f m(t) \tag{3.31}$$

The characteristics of an ideal frequency discriminator are shown in Fig. 3-6.

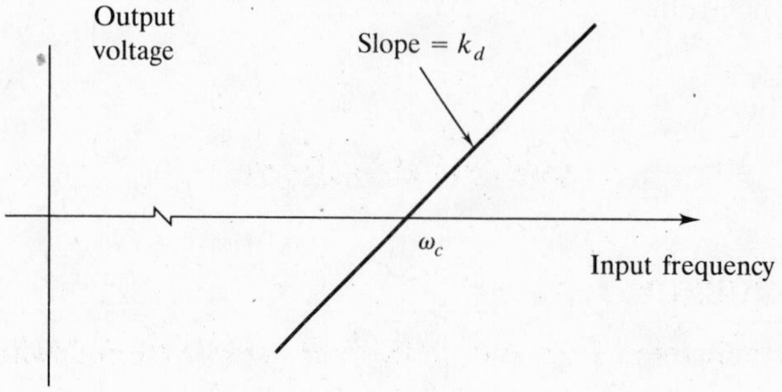

Fig. 3-6 Characteristics of ideal frequency discriminator

The frequency discriminator also can be used to demodulate PM signals. For PM, $\phi(t)$ is given by [Eq. (*3.6*)]

$$\phi(t) = k_p m(t)$$

Then $y_d(t)$, given by Eq. (*3.30*), becomes

$$y_d(t) = k_d k_p \frac{dm(t)}{dt} \tag{3.32}$$

Integration of the discriminator output yields a signal which is proportional to $m(t)$. A demodulator for PM can therefore be implemented as an FM demodulator followed by an integrator.

A simple approximation to the ideal discriminator is an ideal differentiator followed by an envelope detector (Fig. 3-7). If the input to the differentiator is

$$x_c(t) = A\cos\left[\omega_c t + \phi(t)\right]$$

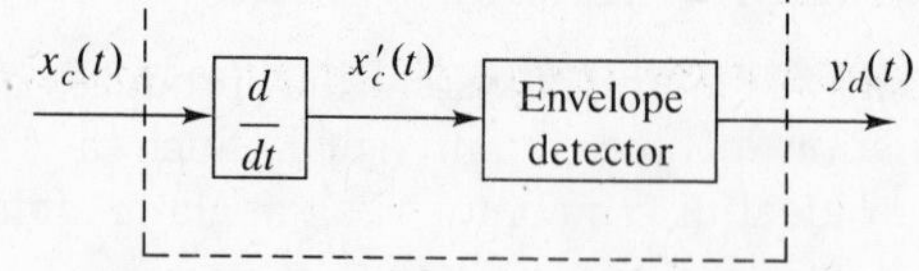

Fig. 3-7 Frequency discriminator

then the output of the differentiator is

$$x'_c(t) = -A\left[\omega_c + \frac{d\phi(t)}{dt}\right]\sin\left[\omega_c t + \phi(t)\right] \tag{3.33}$$

The signal $x'_c(t)$ is both amplitude- and angle-modulated. The envelope of $x'_c(t)$ is

$$A\left[\omega_c + \frac{d\phi(t)}{dt}\right]$$

The output of the envelope detector is, by Eq. (3.4),

$$y_d(t) = \omega_i \tag{3.34}$$

which is the instantaneous frequency of the $x_c(t)$.

There are many other techniques that can be used to implement a frequency discriminator. (See Probs. 3.20, 3.21, and 3.22.)

Solved Problems

INSTANTANEOUS FREQUENCY

3.1. Determine the instantaneous frequency in hertz of each of the following signals:

(a) $10\cos\left(200\pi t + \frac{\pi}{3}\right)$

(b) $10\cos(20\pi t + \pi t^2)$

(c) $\cos 200\pi t \cos(5\sin 2\pi t) + \sin 200\pi t \sin(5\sin 2\pi t)$

(a)
$$\theta(t) = 200\pi t + \frac{\pi}{3}$$

$$\omega_i = \frac{d\theta}{dt} = 200\pi = 2\pi(100)$$

The instantaneous frequency of the signal is 100 Hz, which is constant.

(b)
$$\theta(t) = 20\pi t + \pi t^2$$

$$\omega_i = \frac{d\theta}{dt} = 20\pi + 2\pi t = 2\pi(10 + t)$$

The instantaneous frequency of the signal is 10 Hz at $t = 0$ and increases linearly at a rate of 1 Hz/s.

(c)
$$\cos 200\pi t \cos(5\sin 2\pi t) + \sin 200\pi t \sin(5\sin 2\pi t) = \cos(200\pi t - 5\sin 2\pi t)$$

$$\theta(t) = 200\pi t - 5\sin 2\pi t$$

$$\omega_i = \frac{d\theta}{dt} = 200\pi - 10\pi\cos 2\pi t = 2\pi(100 - 5\cos 2\pi t)$$

The instantaneous frequency of the signal is 95 Hz at $t = 0$ and oscillates sinusoidally between 95 and 105 Hz.

3.2. Consider an angle-modulated signal

$$x_c(t) = 10\cos\left[(10^8)\pi t + 5\sin 2\pi(10^3)t\right]$$

Find the maximum phase deviation and the maximum frequency deviation.

Comparing the given $x_c(t)$ with Eq. (3.1), we have

$$\theta(t) = \omega_c t + \phi(t) = (10^8)\pi t + 5\sin 2\pi(10^3)t$$

and
$$\phi(t) = 5\sin 2\pi(10^3)t$$

Now
$$\phi'(t) = 5(2\pi)(10^3)\cos 2\pi(10^3)t$$

Thus, the maximum phase deviation is

$$|\phi(t)|_{\max} = 5 \text{ rad}$$

and the maximum frequency deviation is

$$\Delta\omega = |\phi'(t)|_{\max} = 5(2\pi)(10^3) \text{ rad/s}$$

or

$$\Delta f = 5 \text{ kHz}$$

PHASE MODULATION AND FREQUENCY MODULATION

3.3. An angle-modulated signal is described by

$$x_c(t) = 10\cos\left[2\pi(10^6)t + 0.1\sin(10^3)\pi t\right]$$

(*a*) Considering $x_c(t)$ as a PM signal with $k_p = 10$, find $m(t)$.

(*b*) Considering $x_c(t)$ as an FM signal with $k_f = 10\pi$, find $m(t)$.

(*a*)

$$\begin{aligned} x_{\text{PM}}(t) &= A\cos\left[\omega_c t + k_p m(t)\right] \\ &= 10\cos\left[2\pi(10^6)t + 10m(t)\right] \\ &= 10\cos\left[2\pi(10^6)t + 0.1\sin(10^3)\pi t\right] \end{aligned}$$

Thus

$$m(t) = 0.01\sin(10^3)\pi t$$

(*b*)

$$\begin{aligned} x_{\text{FM}}(t) &= A\cos\left[\omega_c t + k_f\int_{-\infty}^{t} m(\lambda)\,d\lambda\right] \\ &= 10\cos\left[2\pi(10^6)t + 0.1\sin(10^3)\pi t\right] \end{aligned}$$

Assuming

$$m(t) = a_m\cos(10^3)\pi t$$

we get

$$\begin{aligned} 10\pi\int_{-\infty}^{t} m(\lambda)\,d\lambda &= 10\pi a_m\int_{-\infty}^{t}\cos(10^3)\pi\lambda\,d\lambda \\ &= \frac{a_m}{100}\sin(10^3)\pi t = 0.1\sin(10^3)\pi t \end{aligned}$$

Thus $a_m = 10$, and

$$m(t) = 10\cos(10^3)\pi t$$

3.4. Let $m_1(t)$ and $m_2(t)$ be two message signals, and let $x_{c_1}(t)$ and $x_{c_2}(t)$ be the modulated signals corresponding to $m_1(t)$ and $m_2(t)$, respectively.

(*a*) Show that if the modulation is DSB (AM), then $m_1(t) + m_2(t)$ will produce a modulated signal equal to $x_{c_1}(t) + x_{c_2}(t)$. (This is why AM is sometimes referred to as a *linear* modulation.)

(*b*) Show that if the modulation is PM, then the modulated signal produced by $m_1(t) + m_2(t)$ will not be $x_{c_1}(t) + x_{c_2}(t)$, that is, superposition does not apply to angle-modulated signal. (This is why angle modulation is sometimes referred to as a *nonlinear* modulation.)

(*a*) For DSB (AM), from Eq. (*2.3*) we have

$$m_1(t) \longrightarrow x_{c_1}(t) = m_1(t) \cos \omega_c t$$

$$m_2(t) \longrightarrow x_{c_2}(t) = m_2(t) \cos \omega_c t$$

$$\begin{aligned} m_1(t) + m_2(t) \longrightarrow x_c(t) &= [m_1(t) + m_2(t)] \cos \omega_c t \\ &= m_1(t) \cos \omega_c t + m_2(t) \cos \omega_c t \\ &= x_{c_1}(t) + x_{c_2}(t) \end{aligned}$$

Hence DSB (AM) modulation is a linear modulation.

(*b*) For PM, by Eq. (*3.9*) we have

$$m_1(t) \longrightarrow x_{c_1}(t) = A \cos [\omega_c t + k_p m_1(t)]$$

$$m_2(t) \longrightarrow x_{c_2}(t) = A \cos [\omega_c t + k_p m_2(t)]$$

$$\begin{aligned} m_1(t) + m_2(t) \longrightarrow x_c(t) &= A \cos \{\omega_c t + k_p[m_1(t) + m_2(t)]\} \\ &\neq x_{c_1}(t) + x_{c_2}(t) \end{aligned}$$

Hence PM is a nonlinear modulation.

3.5. Derive Eq. (*3.24*).

In a sinusoidal angle modulation, the modulated signal [Eq. (*3.23*)]

$$x_c(t) = A \cos(\omega_c t + \beta \sin \omega_m t)$$

can be expressed as

$$x_c(t) = A \operatorname{Re}(e^{j\omega_c t} e^{j\beta \sin \omega_m t}) \tag{3.35}$$

The function $e^{j\beta \sin \omega_m t}$ is clearly a periodic function with period $T_m = 2\pi/\omega_m$. It therefore has a Fourier series representation

$$e^{j\beta \sin \omega_m t} = \sum_{n=-\infty}^{\infty} c_n e^{jn\omega_m t}$$

By Eq. (*1.19*), the Fourier coefficients c_n can be found to be

$$c_n = \frac{\omega_m}{2\pi} \int_{-\pi/\omega_m}^{\pi/\omega_m} e^{j\beta \sin \omega_m t} e^{-jn\omega_m t}\, dt$$

Setting $\omega_m t = x$, we have

$$c_n = \frac{1}{2\pi} \int_{-\pi}^{\pi} e^{j(\beta \sin x - nx)}\, dx = J_n(\beta)$$

where $J_n(\beta)$ is the Bessel function of the first kind of order n and argument β (see App. B).
Thus

$$e^{j\beta \sin \omega_m t} = \sum_{n=-\infty}^{\infty} J_n(\beta) e^{jn\omega_m t} \tag{3.36}$$

Substituting Eq. (*3.36*) into Eq. (*3.35*), we obtain

$$\begin{aligned} x_c(t) &= A \operatorname{Re}\left[e^{j\omega_c t} \sum_{n=-\infty}^{\infty} J_n(\beta) e^{jn\omega_m t}\right] \\ &= A \operatorname{Re}\left[\sum_{n=-\infty}^{\infty} J_n(\beta) e^{j(\omega_c + n\omega_m)t}\right] \end{aligned}$$

Taking the real part yields

$$x_c(t) = A \sum_{n=-\infty}^{\infty} J_n(\beta)\cos(\omega_c + n\omega_m)t$$

3.6. Find the normalized average power in an angle-modulated signal with sinusoidal modulation.

From Eq. (*3.24*), an angle-modulated signal with a single-tone modulation can be expressed as

$$x_c(t) = \sum_{n=-\infty}^{\infty} AJ_n(\beta)\cos(\omega_c + n\omega_m)t$$

The normalized average power in $x_c(t)$ is given by

$$P = \sum_{n=-\infty}^{\infty} \frac{1}{2}A^2J_n^2(\beta) = \frac{1}{2}A^2 \sum_{n=-\infty}^{\infty} J_n^2(\beta) = \frac{1}{2}A^2 \tag{3.37}$$

since

$$\sum_{n=-\infty}^{\infty} J_n^2(\beta) = 1 \tag{3.38}$$

FOURIER SPECTRA OF ANGLE-MODULATED SIGNALS

3.7. A carrier is angle-modulated by the sum of two sinusoids

$$x_c(t) = A\cos(\omega_c t + \beta_1 \sin\omega_1 t + \beta_2 \sin\omega_2 t) \tag{3.39}$$

where ω_1 and ω_2 are not harmonically related. Find the spectrum of $x_c(t)$.

In a manner similar to Prob. 3.5, $x_c(t)$ can be expressed as

$$x_c(t) = A\,\mathrm{Re}\left(e^{j\omega_c t}e^{j(\beta_1 \sin\omega_1 t + \beta_2 \sin\omega_2 t)}\right)$$

$$= A\,\mathrm{Re}\left(e^{j\omega_c t}e^{j\beta_1 \sin\omega_1 t}e^{j\beta_2 \sin\omega_2 t}\right) \tag{3.40}$$

Using Eq. (*3.36*), we have

$$e^{j\beta_1 \sin\omega_1 t} = \sum_{n=-\infty}^{\infty} J_n(\beta_1)e^{jn\omega_1 t}$$

$$e^{j\beta_2 \sin\omega_2 t} = \sum_{m=-\infty}^{\infty} J_m(\beta_2)e^{jm\omega_2 t}$$

Substituting these expressions into Eq. (*3.40*) and taking the real part, we obtain

$$x_c(t) = A \sum_{n=-\infty}^{\infty} \sum_{m=-\infty}^{\infty} J_n(\beta_1)J_m(\beta_2)\cos(\omega_c + n\omega_1 + m\omega_2)t \tag{3.41}$$

From Eq. (*3.41*) we see that the spectrum of $x_c(t)$ consists of four categories: (1) the carrier line; (2) sideband lines at $\omega_c + n\omega_1$ due to one tone alone; (3) sideband lines at $\omega_c \pm m\omega_2$ due to the other tone alone; and (4) sideband lines at $\omega_c \pm n\omega_1 \pm m\omega_2$ due to the nonlinear property of angle modulation.

3.8. In a tone-modulated angle modulation, the modulated signal $x_c(t)$ is [see Eq. (*3.23*)]

$$x_c(t) = A\cos(\omega_c t + \beta \sin\omega_m t)$$

When $\beta \ll 1$, we have NB angle modulation.

(*a*) Find the spectrum of this NB angle-modulated signal.

(*b*) Compare the result with that of a tone-modulated AM signal.

(*c*) Discuss the similarities and differences by drawing their phasor representations.

(*a*)
$$x_c(t) = A\cos(\omega_c t + \beta \sin \omega_m t)$$
$$= A\cos\omega_c t \cos(\beta \sin\omega_m t) - A\sin\omega_c t \sin(\beta\sin\omega_m t)$$

When $\beta \ll 1$, we can write

$$\cos(\beta\sin\omega_m t) \approx 1$$
$$\sin(\beta\sin\omega_m t) \approx \beta\sin\omega_m t$$

Then the NB signal can be approximated by

$$x_{\mathrm{NB}c}(t) \approx A\cos\omega_c t - \beta A \sin\omega_m t \sin\omega_c t$$
$$= A\cos\omega_c t - \frac{\beta A}{2}\cos(\omega_c - \omega_m)t + \frac{\beta A}{2}\cos(\omega_c + \omega_m)t \tag{3.42}$$

Note that Eq. (*3.42*) also can be easily obtained from Eq. (*3.16*) by letting $\phi(t) = \beta\sin\omega_m t$.

From Eq. (*3.42*) we see that the spectrum of $x_{\mathrm{NB}c}(t)$ consists of a carrier line and a pair of side lines at $\omega_c \pm \omega_m$.

(*b*) The above result is almost identical to the situation for a tone-modulated AM signal given by [see Eq. (*2.19*) of Prob. 2.4]

$$x_{\mathrm{AM}}(t) = A\cos\omega_c t + \mu A\cos\omega_m t \cos\omega_c t$$
$$= A\cos\omega_c t + \frac{\mu A}{2}\cos(\omega_c - \omega_m)t + \frac{\mu A}{2}\cos(\omega_c + \omega_m)t \tag{3.43}$$

where μ is the modulation index for AM.

Comparison of Eqs. (*3.42*) and (*3.43*) shows that the main difference between NB angle modulation and AM is the phase reversal of the lower sideband component.

(*c*) By using Eq. (*3.35*) and

$$e^{j\beta\sin\omega_m t} \approx 1 + j\beta\sin\omega_m t \qquad \text{for } \beta \ll 1$$

Eq. (*3.42*) can be written in phasor form as

$$x_{\mathrm{NB}c}(t) = \mathrm{Re}\left[Ae^{j\omega_c t}(1 + j\beta\sin\omega_m t)\right]$$
$$= \mathrm{Re}\left[Ae^{j\omega_c t}\left(1 + \frac{\beta}{2}e^{j\omega_m t} - \frac{\beta}{2}e^{-j\omega_m t}\right)\right] \tag{3.44}$$

Similarly, Eq. (*3.43*) can be written in phasor form as

$$x_{\mathrm{AM}}(t) = \mathrm{Re}\left[Ae^{j\omega_c t}(1 + \mu\cos\omega_m t)\right]$$
$$= \mathrm{Re}\left[Ae^{j\omega_c t}\left(1 + \frac{\mu}{2}e^{j\omega_m t} + \frac{\mu}{2}e^{-j\omega_m t}\right)\right] \tag{3.45}$$

By taking the term $Ae^{j\omega_c t}$ as the reference, the phasor representations of Eqs. (*3.44*) and (*3.45*) are shown in Fig. 3-8. From Fig. 3-8, the difference between Eqs. (*3.44*) and (*3.45*) is obvious. In NB angle modulation, the modulation is added in quadrature with the carrier, which results in phase

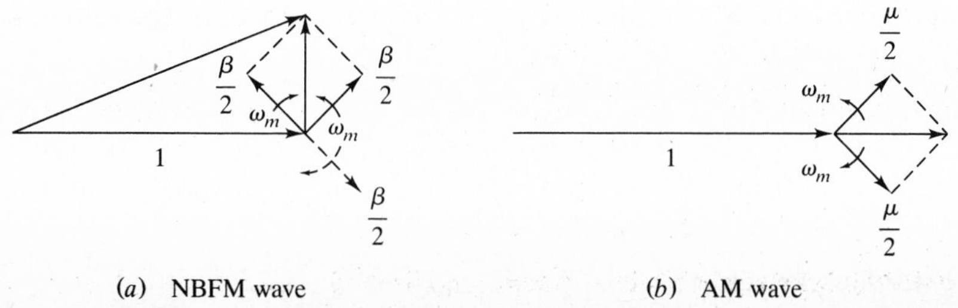

Fig. 3-8 Phasor representation

fluctuation with little amplitude change. In the AM case, the modulation is added in phase with the carrier, producing amplitude fluctuation with no phase deviation.

BANDWIDTH OF ANGLE-MODULATED SIGNALS

3.9. Given the angle-modulated signal

$$x_c(t) = 10\cos(2\pi 10^8 t + 200\cos 2\pi 10^3 t)$$

what is its bandwidth?

The instantaneous frequency is

$$\omega_i = 2\pi(10^8) - 4\pi(10^5)\sin 2\pi(10^3)t$$

So $\Delta\omega = 4\pi(10^5)$, $\omega_m = 2\pi(10^3)$, and

$$\beta = \frac{\Delta\omega}{\omega_m} = \frac{4\pi(10^5)}{2\pi(10^3)} = 200$$

By Eq. (*3.25*),

$$W_B = 2(\beta + 1)\omega_m = 8.04\pi(10^5) \text{ rad/s}$$

Since $\beta \gg 1$,

$$W_B \approx 2\,\Delta\omega = 8\pi(10^5) \text{ rad/s} \qquad \text{or} \qquad f_B = 400 \text{ kHz}$$

3.10. A 20-megahertz (MHz) carrier is frequency-modulated by a sinusoidal signal such that the maximum frequency deviation is 100 kHz. Determine the modulation index and the approximate bandwidth of the FM signal if the frequency of the modulating signal is

(*a*) 1 kHz, (*b*) 100 kHz, and (*c*) 500 kHz.

$$\Delta f = 100 \text{ kHz}, \qquad f_c = 20 \text{ MHz} \gg f_m$$

For sinusoidal modulation, $\beta = \Delta f / f_m$.

(*a*) With $f_m = 1$ kHz, $\beta = 100$. This is a WBFM signal, and $f_B \approx 2\,\Delta f = 200$ kHz.

(*b*) With $f_m = 100$ kHz, $\beta = 1$. Thus, by Eq. (*3.25*),

$$f_B \approx 2(\beta + 1) f_m = 400 \text{ kHz}$$

(*c*) With $f_m = 500$ kHz, $\beta = 0.2$. This is an NBFM signal, and $f_B \approx 2 f_m = 1000$ kHz $= 1$ MHz.

3.11. Consider an angle-modulated signal

$$x_c(t) = 10\cos(\omega_c t + 3\sin\omega_m t)$$

Assume PM and $f_m = 1$ kHz. Calculate the modulation index and find the bandwidth when (*a*) f_m is doubled and (*b*) f_m is decreased by one-half.

$$x_{\text{PM}}(t) = A\cos\left[\omega_c t + k_p m(t)\right] = 10\cos(\omega_c t + 3\sin\omega_m t)$$

Thus, $m(t) = a_m \sin\omega_m t$, and

$$x_{\text{PM}}(t) = 10\cos(\omega_c t + k_p a_m \sin\omega_m t)$$

From Eq. (*3.21*) or Eq. (*3.23*),

$$\beta = k_p a_m = 3$$

We see that the value of β is independent of f_m. By Eq. (3.25), when $f_m = 1$ kHz,

$$f_B = 2(\beta + 1)f_m = 8 \text{ kHz}$$

(a) When f_m is doubled, $\beta = 3$, $f_m = 2$ kHz, and

$$f_B = 2(3 + 1)2 = 16 \text{ kHz}$$

(b) When f_m is decreased by one-half, $\beta = 3$, $f_m = 0.5$ kHz, and

$$f_B = 2(3 + 1)(0.5) = 4 \text{ kHz}$$

3.12. Repeat Prob. 3.11 when FM is assumed.

$$x_{FM}(t) = A\cos\left[\omega_c t + k_f \int_{-\infty}^{t} m(\lambda)\,d\lambda\right] = 10\cos(\omega_c t + 3\sin\omega_m t)$$

Thus $m(t) = a_m \cos\omega_m t$ and

$$x_{FM}(t) = 10\cos\left(\omega_c t + \frac{a_m k_f}{\omega_m}\sin\omega_m t\right)$$

From Eq. (3.21) or Eq. (3.23),

$$\beta = \frac{a_m k_f}{\omega_m} = \frac{a_m k_f}{2\pi f_m} = \frac{a_m k_f}{2\pi(10^3)} = 3$$

We see that the value of β is inversely proportional to f_m. Thus, by Eq. (3.25), when $f_m = 1$ kHz,

$$f_B = 2(\beta + 1)f_m = 2(3 + 1)1 = 8 \text{ kHz}$$

(a) When f_m is doubled, $\beta = 3/2$, $f_m = 2$ kHz, and

$$f_B = 2(\beta + 1)f_m = 2\left(\frac{3}{2} + 1\right)2 = 10 \text{ kHz}$$

(b) When f_m is decreased by one-half, $\beta = 6$, $f_m = 0.5$ kHz, and

$$f_B = 2(\beta + 1)f_m = 2(6 + 1)(0.5) = 7 \text{ kHz}$$

3.13. A carrier is frequency-modulated with a sinusoidal signal of 2 kHz resulting in a maximum frequency deviation of 5 kHz.

(a) Find the bandwidth of the modulated signal.

(b) The amplitude of the modulating sinusoid is increased by a factor of 3, and its frequency is lowered to 1 kHz. Find the maximum frequency deviation and the bandwidth of the new modulated signal.

(a) From Eq. (3.21),

$$\beta = \frac{k_f a_m}{\omega_m} = \frac{\Delta f}{f_m} = \frac{5(10^3)}{2(10^3)} = 2.5$$

By Eq. (3.25), the bandwidth is

$$f_B = 2(\beta + 1)f_m = 2(2.5 + 1)2 = 14 \text{ kHz}$$

(b) Let β_1 be the new modulation index. Then

$$\beta_1 = \frac{k_f 3a_m}{\frac{1}{2}\omega_m} = 6\frac{k_f a_m}{\omega_m} = 6\beta = 6(2.5) = 15$$

Thus

$$\Delta f = \beta_1 f_{m1} = (15)(1) = 15 \text{ kHz}$$

$$f_B = 2(\beta_1 + 1)f_{m1} = 2(15 + 1)(1) = 32 \text{ kHz}$$

3.14. In addition to *Carson's rule* (*3.27*), the following formula is often used to estimate the bandwidth of an FM signal:

$$W_B \approx 2(D+2)\omega_M \qquad \text{for } D > 2$$

where $\omega_M = 2\pi f_M$ and f_M is the highest frequency of the signal in hertz. Compute the bandwidth, using this formula, and compare it to the bandwidth, using Carson's rule for the FM signal with $\Delta f = 75$ kHz and $f_M = 15$ kHz.

Note that commercial FM broadcast stations in the United States are limited to a maximum frequency deviation of 75 kHz, and modulating frequencies typically cover 50 Hz to 15 kHz.

Using Eq. (*3.26*) with $\omega_M = 2\pi f_M$, where $f_M = 15$ kHz, we have

$$D = \frac{\Delta f}{f_M} = \frac{75(10^3)}{15(10^3)} = 5$$

and by using the given formula, the bandwidth is

$$f_B = 2(D+2)f_M = 210 \text{ kHz}$$

Using Carson's rule, Eq. (*3.27*), we see that the bandwidth is

$$f_B = 2(D+1)f_M = 180 \text{ kHz}$$

Note: High-quality FM radios require a bandwidth of at least 200 kHz. Thus it seems that Carson's rule underestimates the bandwidth.

GENERATION OF ANGLE-MODULATED SIGNALS

3.15. Consider the frequency multiplier of Fig. 3-4 and an NBFM signal

$$x_{\text{NBFM}}(t) = A\cos(\omega_c t + \beta \sin \omega_m t)$$

with $\beta < 0.5$ and $f_c = 200$ kHz. Let f_m range from 50 Hz to 15 kHz, and let the maximum frequency deviation Δf at the output be 75 kHz. Find the required frequency multiplication n and the maximum allowed frequency deviation at the input.

From Eq. (*3.22*), $\beta = \Delta f / f_m$. Thus,

$$\beta_{\min} = \frac{75(10^3)}{15(10^3)} = 5 \qquad \beta_{\max} = \frac{75(10^3)}{50} = 1500$$

If $\beta_1 = 0.5$, where β_1 is the input β, then the required frequency multiplication is

$$n = \frac{\beta_{\max}}{\beta_1} = \frac{1500}{0.5} = 3000$$

The maximum allowed frequency deviation at the input, denoted Δf_1, is

$$\Delta f_1 = \frac{\Delta f}{n} = \frac{75(10^3)}{3000} = 25 \text{ Hz}$$

3.16. A block diagram of an indirect (Armstrong) FM transmitter is shown in Fig. 3-9. Compute the maximum frequency deviation Δf of the output of the FM transmitter and the carrier frequency f_c if $f_1 = 200$ kHz, $f_{LO} = 10.8$ MHz, $\Delta f_1 = 25$ Hz, $n_1 = 64$, and $n_2 = 48$.

$$\Delta f = (\Delta f_1)(n_1)(n_2) = (25)(64)(48) \text{ Hz} = 76.8 \text{ kHz}$$

$$f_2 = n_1 f_1 = (64)(200)(10^3) = 12.8(10^6) \text{ Hz} = 12.8 \text{ MHz}$$

$$f_3 = f_2 \pm f_{LO} = (12.8 \pm 10.8)(10^6) \text{ Hz} = \begin{cases} 23.6 & \text{MHz} \\ 2.0 & \text{MHz} \end{cases}$$

Thus, when $f_3 = 23.6$ MHz, then

$$f_c = n_2 f_3 = (48)(23.6) = 1132.8 \text{ MHz}$$

When $f_3 = 2$ MHz, then

$$f_c = n_2 f_3 = (48)(2) = 96 \text{ MHz}$$

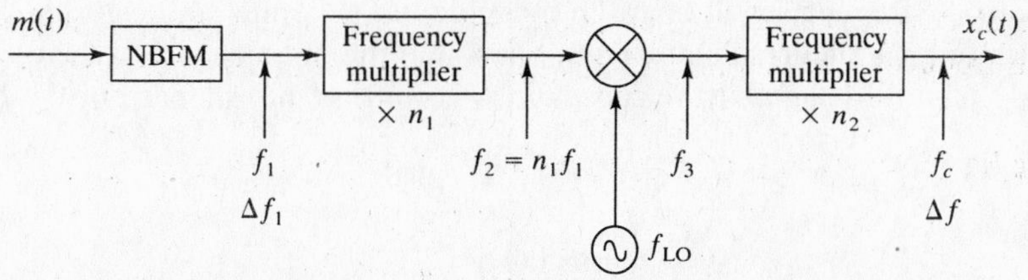

Fig. 3-9 Block diagram of an indirect FM transmitter

3.17. In an Armstrong-type FM generator of Fig. 3-9 (Prob. 3.16), the crystal oscillator frequency is 200 kHz. The maximum phase deviation is limited to 0.2 to avoid distortion. Let f_m range from 50 Hz to 15 kHz. The carrier frequency at the output is 108 MHz, and the maximum frequency deviation is 75 kHz. Select multiplier and mixer oscillator frequencies.

Referring to Fig. 3-9, we have

$$\Delta f_1 = \beta f_m = (0.2)(50) = 10 \text{ Hz}$$

$$\frac{\Delta f}{\Delta f_1} = \frac{75(10^3)}{10} = 7500 = n_1 n_2$$

$$f_2 = n_1 f_1 = n_1(2)(10^5) \qquad \text{Hz}$$

Assuming down conversion, we have

$$f_2 - f_{LO} = \frac{f_c}{n_2}$$

Thus,

$$f_{LO} = n_1 f_1 - \frac{f_c}{n_2} = \frac{7500(2)(10^5) - 108(10^6)}{n_2} = \frac{1392}{n_2}(10^6) \qquad \text{Hz}$$

Letting $n_2 = 150$, we obtain

$$n_1 = 50 \qquad \text{and} \qquad f_{LO} = 9.28 \text{ MHz}$$

3.18. A given angle-modulated signal has a maximum frequency deviation of 50 Hz for an input sinusoid of unit amplitude and a frequency of 120 Hz. Determine the required frequency multiplication factor n to produce a maximum frequency deviation of 20 kHz when the input sinusoid has unit amplitude and a frequency of 240 Hz and the angle modulation used is (*a*) PM and (*b*) FM.

(*a*) From Eqs. (*3.21*) and (*3.22*) we see that in sinusoidal PM, the maximum frequency deviation Δf is proportional to f_m. Thus

$$\Delta f_1 = \left(\frac{240}{120}\right)(50) = 100 \text{ Hz}$$

Hence

$$n = \frac{\Delta f_2}{\Delta f_1} = \frac{20(10^3)}{100} = 200$$

(*b*) Again from Eqs. (*3.21*) and (*3.22*) we see that in sinusoidal FM, the maximum frequency deviation Δf is independent of f_m. Thus

$$n = \frac{\Delta f_2}{\Delta f_1} = \frac{20(10^3)}{50} = 400$$

3.19. At low carrier frequencies it may be possible to generate an FM signal by varying the capacitance of a parallel resonant circuit. Show that the output $x_c(t)$ of the tuned circuit shown in Fig. 3-10 is an FM signal if the capacitance has a time dependence of the form

$$C(t) = C_0 - km(t)$$

and

$$\left|\frac{k}{C_0} m(t)\right| \ll 1$$

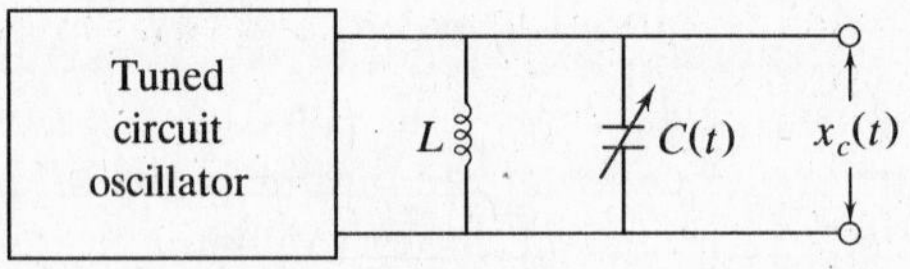

Fig. 3-10

If we assume $km(t)$ is small and slowly varying, then the output frequency ω_i of the oscillator is given by

$$\omega_i = \frac{1}{\sqrt{LC(t)}} = \frac{1}{\sqrt{L[C_0 - km(t)]}} = \frac{1}{\sqrt{LC_0}}\left[1 - \frac{k}{C_0} m(t)\right]^{-1/2}$$

Since $|(k/C_0)m(t)| \ll 1$, we can use the approximation

$$(1-z)^{-1/2} \approx 1 + \frac{1}{2} z$$

and obtain

$$\omega_i \approx \omega_c\left[1 + \frac{1}{2}\frac{k}{C_0} m(t)\right] = \omega_c + k_f m(t)$$

where

$$\omega_c = \frac{1}{\sqrt{LC_0}} \qquad \text{and} \qquad k_f = \frac{1}{2}\frac{\omega_c k}{C_0}$$

Thus, by Eq. (*3.12*), $x_c(t)$ is an FM signal.

3.20. An FM signal

$$x_{FM}(t) = A\cos\left[\omega_c t + k_f \int_{-\infty}^{t} m(\lambda)\, d\lambda\right]$$

is applied to the system shown in Fig. 3-11 consisting of a high-pass *RC* filter and an envelope detector. Assume that $\omega RC \ll 1$ in the frequency band occupied by $x_{FM}(t)$. Determine the output signal $y(t)$, assuming that $k_f|m(t)| < \omega_c$ for all t.

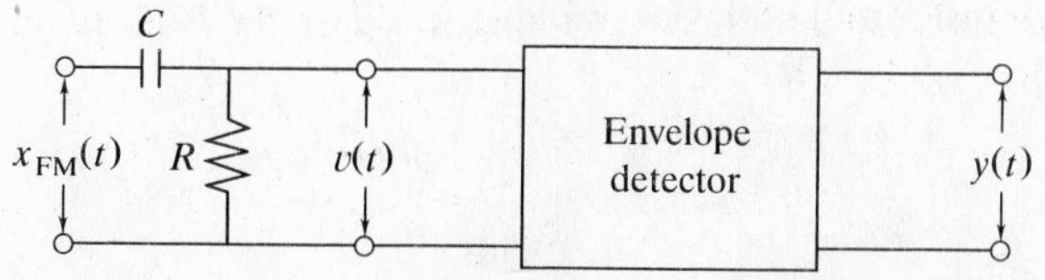

Fig. 3-11

The frequency response $H(\omega)$ of the RC high-pass filter is

$$H(\omega)=\frac{R}{R+1/(j\omega C)}=\frac{j\omega RC}{1+j\omega RC}$$

If $\omega RC \ll 1$, then

$$H(\omega)\approx j\omega RC$$

Since, multiplication by $j\omega$ in the frequency domain is equivalent to differentiation in the time domain [see Eq. (*1.35*)], the output $v(t)$ of the RC filter is

$$v(t)\approx RC\frac{d}{dt}[x_{FM}(t)]$$

$$=-ARC[\omega_c+k_f m(t)]\sin\left[\omega_c t+k_f\int_{-\infty}^{t}m(\lambda)\,d\lambda\right]$$

The corresponding envelope detector output is

$$y(t)=ARC[\omega_c+k_f m(t)]$$

which shows that, except for a dc term $ARC\omega_c$, the output is proportional to $m(t)$.

3.21. Delay lines might be used to approximate the derivative of the signal by realizing that

$$x'(t)\approx\frac{x(t)-x(t-\tau)}{\tau}\tag{3.46}$$

Draw the system, and suggest how small τ must be in order for the right side to be a good approximation of the derivative.

A system to realize Eq. (*3.46*) is shown in Fig. 3-12.

$$y(t)=\frac{1}{\tau}[x(t)-x(t-\tau)]$$

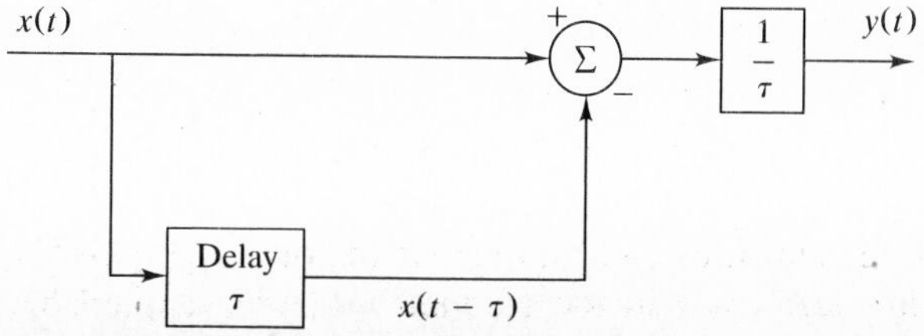

Fig. 3-12

Taking the Fourier transform of both sides yields

$$Y(\omega)=\frac{1}{\tau}\left[X(\omega)-e^{-j\omega\tau}X(\omega)\right]=\frac{1}{\tau}X(\omega)(1-e^{-j\omega\tau})$$

If $\omega\tau \ll 1$, then $1-e^{-j\omega\tau}\approx j\omega\tau$ and

$$Y(\omega)\approx j\omega X(\omega)$$

which indicates that $y(t)$ is approximately equal to the derivative of $x(t)$ and τ must satisfy the following condition:

$$\tau \ll \frac{1}{\omega}=\frac{1}{\omega_c+\Delta\omega}$$

3.22. Consider an FM signal

$$x_{FM}(t)=A\cos\left[\omega_c t+k_f\int_{-\infty}^{t} m(\lambda)\,d\lambda\right]$$

Let t_1 and t_2 $(t_2>t_1)$ denote the times associated with two adjacent zero crossings of $x_{FM}(t)$ (Fig. 3-13). If

$$\int_{t_1}^{t_2} m(\lambda)\,d\lambda \approx m(t)(t_2-t_1) \qquad t_1\le t\le t_2$$

then show that

$$k_f m(t)\approx\frac{\pi}{\Delta t}-\omega_c$$

where $\Delta t=t_2-t_1$.

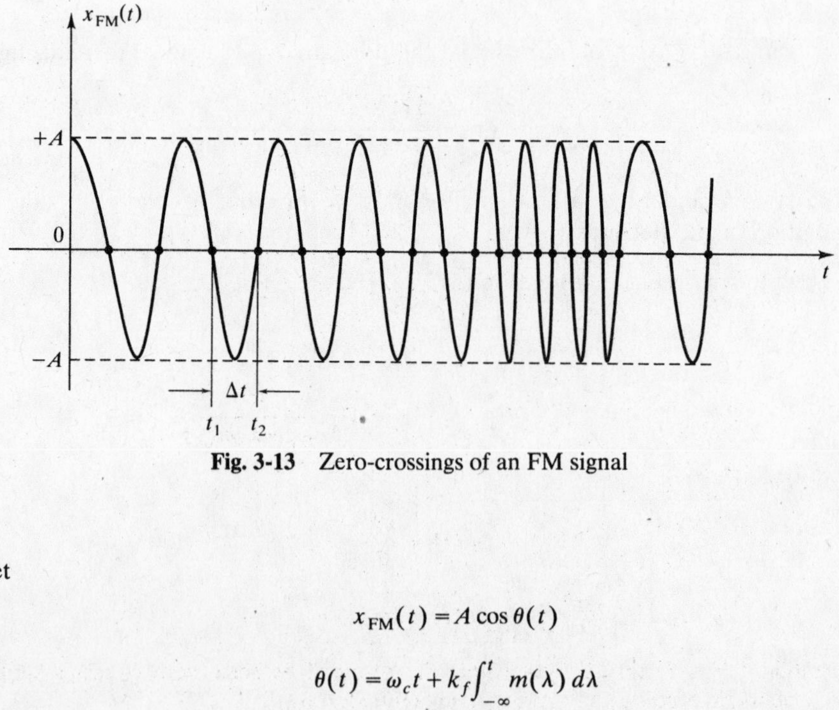

Fig. 3-13 Zero-crossings of an FM signal

Let

$$x_{FM}(t)=A\cos\theta(t)$$

where

$$\theta(t)=\omega_c t+k_f\int_{-\infty}^{t} m(\lambda)\,d\lambda$$

Let t_1 and t_2 $(t_2 > t_1)$ be the times associated with two adjacent zero crossings, that is,

$$x_{\text{FM}}(t_1) = x_{\text{FM}}(t_2) = 0$$

Then
$$\theta(t_2) - \theta(t_1) = \pi = \omega_c(t_2 - t_1) + k_f \int_{t_1}^{t_2} m(\lambda)\, d\lambda$$

The bandwidth of the message $m(t)$ is assumed much less than the bandwidth of the modulated signal. Then $m(t)$ is essentially constant over the interval $[t_1, t_2]$, and we have

$$\left[\omega_c + k_f m(t)\right](t_2 - t_1) = \pi$$

Thus, by Eq. (*3.12*),

$$\omega_i = \omega_c + k_f m(t) = \frac{\pi}{t_2 - t_1}$$

or
$$k_f m(t) = \frac{\pi}{\Delta t} - \omega_c$$

where $\Delta t = t_2 - t_1$.

3.23. The result of Prob. 3.22 indicates that $m(t)$ can be recovered by counting the zero crossings in $x_{\text{FM}}(t)$. Let N denote the number of zero crossings in time T. Show that if T satisfies the condition

$$\frac{1}{f_c} < T \ll \frac{1}{f_M}$$

where f_M is the bandwidth of $m(t)$ in hertz $(\omega_M = 2\pi f_M)$, then

$$km(t) \approx \frac{N}{2T} - f_c$$

Let $t_1, t_2, t_3, \ldots$ denote the times of zero crossings and $T_1 = t_2 - t_1$, $T_2 = t_3 - t_2, \ldots$. Assume that there are N zero crossings in

$$T = T_1 + T_2 + \cdots + T_N$$

From the result of Prob. 3.22, we have

$$k_f m(t) = \frac{\pi}{T_1} - \omega_c$$

or
$$T_1 = \frac{\pi}{\omega_c + k_f m(t)}$$

This is true for $T_2, T_3, \ldots$; that is,

$$T_i = \frac{\pi}{\omega_c + k_f m(t)} \qquad i = 1, 2, 3, \ldots, N$$

Thus,

$$T_1 + T_2 + \cdots + T_N = T = \frac{\pi N}{\omega_c + k_f m(t)}$$

Hence, we obtain

$$k_f m(t) = \frac{\pi N}{T} - \omega_c$$

or
$$\frac{k_f}{2\pi} m(t) = km(t) = \frac{N}{2T} - f_c$$

The condition $1/f_c < T$ ensures that within T there will be some zero crossings, and the condition $T \ll 1/f_M$ offers no excessive averaging (or smoothing) of $m(t)$.

Supplementary Problems

3.24. An angle-modulated signal is given by

$$x_c(t) = 5\cos\left[2\pi(10^6)t + 0.2\cos 200\pi t\right]$$

Can you identify whether $x_c(t)$ is a PM or an FM signal?

Ans. No. It can be either a PM or an FM signal.

3.25. The frequency multiplier is a nonlinear device followed by a bandpass filter, as shown in Fig. 3-14. Suppose that the nonlinear device is an ideal square-law device with input-output characteristics

$$e_o(t) = ae_i^2(t)$$

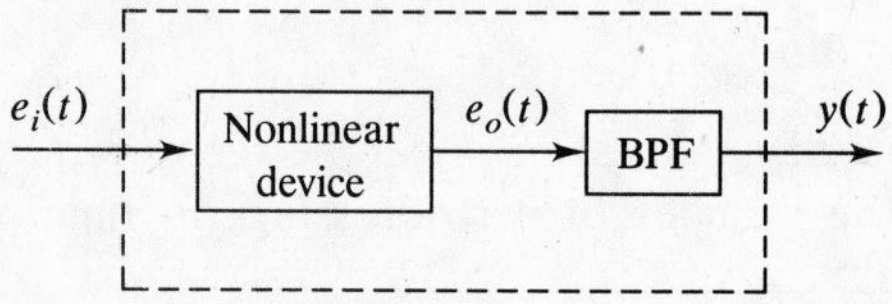

Fig. 3-14

Find the output $y(t)$ if the input is an FM signal given by

$$e_i(t) = A\cos(\omega_c t + \beta \sin \omega_m t)$$

Ans. $y(t) = A'\cos(2\omega_c t + 2\beta \sin \omega_m t)$, where $A' = \frac{1}{2}aA^2$. This result indicates that a square-law device can be used as a frequency doubler.

3.26. Assume that the 10.8-MHz signal in Fig. 3-15 is derived from the 200-kHz oscillator (multiplication by 54) and that the 200-kHz oscillator drift is 0.1 Hz.

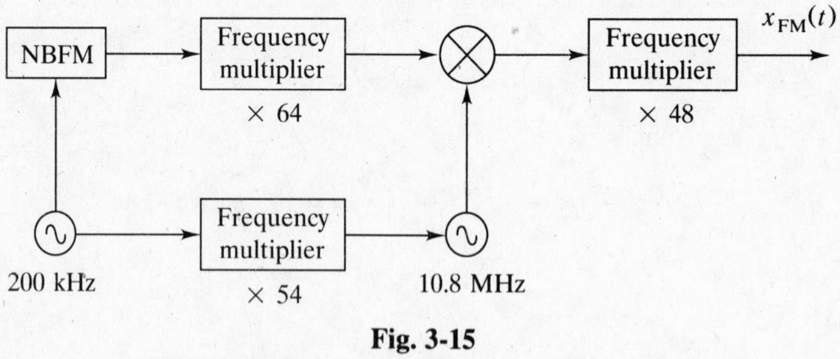

Fig. 3-15

(*a*) Find the drift in the 10.8-MHz signal.

(*b*) Find the drift in the carrier of the resulting FM signal.

Ans. (*a*) ± 5.4 Hz, (*b*) 48 Hz

3.27. A given FM signal has a maximum frequency deviation of 25 Hz for a modulating sinusoid of unit amplitude and a frequency of 100 kHz. Find the required value of frequency multiplication n to produce a maximum frequency deviation of 20 kHz when the modulating sinusoid has unit amplitude and a frequency of 200 Hz.

Ans. $n = 800$

3.28. A block diagram of a typical FM receiver, covering the broadcast range of 88 to 108 MHz, is shown in Fig. 3-16. The IF amplifier frequency is 10.7 MHz. The limiter is used to remove the amplitude fluctuations caused by channel imperfection. The FM receiver is tuned to a carrier frequency of 100 MHz.

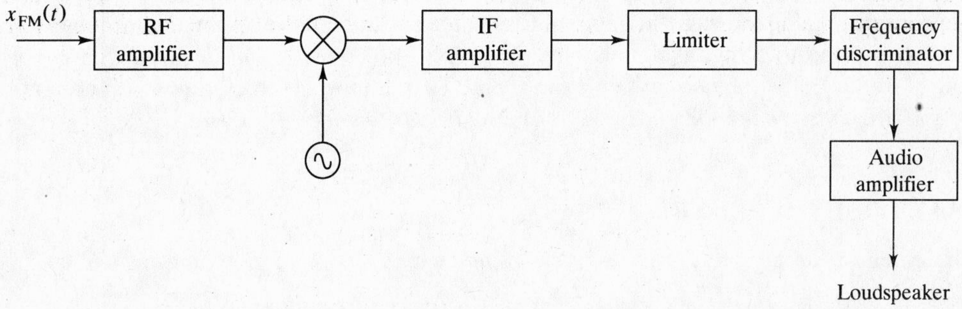

Fig. 3-16 FM receiver

(*a*) A 10-kHz audio signal frequency modulates a 100-MHz carrier, producing $\beta = 0.2$. Find the bandwidths required for the RF and IF amplifiers and for the audio amplifier.

(*b*) Repeat (*a*) if $\beta = 5$.

Ans. (*a*) RF and IF amplifiers: 24 kHz; audio amplifier: 10 kHz

(*b*) RF and IF amplifiers: 120 kHz; audio amplifier: 10 kHz

Chapter 4

Digital Transmission of Analog Signals

4.1 INTRODUCTION

The trend in the design of new communication systems has been toward increasing the use of digital techniques. Digital communications offer several important advantages compared to analog communications, for example, higher performance, greater versatility, and higher security.

To transmit analog message signals, such as voice and video signals, by digital means, the signal has to be converted to a digital signal. This process is known as the *analog-to-digital conversion*, or sometimes referred as *digital pulse modulation*. Two important techniques of analog-to-digital conversion are *pulse code modulation* (PCM) and *delta modulation* (DM).

4.2 PULSE CODE MODULATION

The essential processes of PCM are *sampling*, *quantizing*, and *encoding*, as shown in Fig. 4-1.

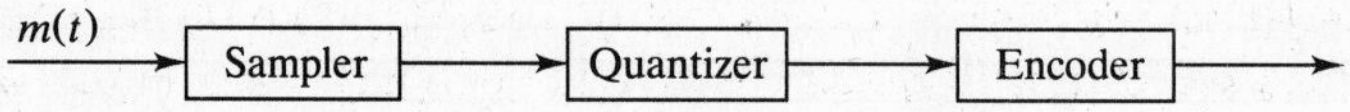

Fig. 4-1 Pulse code modulation

Sampling is the process in which a continuous-time signal is sampled by measuring its amplitude at discrete instants. Representing the sampled values of the amplitude by a finite set of levels is called *quantizing*. Designating each quantized level by a code is called *encoding*.

While sampling converts a continuous-time signal to a discrete-time signal, quantizing converts a continuous-amplitude sample to a discrete-amplitude sample. Thus sampling and quantizing operations transform an analog signal to a digital signal.

The quantizing and encoding operations are usually performed in the same circuit, which is called an analog-to-digital (A/D) converter. The combined use of quantizing and encoding distinguishes PCM from analog pulse modulation techniques.

In the following sections, we discuss the operations of sampling, quantizing, and encoding.

4.3 SAMPLING THEOREM

Digital transmission of analog signals is possible by virtue of the *sampling theorem*, and the sampling operation is performed in accordance with the sampling theorem.

A. Band-Limited Signals:

A *band-limited signal* is a signal $m(t)$ for which the Fourier transform of $m(t)$ is identically zero above a certain frequency ω_M:

$$m(t) \leftrightarrow M(\omega) = 0 \qquad \text{for } |\omega| > \omega_M = 2\pi f_M \tag{4.1}$$

B. Sampling Theorem:

If a signal $m(t)$ is a real-valued band-limited signal satisfying condition (*4.1*), then $m(t)$ can be uniquely determined from its values $m(nT_s)$ sampled at uniform intervals $T_s[\leq 1/(2f_M)]$. In fact,

$m(t)$ is given by

$$m(t) = \sum_{n=-\infty}^{\infty} m(nT_s) \frac{\sin \omega_M(t - nT_s)}{\omega_M(t - nT_s)} \tag{4.2}$$

We refer to T_s as the *sampling period* and to its reciprocal $f_s = 1/T_s$ as the *sampling rate*.

Thus, the sampling theorem states that a band-limited signal which has no frequency components higher than f_M Hz can be recovered completely from a set of samples taken at the rate of f_s ($\geq 2f_M$) samples per second.

The above sampling theorem is often called the *uniform sampling theorem* for baseband or low-pass signals.

The minimum sampling rate, $2f_M$ samples per second, is called the *Nyquist rate*; its reciprocal $1/(2f_M)$ (measured in seconds) is called the *Nyquist interval*. For the proof of the sampling theorem see Prob. 4.2.

The requirement that the sampling rate be equal to or greater than twice the highest frequency applies to baseband or low-pass signals. However, when bandpass signals are to be sampled, lower sampling rates can sometimes be used (see Prob. 4.7).

4.4 SAMPLING

A. Instantaneous Sampling:

Suppose we sample an arbitrary signal $m(t)$ [Fig. 4-2(a)] instantaneously and at a uniform rate, once every T_s s. Then we obtain an infinite sequence of samples $\{m(nT_s)\}$, where n takes on all possible integer values. This ideal form of sampling is called *instantaneous sampling*.

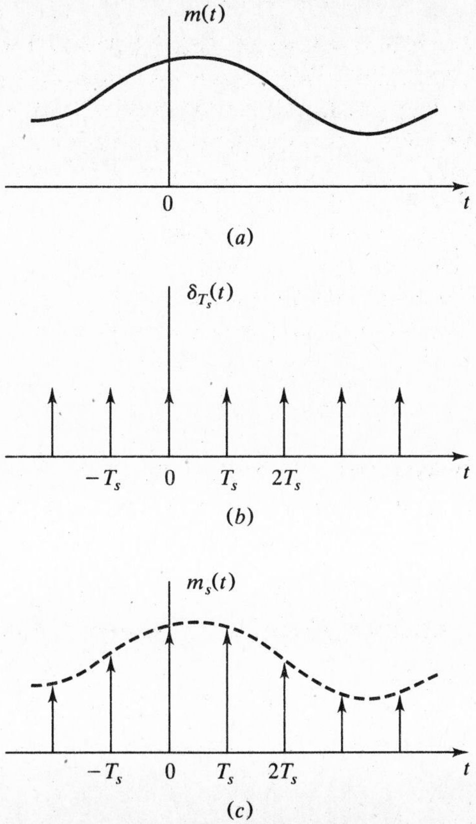

Fig. 4-2 Ideal signal sampling

B. Ideal Sampled Signal:

Multiplication of $m(t)$ by a unit impulse train $\delta_T(t)$ [Fig. 4-2(b), Eq. (1.105)] yields

$$m_s(t) = m(t)\delta_{T_s}(t) = \sum_{n=-\infty}^{\infty} m(nT_s)\delta(t - nT_s) \tag{4.3}$$

The signal $m_s(t)$ [Fig. 4-2(c)] is referred to as the *ideal sampled signal.*

C. Practical Sampling:

1. Natural Sampling:

Although instantaneous sampling is a convenient model, a more practical way of sampling a band-limited analog signal $m(t)$ is performed by high-speed switching circuits. An equivalent circuit employing a mechanical switch and the resulting sampled signal are shown in Fig. 4-3(a) and (b), respectively.

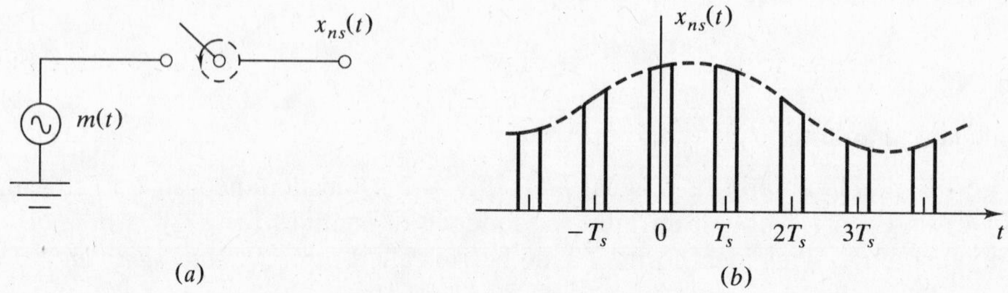

Fig. 4-3 Natural sampling

The sampled signal $x_{ns}(t)$ can be written as

$$x_{ns}(t) = m(t)x_p(t) \tag{4.4}$$

where $x_p(t)$ is the periodic train of rectangular pulses with period T_s, and each rectangular pulse in $x_p(t)$ has width d and unit amplitude.

The sampling here is termed *natural sampling*, since the top of each pulse in $x_{ns}(t)$ retains the shape of its corresponding analog segment during the pulse interval. The effect of the finite width of the sampling pulses is investigated in Prob. 4.9.

2. Flat-Top Sampling:

The simplest and thus most popular practical sampling method is actually performed by a functional block termed the *sample-and-hold* (S/H) *circuit* [Fig. 4-4(a)]. This circuit produces a

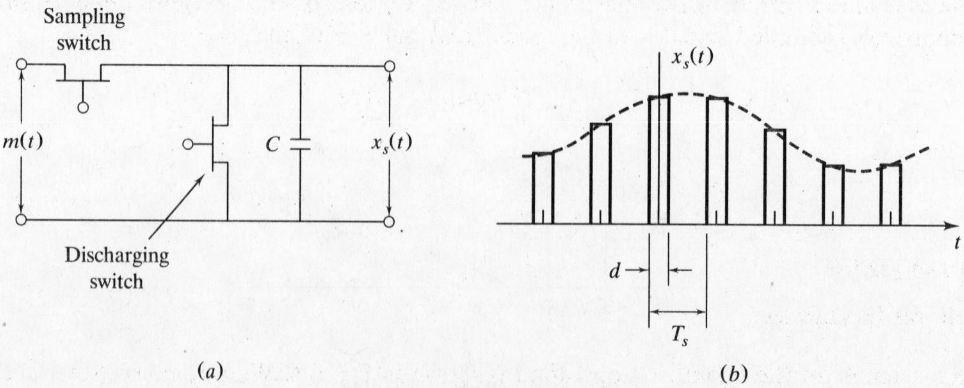

Fig. 4-4 Flat-top sampling

flat-top sampled signal $x_s(t)$ [Fig. 4-4(*b*)]. The effect of the flat-top sampling is discussed in Prob. 4.11.

4.5 PULSE AMPLITUDE MODULATION

The signal $x_s(t)$ depicted in Fig. 4-4(*b*) represents a pulse amplitude-modulated signal. In pulse amplitude modulation (PAM), the carrier signal consists of a periodic train of rectangular pulses, and the amplitudes of rectangular pulses vary with the instantaneous sampled values of an analog message signal. Note that the carrier frequency (that is, the pulse repetition frequency) is the same as the sampling rate.

The PAM signal $x_s(t)$ can be expressed as

$$x_s(t) = \sum_{n=-\infty}^{\infty} m(nT_s)\,p(t-nT_s) \tag{4.5}$$

where $p(t)$ is a rectangular pulse of unit amplitude and duration d, defined as (see Fig. 4-5)

$$p(t) = \begin{cases} 1 & |t| < \dfrac{d}{2} \\ 0 & \text{otherwise} \end{cases} \tag{4.6}$$

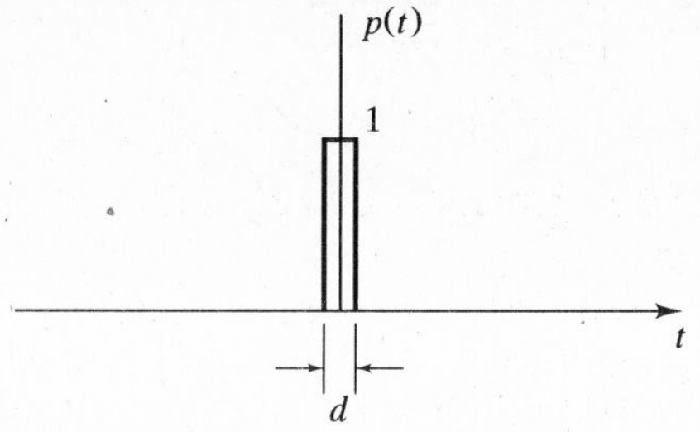

Fig. 4-5 Rectangular pulse

It can be shown (Prob. 4.10) that $x_s(t)$ can be expressed as the convolution of $m_s(t)$, the instantaneously sampled signal, and the rectangular pulse $p(t)$, that is,

$$x_s(t) = m_s(t) * p(t) \tag{4.7}$$

4.6 QUANTIZING

A. Uniform Quantizing:

An example of the quantizing operation is shown in Fig. 4-6. We assume that the amplitude of $m(t)$ is confined to the range $(-m_p, m_p)$. As illustrated in Fig. 4-6(*a*), this range is divided in L

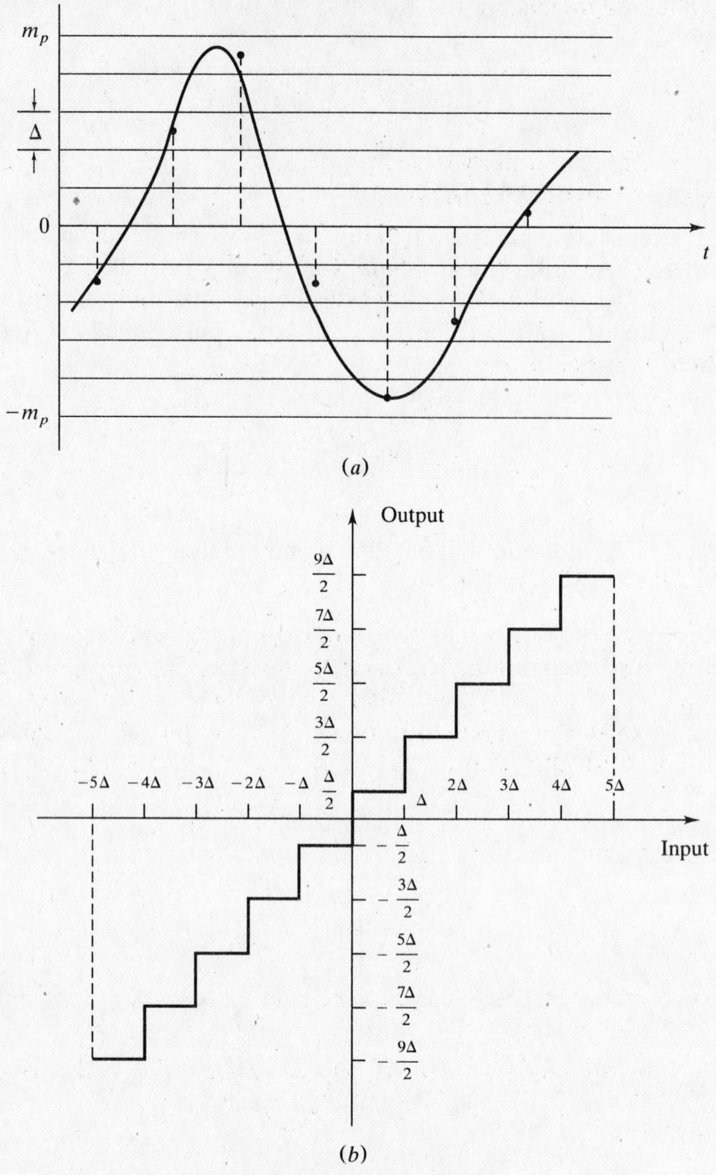

Fig. 4-6 Uniform quantizing

zones, each of step size Δ, given by

$$\Delta = \frac{2m_p}{L} \tag{4.8}$$

A sample amplitude value is approximated by the midpoint of the interval in which it lies.

The input-output characteristics of a uniform quantizer are shown in Fig. 4-6(*b*).

B. Quantizing Noise:

The difference between the input and output signals of the quantizer becomes the *quantizing error*, or *quantizing noise*. It is apparent that with a random input signal, the quantizing error q_e

varies randomly within the interval

$$-\frac{\Delta}{2} \leq q_e \leq \frac{\Delta}{2} \tag{4.9}$$

Assuming that the error is equally likely to lie anywhere in the range $(-\Delta/2, \Delta/2)$, the mean-square quantizing error $\langle q_e^2 \rangle$ is given by (see Chap. 5)

$$\langle q_e^2 \rangle = \frac{1}{\Delta} \int_{-\Delta/2}^{\Delta/2} q_e^2 \, dq_e = \frac{\Delta^2}{12} \tag{4.10}$$

Substituting Eq. (*4.8*) into Eq. (*4.10*), we have

$$\langle q_e^2 \rangle = \frac{m_p^2}{3L^2} \tag{4.11}$$

C. Nonuniform Quantizing and Companding:

For many classes of signals the uniform quantizing is not efficient. For example, in speech communication it is found (statistically) that smaller amplitudes predominate in speech and that larger amplitudes are relatively rare. The uniform quantizing scheme is thus wasteful for speech signals; many of the quantizing levels are rarely used. An efficient scheme is to employ a nonuniform quantizing method in which smaller steps for small amplitudes are used (Fig. 4-7).

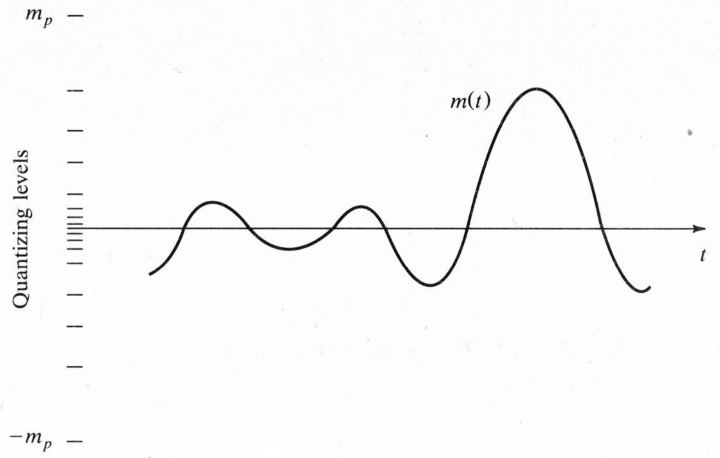

Fig. 4-7 Nonuniform quantizing

The same result can be achieved by first compressing signal samples and then using a uniform quantizing. The input-output characteristics of a compressor are shown in Fig. 4-8. The horizontal axis is the normalized input signal (that is, m/m_p), and the vertical axis is the output signal y. The compressor maps input signal increment Δm into large increment Δy for small input signals and small increments for large input signals. Hence by applying the compressed signal to a uniform quantizer, a given interval Δm contains a larger number of steps (or smaller step size) when m is small.

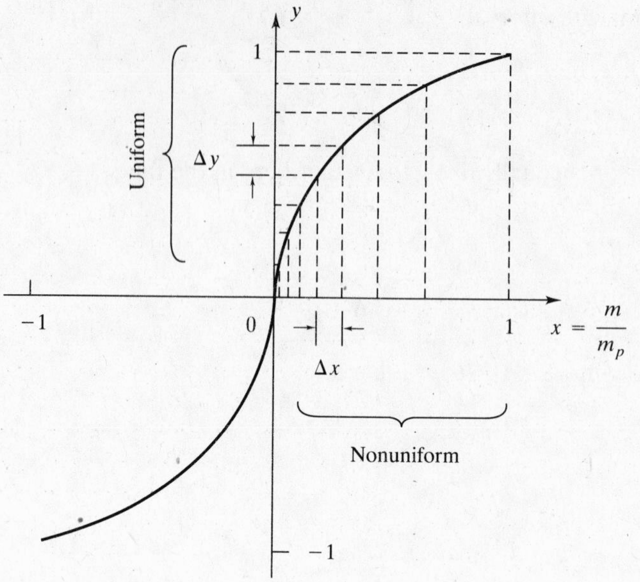

Fig. 4-8 Characteristics of a compressor

A particular form of compression law that is used in practice (in North America and Japan) is the so-called μ *law*, defined by

$$y = \frac{\ln(1 + \mu|m/m_p|)}{\ln(1+\mu)} \operatorname{sgn}(m) \qquad \left|\frac{m}{m_p}\right| \le 1 \tag{4.12}$$

where μ is a positive constant and

$$\operatorname{sgn}(m) = \begin{cases} +1 & m > 0 \\ -1 & m < 0 \end{cases}$$

Another compression law that is used in practice (in Europe) is the so-called A *law*, defined by

$$y = \begin{cases} \dfrac{A}{1 + \ln A}\left(\dfrac{m}{m_p}\right) & \left|\dfrac{m}{m_p}\right| \le \dfrac{1}{A} \\[2ex] \dfrac{(1 + \ln A|m/m_p|)}{1 + \ln A} \operatorname{sgn}(m) & \dfrac{1}{A} \le \left|\dfrac{m}{m_p}\right| \le 1 \end{cases} \tag{4.13}$$

For the μ law, $\mu = 255$ is used in digital telephone systems in North America. For the A law, $A = 87.6$ is used in European systems. These values are selected to obtain a nearly constant output signal-to-quantizing noise ratio over an input signal power dynamic range of 40 decibels (dB).

To restore the signal samples to their correct relative level, an expander with a characteristic complementary to that of the compressor is used in the receiver. The combination of compression and expansion is called *companding*.

4.7 ENCODING

An encoder in PCM translates the quantized sample into a code number. Usually the code number is converted to its representation in binary sequence. The binary sequence is converted to a sequential string of pulses for transmission (see Sec. 4.10). In this case the system is referred to as *binary PCM*. The essential features of binary PCM are shown in Fig. 4-9. Assume that an analog

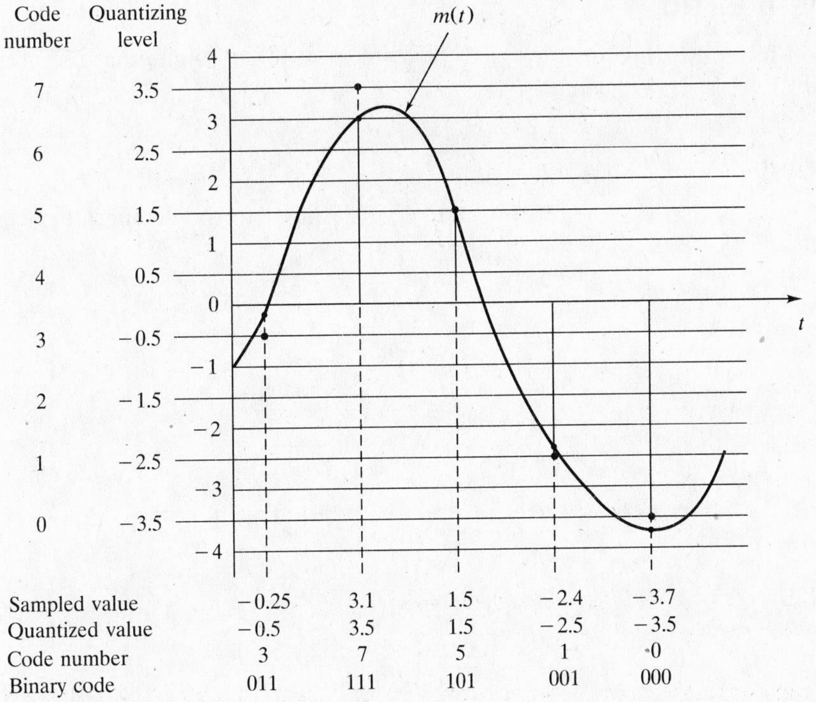

Fig. 4-9 PCM

signal $m(t)$ is confined to the range -4 to 4 volts (V). The step size Δ is set to 1 V. Thus, eight quantizing levels are employed; these are located at $-3.5, -2.5, \ldots, +3.5$ V. We assign the code number 0 to the level at -3.5 V, the code number 1 to the level at -2.5 V, and so on, until the level at $+3.5$ V, which is assigned the code number 7. Each code number has its binary code representation, ranging from 000 for code number 0 to 111 for code number 7. Each sample of $m(t)$ is assigned to the quantizing level closest to the sampled value.

4.8 BANDWIDTH REQUIREMENTS OF PCM

Suppose that in a binary PCM, L quantizing levels are used, satisfying

$$L = 2^n \qquad n = \log_2 L \tag{4.14}$$

where n is an integer. For this case, $n = \log_2 L$ binary pulses must be transmitted for each sample of the message signal. If the message bandwidth is f_m and the sampling rate is f_s ($\geq 2f_m$), then nf_s binary pulses must be transmitted per second.

Assuming the PCM signal is a low-pass signal of bandwidth f_{PCM}, the required minimum sampling rate is $2f_{PCM}$. Thus

$$2f_{PCM} = nf_s$$

or

$$f_{PCM} = \frac{n}{2} f_s \geq nf_m \qquad \text{Hz} \tag{4.15}$$

Equation (*4.15*) shows that the minimum required bandwidth for PCM is proportional to the message signal bandwidth and the number of bits per symbol. Note that the actual bandwidth required for a PCM system depends on the PCM representation.

4.9 DELTA MODULATION

A method for converting analog signals to a string of binary digits that requires much simpler circuitry than PCM is *delta modulation*.

A. Delta Modulator:

A simple DM system is shown in Fig. 4-10(*a*). The input to the comparator is

$$e(t) = m(t) - \tilde{m}(t) \tag{4.16}$$

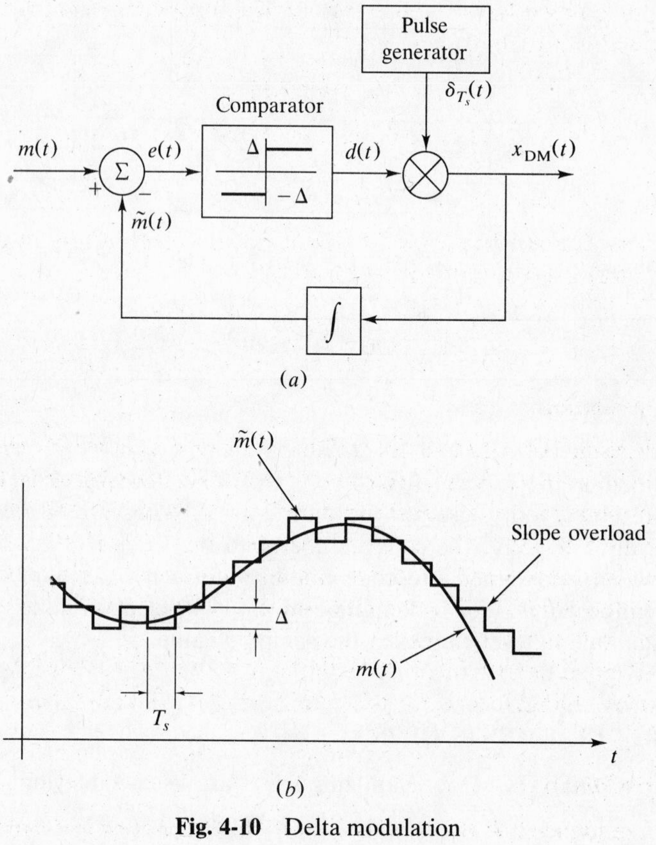

Fig. 4-10 Delta modulation

where $m(t)$ is the message signal and $\tilde{m}(t)$ is a reference signal. The output of the comparator is

$$d(t) = \Delta \operatorname{sgn}[e(t)] = \begin{cases} \Delta & e(t) > 0 \\ -\Delta & e(t) < 0 \end{cases} \tag{4.17}$$

Then the output of the delta modulator is

$$x_{\mathrm{DM}}(t) = \Delta \operatorname{sgn}[e(t)] \sum_{n=-\infty}^{\infty} \delta(t - nT_s)$$

$$= \Delta \sum_{n=-\infty}^{\infty} \operatorname{sgn}[e(nT_s)]\delta(t - nT_s) \tag{4.18}$$

Thus the output of the delta modulator is a series of impulses, each having positive or negative polarity depending on the sign of $e(t)$ at the sampling instants. Integrating $x_{\mathrm{DM}}(t)$, we obtain

$$\tilde{m}(t) = \sum_{n=-\infty}^{\infty} \Delta \operatorname{sgn}[e(nT_s)] \tag{4.19}$$

which is a staircase approximation of $m(t)$, as shown in Fig. 4-10(b).

B. Demodulation:

Demodulation of DM is accomplished by integrating $x_{\mathrm{DM}}(t)$ to form the staircase approximation $\tilde{m}(t)$ and then passing through a low-pass filter to eliminate the discrete jumps in $\tilde{m}(t)$.

Small step size is desirable to reproduce the input waveform accurately. However, small step size must be accompanied by a fast sampling rate to avoid slope overload, as shown in Fig. 4-10(b). To avoid slope overload, we require that

$$\frac{\Delta}{T_s} > \left|\frac{dm(t)}{dt}\right|_{\max} \tag{4.20}$$

C. Quantizing Error:

We assume that the quantizing error in DM is equally likely to lie anywhere in the interval $(-\Delta, \Delta)$, so that the mean-square quantizing error is

$$\langle q_e^2 \rangle = \frac{1}{2\Delta}\int_{-\Delta}^{\Delta} q_e^2 \, dq_e = \frac{\Delta^2}{3} \tag{4.21}$$

D. Adaptive Delta Modulation:

The delta modulation discussed so far suffers from one serious disadvantage: The dynamic range of amplitude of $m(t)$ is too small, because of the threshold and overload effects. This problem can be overcome by making a delta modulator adaptive. In adaptive DM, the step size Δ is varied according to the level of the input signal. For example, in Fig. 4-10(b), when the signal $m(t)$ is falling rapidly, slope overload occurs. If the step size is increased during this period, the overload could be avoided. On the other hand, if the slope of $m(t)$ is small, reduction of step size will reduce the threshold level as well as the quantizing noise.

The use of an adaptive delta modulator requires that the receiver be adaptive also, so that the step size at the receiver changes to match the change in Δ at the modulator.

4.10 SIGNALING FORMATS

Digital data (a sequence of binary digits) can be transmitted by various pulse waveforms. Sometimes these pulse waveforms have been called *line codes*. Figure 4-11 shows a number of signal formats for transmission of binary data 10110001.

(a) Unipolar Nonreturn-to-Zero (NRZ) Signaling:

Symbol 1 is represented by transmitting a pulse of constant amplitude for the entire duration of the bit interval, and symbol 0 is represented by no pulse. NRZ indicates that the assigned amplitude level is maintained throughout the entire bit interval.

(b) Bipolar NRZ Signaling:

Symbols 1 and 0 are represented by pulses of equal positive and negative amplitudes. In either case, the assigned pulse amplitude level is maintained throughout the bit interval.

(c) Unipolar Return-to-Zero (RZ) Signaling:

Symbol 1 is represented by a positive pulse that returns to zero before the end of the bit interval, and symbol 0 is represented by the absence of pulse.

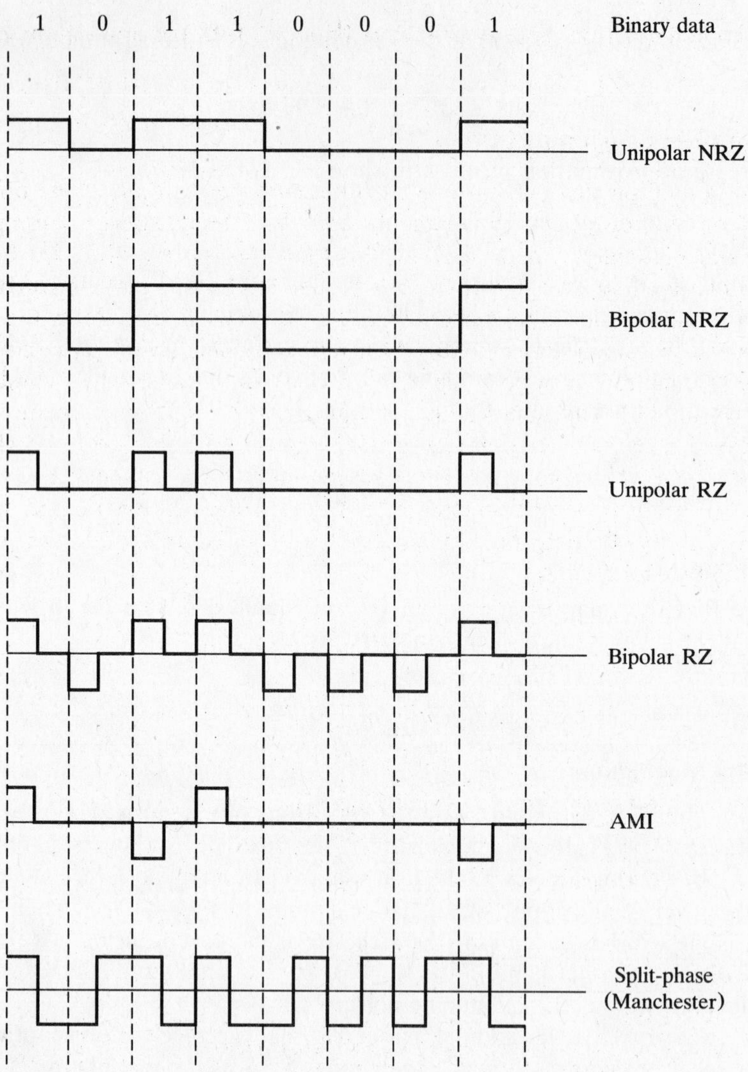

Fig. 4-11 Binary signaling formats

(*d*) **Bipolar RZ Signaling:**

Positive and negative pulses of equal amplitude are used for symbols 1 and 0, respectively. In either case, the pulse returns to 0 before the end of the bit interval.

(*e*) **Alternate Mark Inversion (AMI) RZ Signaling:**

Positive and negative pulses (of equal amplitude) are used alternately for symbol 1, and no pulse is used for symbol 0. In either case the pulse returns to 0 before the end of the bit interval.

(*f*) **Split-Phase (Manchester) Signaling:**

Symbol 1 is represented by a positive pulse followed by a negative pulse, with both pulses being of equal amplitude and half-bit duration; for symbol 0, the polarities of these pulses are reversed.

Many other signaling formats and variations are discussed in the literature. There are many formats because the channel characteristics vary from application to application. For example, if the channel is alternating current (ac) coupled, a format with a large dc component should not be chosen. Some of the important parameters to consider in selecting a signaling format are the spectral characteristics, immunity of the format to noise, bit synchronization capability, cost and complexity of implementation, and other factors that vary with the application.

4.11 TIME-DIVISION MULTIPLEXING

Time-division multiplexing (TDM) is one of the many applications of the principle of sampling in communication systems. TDM is commonly used to simultaneously transmit several different signals over a single channel. Figure 4-12(*a*) illustrates the scheme of TDM. Each input message signal is first restricted in bandwidth by a low-pass filter, to remove frequencies that are not essential to an adequate signal representation. The low-pass filter outputs are then applied to a commutator which is usually implemented by using electronic switching circuitry. Each signal is sampled at the Nyquist rate or higher. Usually 1.1 times the Nyquist rate is employed in practice to avoid aliasing (Prob. 4.2). The samples are interleaved, and a single composite signal consisting of all the interleaved pulses is transmitted over the channel. Figure 4-12(*b*) shows time-division multiplexing of two PAM signals. At the receiving end of the channel, the composite signal is demultiplexed by using a second commutator whose output is distributed to the appropriate low-pass filter for demodulation. Proper operation of this system depends on proper synchronization between the two commutators.

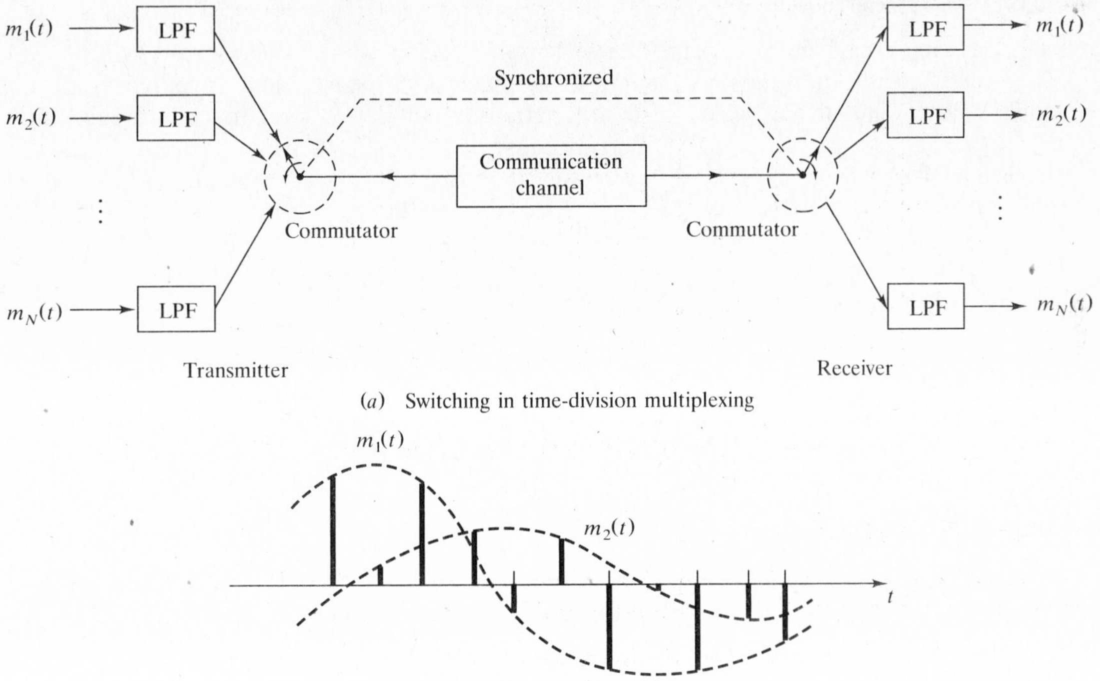

(*a*) Switching in time-division multiplexing

(*b*) Time-division multiplexing of two signals

Fig. 4-12 Time-division multiplexing

If all message signals have equal bandwidths, then the samples are transmitted sequentially, as shown in Fig. 4-12(*a*). If the sampled data signals have unequal bandwidths, more samples must be transmitted per unit time from the wideband signals. This is easily accomplished if the bandwidths are harmonically related. (See Prob. 4.31.)

Although Fig. 4-12 refers to TDM of PAM signals, the same general concepts apply to TDM of PCM or any other pulse signals. TDM is widely used in telephony, telemetry, radio, and data processing. The Bell system, for example, time-multiplexes 24 PCM signals on a telephone channel in its T1 system. In telephony, both frequency-division multiplexing (FDM) and TDM are jointly used to permit many hundreds of different conversations to utilize the same microwave link.

4.12 BANDWIDTH REQUIREMENTS FOR TDM

First, let us define T as the time spacing between adjacent samples in the time-multiplexed signal waveform [see Fig. 4-12(*b*)]. If all input signals have the same bandwidths f_m and are sampled equally, then $T = T_s/n$ [where n is the number of input signals and $T_s = 1/f_s \leq 1/(2f_m)$] is the sampling interval for each signal. Assuming the resultant time-multiplexed signal is a low-pass signal of bandwidth f_{TDM}, then the required minimum sampling rate is $2f_{\text{TDM}}$.

Thus

$$f_{\text{TDM}} = \frac{1}{2T} = \frac{n}{2T_s} = \frac{1}{2}nf_s \geq nf_m \qquad \text{Hz} \tag{4.22}$$

Equation (*4.22*) shows that the minimum required bandwidth for TDM transmission is proportional to the message signal bandwidth and the number of the multiplexed signals.

4.13 PULSE SHAPING AND INTERSYMBOL INTERFERENCE

A. Intersymbol Interference:

In discussing digital signal transmission so far, we have shown the digital pulse as rectangular and have assumed the transmission channel to be linear and distortionless. In practice, however, channels have a limited bandwidth, and hence transmitted pulses tend to be "spread" during transmission. This pulse spreading or dispersion causes overlap of pulses into adjacent time slots, as shown in Fig. 4-13. The signal overlap may result in an error at the receiver. This phenomenon of pulse overlap and the resultant difficulty of discriminating between symbols at the receiver are termed *intersymbol interference* (ISI).

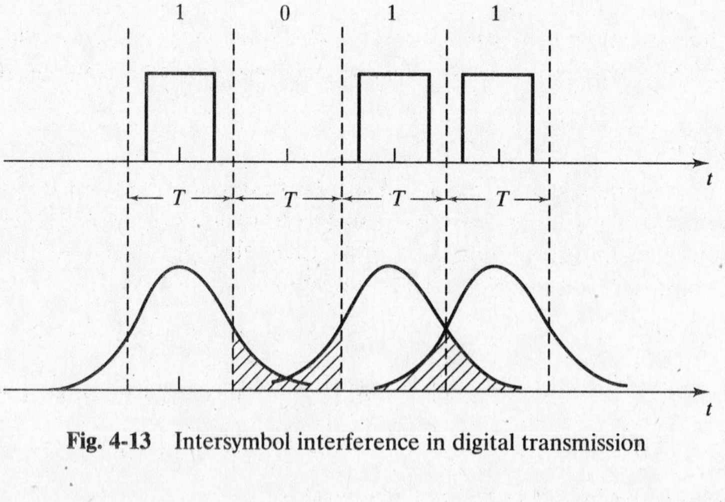

Fig. 4-13 Intersymbol interference in digital transmission

B. Pulse Shaping:

One method of controlling ISI is to shape the transmitted pulses properly. One pulse shape that produces zero ISI is given by [Fig. 4-14(*a*)]

$$h(t) = \frac{1}{T_s}\frac{\sin(\pi t/T_s)}{\pi t/T_s} \tag{4.23}$$

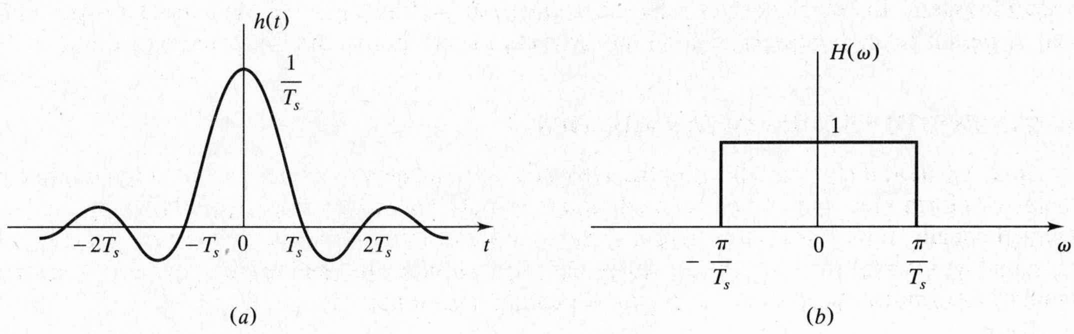

Fig. 4-14 Pulse providing zero ISI

This is the impulse response of an ideal low-pass filter whose frequency response is shown in Fig. 4-14(*b*).

Note that $h(t)$ goes through zero at equally spaced intervals which are multiples of T_s except at the center. Thus if $T_s = 1/(2f_B)$ (Nyquist interval), it is clear that pulses of the same shape that are spaced T_s or an integer multiple of T_s will not interfere (Fig. 4-15).

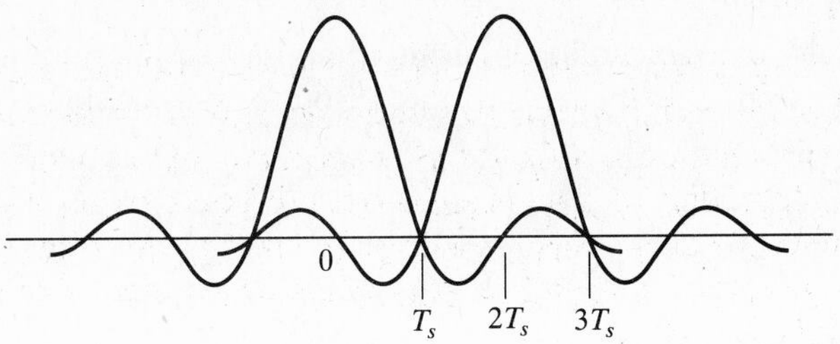

Fig. 4-15 Pulses with zero ISI

In practice, however, there are difficulties with this filter shape. First, an ideal low-pass filter is not causal or physically realizable. Second, this waveform depends critically on timing precision.

C. Raised-Cosine Filter:

An example of a commonly used filter which meets the requirement that the impulse response have zeros at uniformly spaced time intervals (except one at the center) and in which the frequency response decreases toward zero gradually rather than abruptly is the raised-cosine filter. The frequency response of the raised-cosine filter is given by

$$H(\omega) = \begin{cases} 1 & 0 \le |\omega| \le (1-\alpha)W \\ \dfrac{1}{2}\left\{1 - \sin\left[\dfrac{\pi}{2\alpha W}(|\omega| - W)\right]\right\} & (1-\alpha)W \le |\omega| \le (1+\alpha)W \\ 0 & |\omega| > (1+\alpha)W \end{cases} \tag{4.24}$$

where $W = \pi T_s$. The corresponding impulse response is

$$h(t) = \frac{1}{T_s}\left(\frac{\sin Wt}{Wt}\right)\left[\frac{\cos \alpha Wt}{1 - (2\alpha Wt/\pi)^2}\right] \tag{4.25}$$

Note that the second term on the right-hand side of Eq. (*4.25*) is of the form $(\sin x)/x$ which we encountered in the ideal low-pass filter; it retains the original zero crossings of that waveform.

Plots of $H(\omega)$ and $h(t)$ are shown in Fig. 4-16 for three values of α. The parameter α is called the *roll-off* factor. Note that the case for $\alpha = 0$ coincides with the ideal low-pass filter. The case for

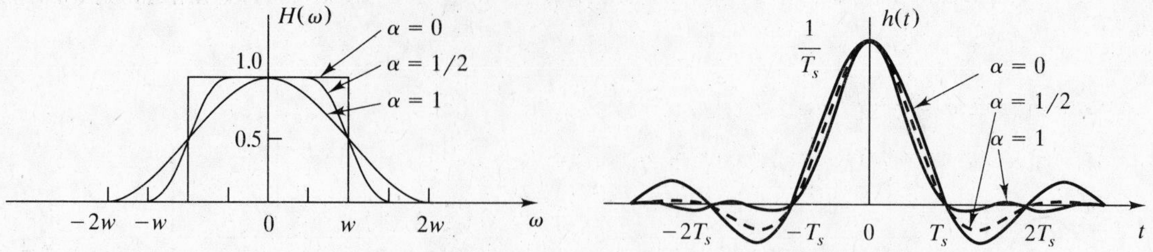

Fig. 4-16 Raised-cosine pulse shaping and resulting time response

$\alpha = 1$ is known as the *full-cosine roll-off characteristic*; its frequency response function is

$$H(\omega) = \begin{cases} \dfrac{1}{2}\left(1 + \cos \dfrac{\pi\omega}{2W}\right) & |\omega| \leq 2W \\ 0 & \text{elsewhere} \end{cases} \tag{4.26}$$

From Fig. 4-16 we see that the bandwidth occupied by a raised-cosine type of transmission characteristic varies from a minimum of $f_B = 1/(2T_s)$ Hz ($\alpha = 0$) to a maximum of $f_B = 1/T_s$ Hz ($\alpha = 1$).

The bandwidth/pulse-rate tradeoff from using the raised-cosine type of transmission characteristics depends, of course, on α. If the desired rate of pulse transmission is $1/T$ pulses per second, the bandwidth f_B required is

$$f_B = \frac{1+\alpha}{2T} \quad \text{Hz} \tag{4.27}$$

Alternatively, with f_B specified, the allowable pulse rate is given by

$$\frac{1}{T} = \frac{2f_B}{1+\alpha} \tag{4.28}$$

4.14 DIGITAL CARRIER MODULATION SYSTEMS

Because baseband digital signals have sizable power at low frequencies, they are suitable for transmission over a pair of wires or coaxial cables. Baseband digital signals cannot be transmitted over a radio link because this would require impractically large antennas to efficiently radiate the low-frequency spectrum of the signal. Hence for such purposes, we use analog modulation techniques in which the digital messages are used to modulate a high-frequency continuous-wave (CW) carrier.

In binary modulation schemes, the modulation process corresponds to switching (or keying) the amplitude, frequency, or phase of the CW carrier between either of two values corresponding to binary symbols 0 and 1. The three types of digital modulation are *amplitude-shift keying*, *frequency-shift keying*, and *phase-shift keying*.

A. Amplitude-Shift Keying (ASK):

In ASK, the modulated signal can be expressed as

$$x_c(t) = \begin{cases} A \cos \omega_c t & \text{symbol } 1 \\ 0 & \text{symbol } 0 \end{cases} \tag{4.29}$$

Note that the modulated signal is still an on-off signal. Thus ASK is also known as *on-off keying* (OOK).

B. Frequency-Shift Keying (FSK):

In FSK, the modulated signal can be expressed as

$$x_c(t) = \begin{cases} A\cos\omega_1 t & \text{symbol 1} \\ A\cos\omega_2 t & \text{symbol 0} \end{cases} \tag{4.30}$$

C. Phase-Shift Keying (PSK):

In PSK, the modulated signal can be expressed as

$$x_c(t) = \begin{cases} A\cos\omega_c t & \text{symbol 1} \\ A\cos(\omega_c t + \pi) & \text{symbol 0} \end{cases} \tag{4.31}$$

Figure 4-17 illustrates these digital modulation schemes for the case in which the data bits are represented by the bipolar NRZ waveform. The performance of these modulation methods in noisy channels is discussed in Chap. 7.

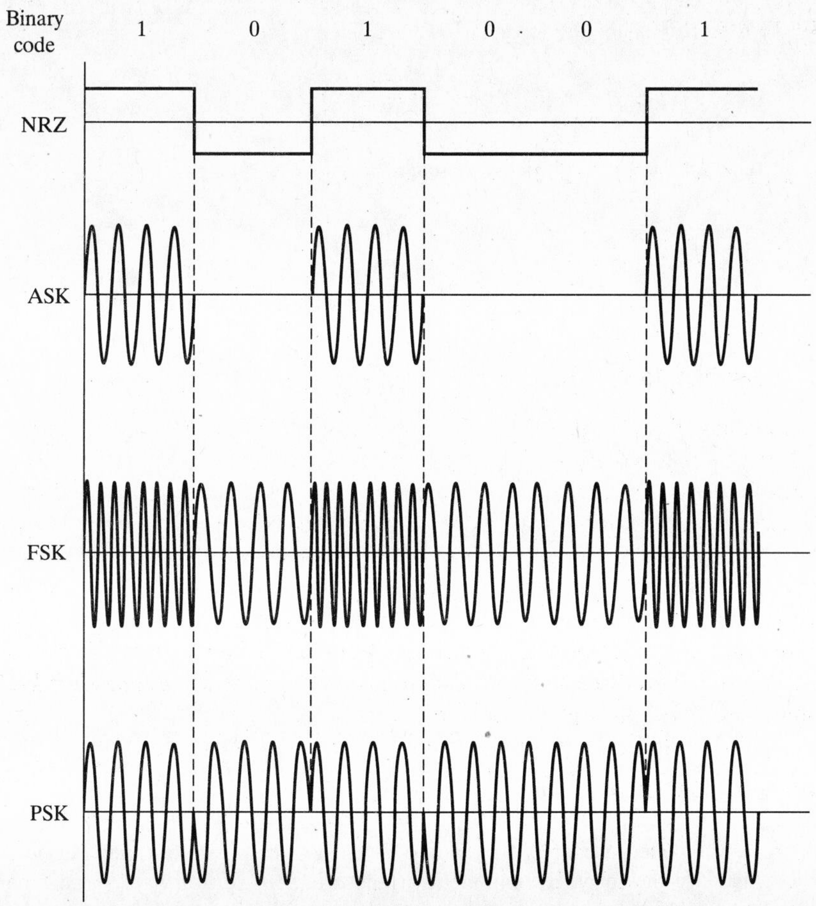

Fig. 4-17 Digital carrier modulation

Solved Problems

SAMPLING AND SAMPLING THEOREM

4.1. Verify Eq. (*4.3*), that is,

$$m_s(t) = m(t)\delta_{T_s}(t) = \sum_{n=-\infty}^{\infty} m(nT_s)\delta(t - nT_s)$$

From Eqs. (*1.106*) and (*1.12*), we have

$$\begin{aligned} m_s(t) = m(t)\delta_{T_s}(t) &= m(t) \sum_{n=-\infty}^{\infty} \delta(t - nT_s) \\ &= \sum_{n=-\infty}^{\infty} m(t)\delta(t - nT_s) \\ &= \sum_{n=-\infty}^{\infty} m(nT_s)\delta(t - nT_s) \end{aligned}$$

4.2. Verify the sampling theorem (4.2).

Let $m(t)$ be a band-limited signal defined by Eq. (*4.1*) [Fig. 4-18(*a*) and (*b*)]. From Eq. (*1.125*) of Prob. 1.22, we have [Fig. 4-18(*c*) and (*d*)]

$$\mathscr{F}\left[\delta_{T_s}(t)\right] = \omega_s \sum_{n=-\infty}^{\infty} \delta(\omega - n\omega_s) \qquad \omega_s = \frac{2\pi}{T_s}$$

According to the frequency convolution theorem (*1.54*), we have

$$\begin{aligned} M_s(\omega) &= \mathscr{F}\left[m(t)\delta_{T_s}(t)\right] \\ &= \frac{1}{2\pi}\left[M(\omega) * \omega_s \sum_{n=-\infty}^{\infty} \delta(\omega - n\omega_s)\right] \\ &= \frac{1}{T_s} \sum_{n=-\infty}^{\infty} M(\omega) * \delta(\omega - n\omega_s) \end{aligned}$$

Using Eq. (*1.52*), we obtain

$$M_s(\omega) = \frac{1}{T_s} \sum_{n=-\infty}^{\infty} M(\omega - n\omega_s) \tag{4.32}$$

Note that $M_s(\omega)$ will repeat periodically without overlap as long as $\omega_s \geq 2\omega_M$, or $2\pi/T_s \geq 2\omega_M$, that is,

$$T_s \leq \frac{\pi}{\omega_M} \qquad \text{or} \qquad T_s \leq \frac{1}{2f_M} \tag{4.33}$$

where $\omega_M = 2\pi f_M$. Therefore, as long as we sample $m(t)$ at uniform intervals less than π/ω_M, or $1/(2f_M)$ s apart [Fig. 4-18(*e*)], $M_s(\omega)$ will be a periodic replica of $M(\omega)$ without overlap [Fig. 4-18(*f*)] and $m(t)$ can be recovered by low-pass filtering of $m_s(t)$. On the other hand, if we sample $m(t)$ at a rate less than the Nyquist rate, that is, if $\omega_s < 2\omega_M$ [Fig. 4-18(*i*)], then the shifted components of $M(\omega)$ overlap as shown in Fig. 4-18(*j*). Because of this overlap, it is no longer possible to recover $m(t)$ from $m_s(t)$ by low-pass filtering since the spectral components in these regions of overlap add, and therefore the signal is distorted. The distortion that occurs when a signal is sampled too slowly is called *aliasing*.

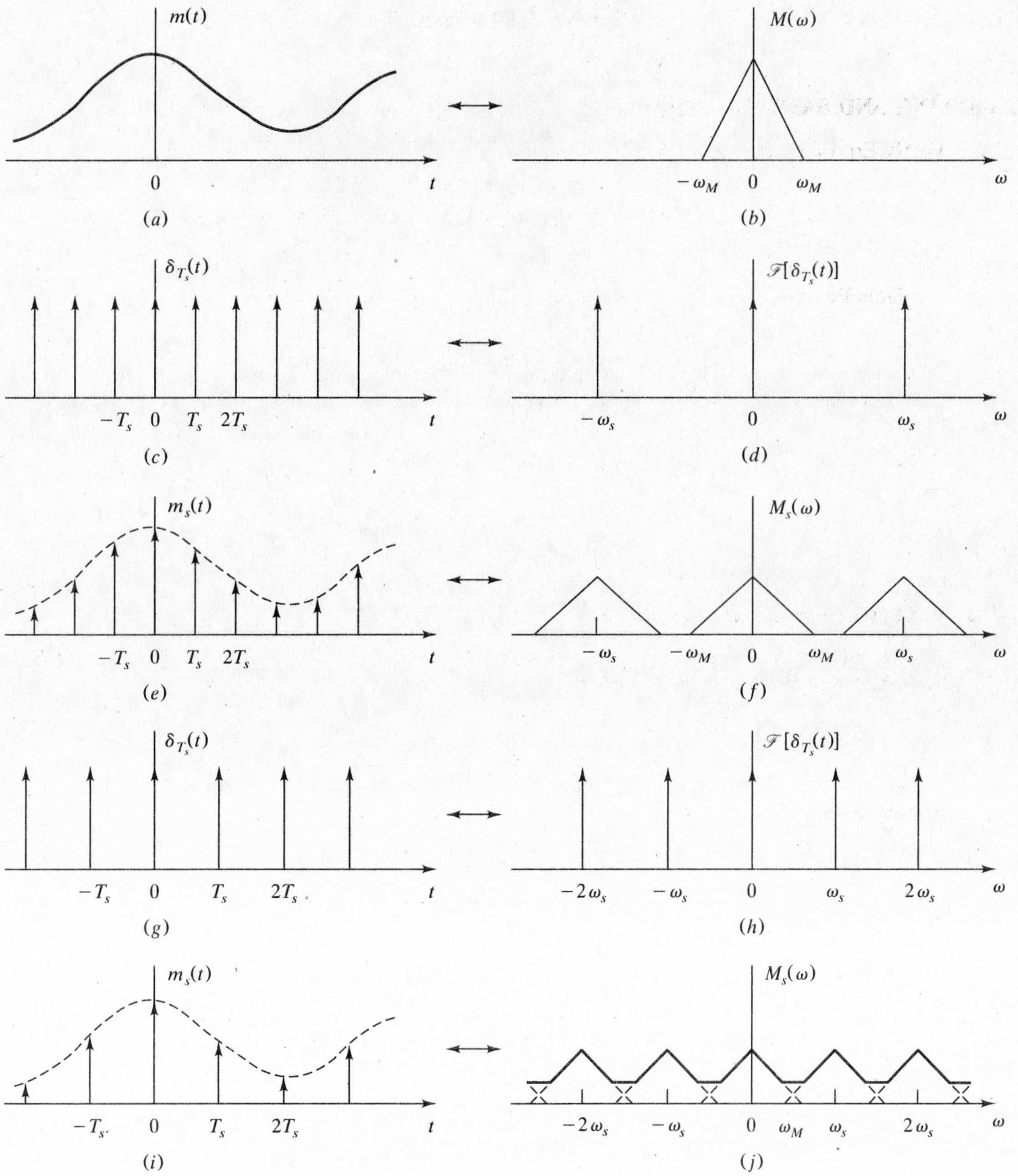

Fig. 4-18 Sampling theorem

Next, from Eq. (*4.32*),

$$T_s M_s(\omega) = \sum_{n=-\infty}^{\infty} M(\omega - n\omega_s) \tag{4.34}$$

Hence, under the following two conditions

1. $M(\omega) = 0 \quad \text{for } |\omega| > \omega_M$
2. $T_s = \dfrac{\pi}{\omega_M}$

we see from Eq. (4.34) that

$$M(\omega) = \frac{\pi}{\omega_M} M_s(\omega) \qquad \text{for } |\omega| < \omega_M \tag{4.35}$$

Next, taking the Fourier transform of Eq. (4.3) and using Eq. (1.39), we have

$$M_s(\omega) = \sum_{n=-\infty}^{\infty} m(nT_s)e^{-jnT_s\omega} \tag{4.36}$$

Substituting Eq. (4.36) into Eq. (4.35), we obtain

$$M(\omega) = \frac{\pi}{\omega_M} \sum_{n=-\infty}^{\infty} m(nT_s)e^{-jnT_s\omega} \qquad \text{for } |\omega| < \omega_M \tag{4.37}$$

Equation (4.37) shows that the Fourier transform $M(\omega)$ of the band-limited signal $m(t)$ is uniquely determined by its sample values $m(nT_s)$. Taking the inverse Fourier transform of Eq. (4.37), we obtain

$$\begin{aligned} m(t) &= \frac{1}{2\pi}\int_{-\infty}^{\infty} M(\omega)e^{j\omega t}\,d\omega \\ &= \frac{1}{2\omega_M}\int_{-\omega_M}^{\omega_M} \sum_{n=-\infty}^{\infty} m(nT_s)e^{j(t-nT_s)\omega}\,d\omega \\ &= \sum_{n=-\infty}^{\infty} m(nT_s)\frac{1}{2\omega_M}\int_{-\omega_M}^{\omega_M} e^{j(t-nT_s)\omega}\,d\omega \\ &= \sum_{n=-\infty}^{\infty} m(nT_s)\frac{\sin \omega_M(t-nT_s)}{\omega_M(t-nT_s)} \end{aligned}$$

4.3. Consider a signal $m(t) = \cos \omega_0 t$, where $\omega_0 = 2\pi f_0$. Illustrate the effect of undersampling of $m(t)$ for a sampling rate of $f_s = (3/2)f_0$.

The spectrum of $m(t)$ is given by [Eq. (1.42)]

$$M(\omega) = \pi\delta(\omega - \omega_0) + \pi\delta(\omega + \omega_0)$$

which is shown in Fig. 4-19(a). Figure 4-19(b) shows the spectrum $M_s(\omega)$ of the ideal sampled signal $m_s(t)$ with $\omega_s = (3/2)\omega_0$, where $\omega_s = 2\pi f_s$. Also indicated by a dashed line is the passband of the

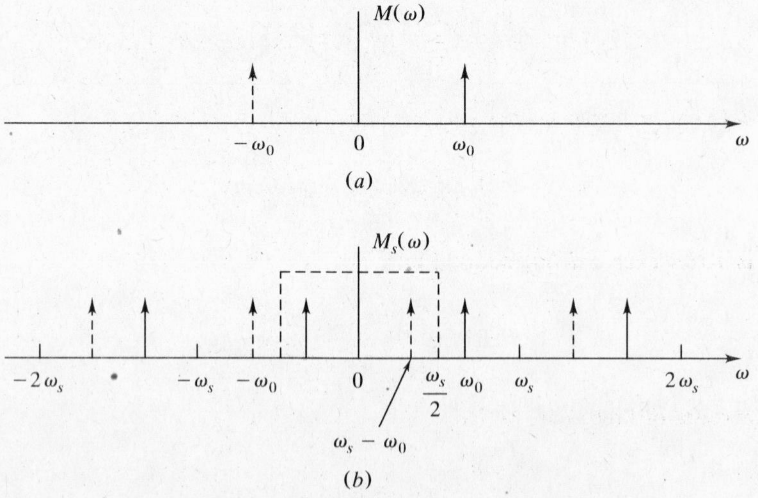

Fig. 4-19 Effect in frequency domain of undersampling

low-pass filter with $\omega_c = \omega_s/2$. Note that aliasing does occur, and the low-pass filtered output $x_r(t)$ is given by

$$x_r(t) = \cos(\omega_s - \omega_0)t = \cos\frac{1}{2}\omega_0 t \neq m(t)$$

In Fig. 4-20 we have depicted the signal $m(t)$, its samples, and the reconstructed signal $x_r(t)$.

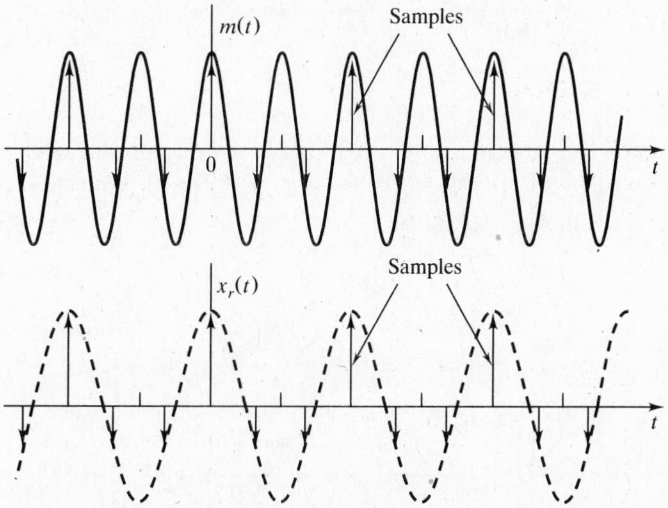

Fig. 4-20 Effect of aliasing on a sinusoidal signal

4.4. Consider the sampling theorem (*4.2*) with $T_s = \pi/\omega_M$, that is,

$$m(t) = \sum_{n=-\infty}^{\infty} m(nT_s)\phi_n(t) \tag{4.38}$$

where

$$\phi_n(t) = \frac{\sin\omega_M(t - nT_s)}{\omega_M(t - nT_s)} \qquad n = 0, \pm 1, \pm 2, \ldots \tag{4.39}$$

Show that $\phi_n(t)$ is orthogonal over the interval $-\infty < t < \infty$, and

$$\int_{-\infty}^{\infty} \phi_n(t)\phi_k(t)\,dt = T_s\delta_{nk} \tag{4.40}$$

where δ_{nk} is *Kronecker's delta*, defined by

$$\delta_{nk} = \begin{cases} 1 & n = k \\ 0 & n \neq k \end{cases}$$

From the result of Prob. 1.14, we have

$$\frac{\sin at}{\pi t} \longleftrightarrow p_a(\omega) = \begin{cases} 1 & |\omega| < a \\ 0 & |\omega| > a \end{cases}$$

Thus

$$\frac{\sin \omega_M t}{\omega_M t} \longleftrightarrow \frac{\pi}{\omega_M} p_{\omega_M}(\omega) = \begin{cases} \dfrac{\pi}{\omega_M} & |\omega| < \omega_M \\ 0 & |\omega| > \omega_M \end{cases}$$

Then by Eq. (*1.30*), we have

$$\phi_n(t) = \frac{\sin \omega_M (t - nT_s)}{\omega_M (t - nT_s)} \longleftrightarrow \Phi_n(\omega) = \begin{cases} \dfrac{\pi}{\omega_M} e^{-j\omega n T_s} & |\omega| < \omega_M \\ 0 & |\omega| > \omega_M \end{cases}$$

$$\phi_k(t) = \frac{\sin \omega_M (t - kT_s)}{\omega_M (t - kT_s)} \longleftrightarrow \Phi_k(\omega) = \begin{cases} \dfrac{\pi}{\omega_M} e^{-j\omega k T_s} & |\omega| < \omega_M \\ 0 & |\omega| > \omega_M \end{cases}$$

Then, by Eq. (*1.128*) of Prob. 1.29, that is,

$$\int_{-\infty}^{\infty} x_1(t) x_2(t)\, dt = \frac{1}{2\pi} \int_{-\infty}^{\infty} X_1(\omega) X_2(-\omega)\, d\omega$$

we obtain

$$\begin{aligned} \int_{-\infty}^{\infty} \phi_n(t)\phi_k(t)\, dt &= \frac{1}{2\pi} \int_{-\omega_M}^{\omega_M} \left(\frac{\pi}{\omega_M}\right)^2 e^{-j\omega n T_s} e^{j\omega k T_s}\, d\omega \\ &= \frac{\pi}{2\omega_M^2} \int_{-\omega_M}^{\omega_M} e^{-j\omega(n-k)T_s}\, d\omega \\ &= \begin{cases} \dfrac{\pi}{\omega_M} = T_s & n = k \\ 0 & n \neq k \end{cases} \end{aligned}$$

Hence, we conclude that

$$\int_{-\infty}^{\infty} \phi_n(t)\phi_k(t)\, dt = T_s \delta_{nk}$$

4.5. If $m(t)$ is band-limited, that is, $M(\omega) = 0$ for $|\omega| > \omega_M$, then show that

$$\int_{-\infty}^{\infty} [m(t)]^2\, dt = T_s \sum_{n=-\infty}^{\infty} [m(nT_s)]^2 \tag{4.41}$$

where $T_s = \pi/\omega_M$.

Using Eqs. (*4.39*) and (*4.40*), we have

$$\begin{aligned} \int_{-\infty}^{\infty} [m(t)]^2\, dt &= \int_{-\infty}^{\infty} \left[\sum_{n=-\infty}^{\infty} m(nT_s)\phi_n(t)\right]\left[\sum_{k=-\infty}^{\infty} m(kT_s)\phi_k(t)\right] dt \\ &= \sum_{n=-\infty}^{\infty} m(nT_s)\left[\sum_{k=-\infty}^{\infty} m(kT_s) \int_{-\infty}^{\infty} \phi_n(t)\phi_k(t)\, dt\right] \\ &= \sum_{n=-\infty}^{\infty} m(nT_s)\left[\sum_{k=-\infty}^{\infty} m(kT_s) T_s \delta_{nk}\right] \\ &= T_s \sum_{n=-\infty}^{\infty} [m(nT_s)]^2 \end{aligned}$$

4.6. Find the Nyquist rate and the Nyquist interval for each of the following signals:

(*a*) $m(t) = 5\cos 1000\pi t \cos 4000\pi t$

(*b*) $m(t) = \dfrac{\sin 200\pi t}{\pi t}$

(*c*) $m(t) = \left(\dfrac{\sin 200\pi t}{\pi t}\right)^2$

(*a*)

$$\begin{aligned} m(t) &= 5\cos 1000\pi t \cos 4000\pi t \\ &= 2.5(\cos 3000\pi t + \cos 5000\pi t) \end{aligned}$$

Thus, $m(t)$ is band-limited with $f_M = 2500$ Hz. Hence, the Nyquist rate is 5000 Hz, and the Nyquist interval is $1/5000$ s = 0.2 ms.

(*b*) From the result of Prob. 1.14, we have

$$\frac{\sin at}{\pi t} \longleftrightarrow p_a(\omega) = \begin{cases} 1 & |\omega| < a \\ 0 & |\omega| > a \end{cases}$$

Thus we see that $m(t)$ is a band-limited signal with $f_M = 100$ Hz. Hence, the Nyquist rate is 200 Hz, and the Nyquist interval is $1/200$ s.

(*c*) From the frequency convolution theorem (*1.54*), we find that the signal $m(t)$ is also band-limited and its bandwidth is twice that of the signal of part (*b*), that is, 200 Hz. Thus, the Nyquist rate is 400 Hz, and the Nyquist interval is $1/400$ s.

4.7. The bandpass sampling theorem states that if a bandpass signal $m(t)$ has a spectrum of bandwidth ω_B $(= 2\pi f_B)$ and an upper frequency limit ω_u $(= 2\pi f_u)$, then $m(t)$ can be recovered from $m_s(t)$ by bandpass filtering if $f_s = 2f_u/k$, where k is the largest integer not exceeding f_u/f_B. All higher sampling rates are not necessarily usable unless they exceed $2f_u$.

Consider the bandpass signal $m(t)$ with a spectrum shown in Fig. 4-21. Check the bandpass sampling theorem by sketching the spectrum of the ideally sampled signal $m_s(t)$ when $f_s = 25$, 45, and 50 kHz. Indicate if and how the signal can be recovered.

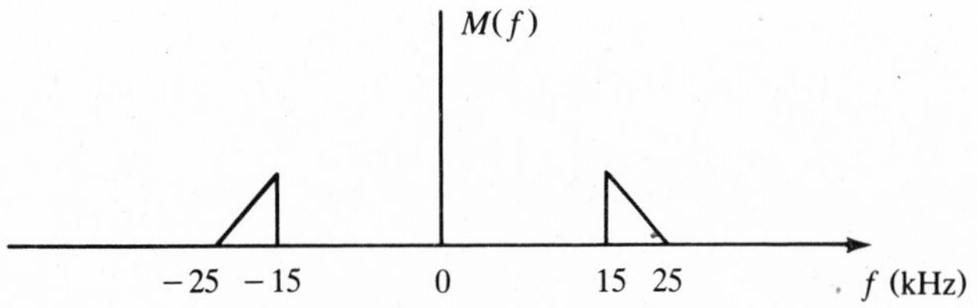

Fig. 4-21 Bandpass signal spectrum

From Fig. 4-21, $f_u = 25$ kHz and $f_B = 10$ kHz. Then $f_u/f_B = 2.5$ and $k = 2$. Hence we have $f_s = 2f_u/k = 25$ kHz.

For $f_s = 25$ kHz: From Fig. 4-22(*a*), we see that $m(t)$ can be recovered from the sampled signal by using a bandpass filter over

$$f_{c_1} \le f \le 25 \text{ kHz} \qquad \text{with} \qquad 10 \text{ kHz} \le f_{c_1} < 15 \text{ kHz}$$

For $f_s = 45$ kHz: From Fig. 4-22(*b*), it is not possible to recover $m(t)$ by filtering.

For $f_s = 50$ kHz: From Fig. 4-22(*c*), we see that $m(t)$ can be recovered by using a low-pass filter with cutoff frequency $f_c = 25$ kHz.

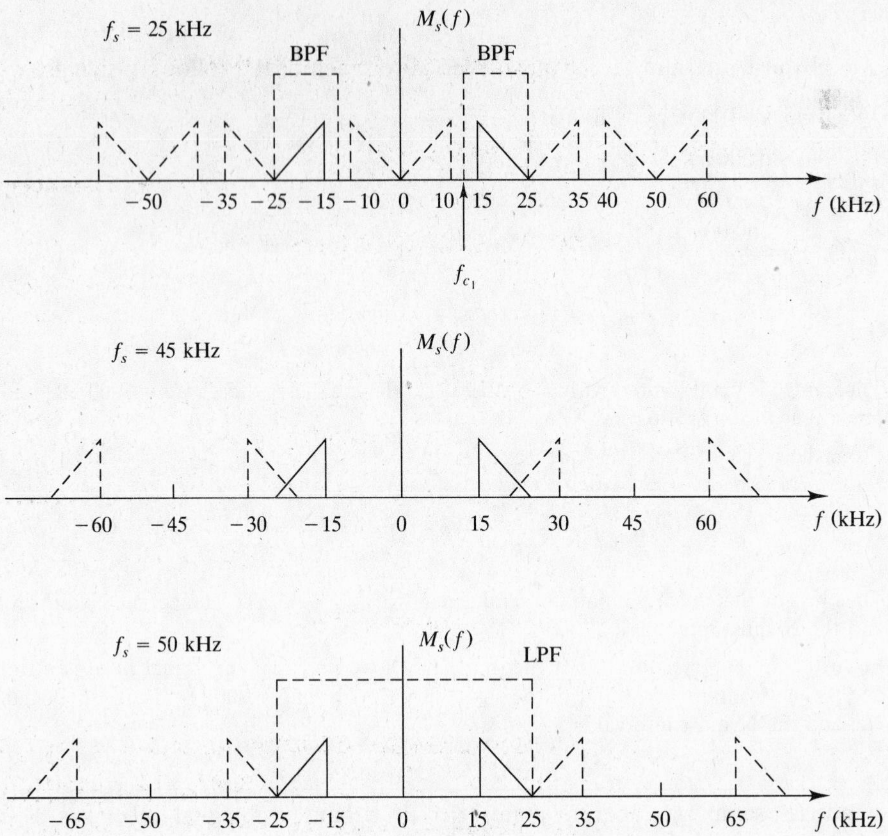

Fig. 4-22

4.8. Given the signal

$$m(t) = 10\cos 2000\pi t \cos 8000\pi t$$

(*a*) What is the minimum sampling rate based on the low-pass uniform sampling theorem?

(*b*) Repeat (*a*) based on the bandpass sampling theorem.

(*a*)
$$m(t) = 10\cos 2000\pi t \cos 8000\pi t$$
$$= 5\cos 6000\pi t + 5\cos 10\,000\pi t$$
$$f_M = 5000 \text{ Hz} = 5 \text{ kHz}$$

Thus, $f_s = 2f_M = 10$ kHz.

(*b*) $f_u = f_M = 5$ kHz and $f_B = (5 - 3) = 2$ kHz.

$$\frac{f_u}{f_B} = \frac{5}{2} = 2.5 \longrightarrow k = 2$$

Based on the bandpass sampling theorem,

$$f_s = \frac{2f_u}{k} = 5 \text{ kHz}$$

4.9. Show that if the sampling rate is equal to or greater than twice the highest message frequency, the message $m(t)$ can be recovered from the natural sampled signal $x_{ns}(t)$ by low-pass filtering.

As shown in Eq. (*4.4*) and Fig. 4-23, the natural sampled signal $x_{ns}(t)$ is equal to $m(t)$ multiplied by a rectangular pulse train $x_p(t)$ whose Fourier series is given by (see Prob. 1.10)

$$x_p(t) = \sum_{n=-\infty}^{\infty} c_n e^{jn\omega_s t} \qquad \omega_s = \frac{2\pi}{T_s}$$

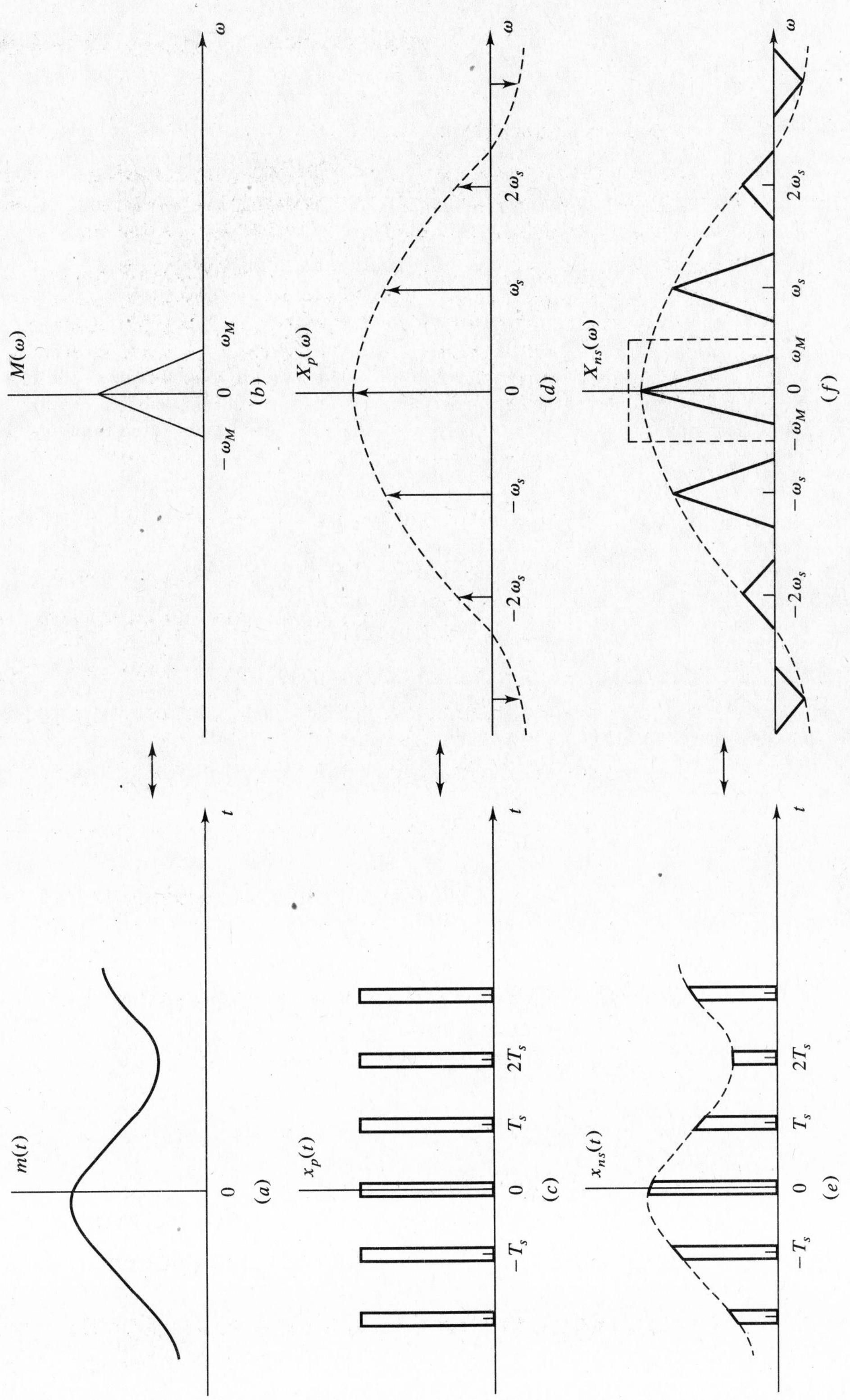

Fig. 4-23 Natural sampling

where

$$c_n = \frac{d}{T_s} \frac{\sin(n\omega_s d/2)}{n\omega_s d/2}$$

Then

$$x_{ns}(t) = m(t)x_p(t) = m(t) \sum_{n=-\infty}^{\infty} c_n e^{jn\omega_s t} = \sum_{n=-\infty}^{\infty} c_n m(t) e^{jn\omega_s t} \tag{4.42}$$

Hence, by the frequency-shifting property of the Fourier transform (*1.31*), we have

$$X_{ns}(\omega) = \sum_{n=-\infty}^{\infty} c_n M(\omega - n\omega_s) \tag{4.43}$$

Equation (*4.43*) indicates that the spectrum $X_{ns}(\omega)$ consists of a weighted version of the spectrum $M(\omega)$ centered on integer multiples of the sampling frequency. The spectral component at $\omega = n\omega_s$ is multiplied by c_n. It is clear from Fig. 4-23(f) that if $\omega_s \geq 2\omega_M$, then $m(t)$ can be recovered from $x_{ns}(t)$ by low-pass filtering of $x_{ns}(t)$.

4.10. Verify Eq. (*4.7*), that is,

$$x_s(t) = m_s(t) * p(t)$$

From Eq. (*4.3*),

$$m_s(t) = \sum_{n=-\infty}^{\infty} m(nT_s)\delta(t - nT_s)$$

Then using Eqs. (*1.48*) and (*1.52*), we obtain

$$\begin{aligned} m_s(t) * p(t) &= \sum_{n=-\infty}^{\infty} m(nT_s)\delta(t - nT_s) * p(t) \\ &= \sum_{n=-\infty}^{\infty} m(nT_s)p(t) * \delta(t - nT_s) \\ &= \sum_{n=-\infty}^{\infty} m(nT_s)p(t - nT_s) = x_s(t) \end{aligned}$$

4.11. Find the spectrum of $x_s(t)$ [Eq. (*4.7*)] in terms of the spectrum of the message $M(\omega)$ and discuss the distortion in the recovered waveform if flat-top sampling is used.

Using Eq. (*1.112*) of Prob. 1.13, we have

$$p(t) \longleftrightarrow P(\omega) = d\frac{\sin(\omega d/2)}{\omega d/2} \tag{4.44}$$

Next, applying the convolution theorem (*1.53*) to Eq. (*4.7*) and using Eq. (*4.32*), we obtain

$$X_s(\omega) = M_s(\omega)P(\omega) = \frac{1}{T_s} \sum_{n=-\infty}^{\infty} M(\omega - n\omega_s)P(\omega) \tag{4.45}$$

Figure 4-24 illustrates a graphical interpretation of Eq. (*4.45*), with the assumed $M(\omega)$. We see that flat-top sampling is equivalent to passing an ideal sampled signal through a filter having the frequency response $H(\omega) = P(\omega)$. The high-frequency roll-off characteristic of $P(\omega)$ acts as a low-pass filter and attenuates the upper portion of the message spectrum. This loss of high-frequency content is known as the *aperture effect*. The larger the pulse duration of aperture d, the larger the effect. Since T_s does not depend on d, the ratio d/T_s is a measure of the flatness of $P(\omega)$ within the bandwidth of the low-pass filter. In practice, this aperture effect can be neglected as long as $d/T_s \leq 0.1$.

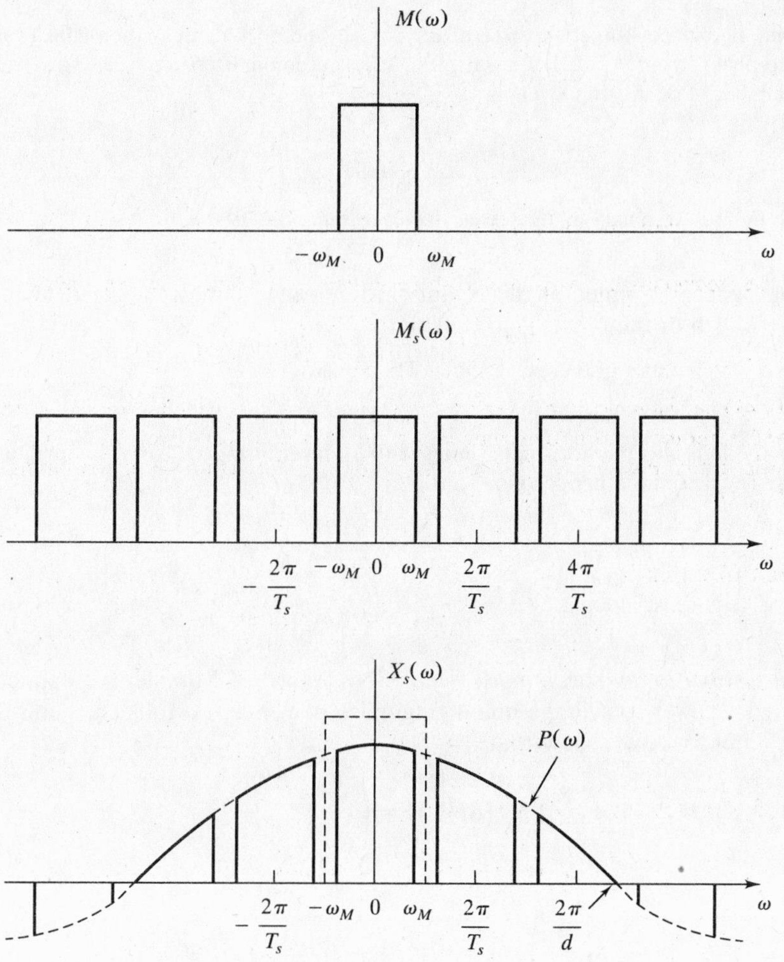

Fig. 4-24 Aperture effect in flat-top sampling

QUANTIZING

4.12. A binary channel with bit rate $R_b = 36\,000$ bits per second (b/s) is available for PCM voice transmission. Find appropriate values of the sampling rate f_s, the quantizing level L, and the binary digits n, assuming $f_M = 3.2$ kHz.

Since we require

$$f_s \geq 2f_M = 6400 \qquad \text{and} \qquad nf_s \leq R_b = 36\,000$$

then

$$n \leq \frac{R_b}{f_s} \leq \frac{36\,000}{6400} = 5.6$$

So $n = 5$, $L = 2^5 = 32$, and

$$f_s = \frac{36\,000}{5} = 7200 \text{ Hz} = 7.2 \text{ kHz}$$

4.13. An analog signal is quantized and transmitted by using a PCM system. If each sample at the receiving end of the system must be known to within ± 0.5 percent of the peak-to-peak full-scale value, how many binary digits must each sample contain?

Let $2m_p$ be the peak-to-peak value of the signal. The peak error is then $0.005(2m_p) = 0.01m_p$, and the peak-to-peak error is $2(0.01m_p) = 0.02m_p$ (the maximum step size Δ). Thus from Eq. (*4.8*) the required number of quantizing levels is

$$L = \frac{2m_p}{\Delta} = \frac{2m_p}{0.02m_p} = 100 \le 2^n$$

Hence, the number of binary digits needed for each sample is $n = 7$.

4.14. An analog signal is sampled at the Nyquist rate f_s and quantized into L levels. Find the time duration τ of 1 b of the binary-encoded signal.

Let n be the number of bits per sample. Then by Eq. (*4.14*)

$$n = [\log_2 L]$$

where $[\log_2 L]$ indicates the next higher integer to be taken if $\log_2 L$ is not an integer value; nf_s binary pulses must be transmitted per second. Thus

$$\tau = \frac{1}{nf_s} = \frac{T_s}{n} = \frac{T_s}{[\log_2 L]}$$

where T_s is the Nyquist interval.

4.15. The output *signal-to-quantizing-noise ratio* $(\text{SNR})_o$ in a PCM system is defined as the ratio of average signal power to average quantizing noise power. For a full-scale sinusoidal modulating signal with amplitude A, show that

$$(\text{SNR})_o = \left(\frac{S}{N_q}\right)_o = \frac{3}{2}L^2 \tag{4.46}$$

or

$$\left(\frac{S}{N_q}\right)_{0\text{ dB}} = 10\log\left(\frac{S}{N_q}\right)_o = 1.76 + 20\log L \tag{4.47}$$

where L is the number of quantizing levels.

The peak-to-peak excursion of the quantizer input is $2A$. From Eq. (*4.8*), the quantizer step size is

$$\Delta = \frac{2A}{L}$$

Then from Eq. (*4.10*) or (*4.11*), the average quantizing noise power is

$$N_q = \langle q_e^2 \rangle = \frac{\Delta^2}{12} = \frac{A^2}{3L^2}$$

The output signal-to-quantizing-noise ratio of a PCM system for a full-scale test tone is therefore

$$(\text{SNR})_o = \left(\frac{S}{N_q}\right)_o = \frac{A^2/2}{A^2/(3L^2)} = \frac{3}{2}L^2$$

Expressing this in decibels, we have

$$\left(\frac{S}{N_q}\right)_{0\text{ dB}} = 10\log\left(\frac{S}{N_q}\right)_o = 1.76 + 20\log L$$

4.16. In a binary PCM system, the output signal-to-quantizing-noise ratio is to be held to a minimum of 40 dB. Determine the number of required levels, and find the corresponding output signal-to-quantizing-noise ratio.

In a binary PCM system, $L = 2^n$, where n is the number of binary digits. Then Eq. (*4.47*) becomes

$$\left(\frac{S}{N_q}\right)_{0\,\text{dB}} = 1.76 + 20\log 2^n = 1.76 + 6.02n \quad \text{dB} \tag{4.48}$$

Now

$$\left(\frac{S}{N_q}\right)_{0\,\text{dB}} = 40\text{ dB} \longrightarrow \left(\frac{S}{N_q}\right)_o = 10\,000$$

Thus, from Eq. (*4.46*),

$$L = \sqrt{\frac{2}{3}\left(\frac{S}{N_q}\right)_o} = \sqrt{\frac{2}{3}(10\,000)} = [81.6] = 82$$

and the number of binary digits n is

$$n = [\log_2 82] = [6.36] = 7$$

Then the number of levels required is $L = 2^7 = 128$, and the corresponding output signal-to-quantizing-noise ratio is

$$\left(\frac{S}{N_q}\right)_{0\,\text{dB}} = 1.76 + 6.02 \times 7 = 43.9 \text{ dB}$$

Note: Equation (*4.48*) indicates that each bit in the code word of a binary PCM system contributes 6 dB to the output signal-to-quantizing-noise ratio. This is called the *6-dB rule*.

4.17. A compact disk (CD) recording system samples each of two stereo signals with a 16-bit analog-to-digital converter (ADC) at 44.1 kb/s.

(*a*) Determine the output signal-to-quantizing-noise ratio for a full-scale sinusoid.

(*b*) The bit stream of digitized data is augmented by the addition of error-correcting bits, clock extraction bits, and display and control bit fields. These additional bits represent 100 percent overhead. Determine the output bit rate of the CD recording system.

(*c*) The CD can record an hour's worth of music. Determine the number of bits recorded on a CD.

(*d*) For a comparison, a high-grade collegiate dictionary may contain 1500 pages, 2 columns per page, 100 lines per column, 8 words per line, 6 letters per word, and 7 b per letter on average. Determine the number of bits required to describe the dictionary, and estimate the number of comparable books that can be stored on a CD.

(*a*) From Eq. (*4.47*),

$$\left(\frac{S}{N_q}\right)_{0\,\text{dB}} = 1.76 + 6.02 \times 16 = 98.08 \text{ dB}$$

The very high SNR of the disk has the effect of increasing the dynamic range of recording, resulting in the excellent clarity of sound from a CD.

(*b*) The input bit rate is

$$2(44.1)(10^3)(16) = 1.411(10^6)\text{ b/s} = 1.411\text{ Mb/s}$$

Including the additional 100 percent overhead, the output bit rate is

$$2(1.411)(10^6)\text{ b/s} = 2.822\text{ Mb/s}$$

(*c*) The number of bits recorded on a CD is

$$2.822(10^6)(3600) = 10.16(10^9)\text{ b} = 10.16\text{ gigabits (GB)}$$

(*d*) The number of bits required to describe the dictionary is

$$1500(2)(100)(8)(6)(7) = 100.8(10^6)\text{ b} = 100.8\text{ Mb}$$

Including the additional 100 percent overhead, then,

$$\frac{10.16(10^9)}{2(100.8)(10^6)} = 50.4$$

Thus a disk contains the equivalent of about 50 comparable-books storage capacity.

4.18. (*a*) Plot the μ law compression characteristic for $\mu = 255$.

(*b*) If $m_p = 20$ V and 256 quantizing levels are employed, what is the voltage between levels when there is no compression? For $\mu = 255$, what is the smallest and what is the largest effective separation between levels?

(*a*) From Eq. (*4.12*), for $\mu = 255$ we have

$$y = \pm \frac{\ln(1 + 255|x|)}{\ln 256} \qquad |x| < 1$$

where $x = m/m_p$. The plot of the μ law compression characteristic for $\mu = 255$ is shown in Fig. 4-25.

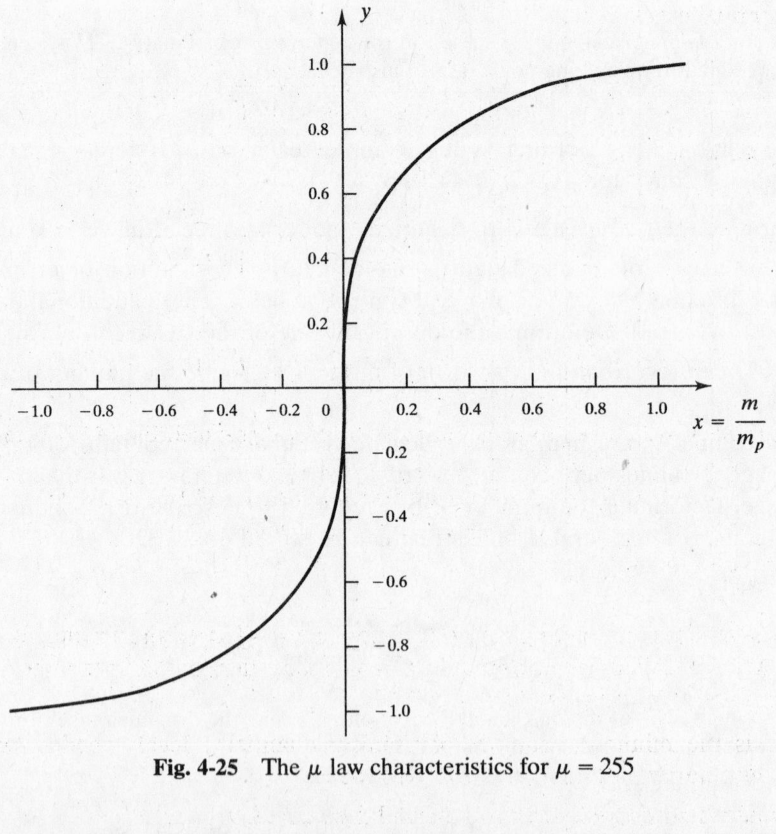

Fig. 4-25 The μ law characteristics for $\mu = 255$

(*b*) With no compression (that is, a uniform quantizing), from Eq. (*4.8*) the step size Δ is

$$\Delta = \frac{2m_p}{L} = \frac{40}{256} = 0.156\text{ V}$$

With compression (that is, a nonuniform quantizing), the smallest effective separation between levels will be the one closest to the origin, and the largest effective separation between levels will be the one closest to $|x| = 1$.

Let x_1 be the value of x corresponding to $y = 1/127$, that is,

$$\frac{\ln(1 + 255|x_1|)}{\ln 256} = \frac{1}{127}$$

Solving for $|x_1|$, we obtain

$$|x_1| = 1.75(10^{-4})$$

Thus the smallest effective separation between levels is given by

$$\Delta_{\min} = m_p|x_1| = 20(1.75)(10^{-4}) = 3.5(10^{-3})\text{ V} = 3.5\text{ mV}$$

Next, let x_{127} be the value of x corresponding to $y = 1 - 1/127$, that is,

$$\frac{\ln(1 + 255|x_{127}|)}{\ln 256} = \frac{126}{127}$$

Solving for $|x_{127}|$, we obtain

$$|x_{127}| = 0.957$$

Thus the largest effective separation between levels is given by

$$\Delta_{\max} = m_p(1 - |x_{127}|) = 20(1 - 0.957) = 0.86\text{ V}$$

4.19. When a μ law compander is used in PCM, the output signal-to-quantizing-noise ratio for $\mu \gg 1$ is approximated by

$$\left(\frac{S}{N_q}\right)_o \approx \frac{3L^2}{[\ln(1+\mu)]^2} \tag{4.49}$$

Derive the 6-dB rule for $\mu = 255$.

$$\left(\frac{S}{N_q}\right)_{0\text{ dB}} = 10\log\left(\frac{S}{N_q}\right)_o = 10\log\frac{3L^2}{[\ln(1+\mu)]^2}$$

For $\mu = 255$,

$$\left(\frac{S}{N_q}\right)_{0\text{ dB}} = 10\log\frac{3L^2}{(\ln 256)^2} = 20\log L - 10.1 \quad \text{dB} \tag{4.50}$$

In a binary PCM, $L = 2^n$, where n is the number of binary digits; then Eq. (*4.50*) becomes

$$\left(\frac{S}{N_q}\right)_{0\text{ dB}} = 20\log 2^n - 10.1 = 6.02n - 10.1 \quad \text{dB} \tag{4.51}$$

which is the 6-dB rule for $\mu = 255$.

4.20. Consider an audio signal with spectral components limited to the frequency band of 300 to 3300 Hz. A PCM signal is generated with a sampling rate of 8000 samples/s. The required output signal-to-quantizing-noise ratio is 30 dB.

(*a*) What is the minimum number of uniform quantizing levels needed, and what is the minimum number of bits per sample needed?

(*b*) Calculate the minimum system bandwidth required.

(*c*) Repeat parts (*a*) and (*b*) when a μ law compander is used with $\mu = 255$.

(*a*) Using Eq. (*4.47*), we have

$$\left(\frac{S}{N_q}\right)_{0\text{ dB}} = 1.76 + 20\log L \geq 30$$

$$\log L \geq \frac{1}{20}(30 - 1.76) = 1.412 \longrightarrow L \geq 25.82$$

Thus the minimum number of uniform quantizing levels needed is 26.

$$n = [\log_2 L] = [\log_2 26] = [4.7] = 5 \text{ b per sample}$$

The minimum number of bits per sample is 5.

(*b*) From Eq. (*4.15*), the minimum required system bandwidth is

$$f_{\text{PCM}} = \frac{n}{2} f_s = \frac{5}{2}(8000) = 20\,000 \text{ Hz} = 20 \text{ kHz}$$

(*c*) Using Eq. (*4.50*),

$$\left(\frac{S}{N_q}\right)_{0 \text{ dB}} = 20 \log L - 10.1 \geq 30$$

$$\log L \geq \frac{1}{20}(30 + 10.1) = 2.005 \longrightarrow L \geq 101.2$$

Thus the minimum number of quantizing levels needed is 102.

$$n = [\log_2 L] = [6.67] = 7$$

The minimum number of bits per sample is 7. The minimum bandwidth required for this case is

$$f_{\text{PCM}} = \frac{n}{2} f_s = \frac{7}{2}(8000) = 28\,000 \text{ Hz} = 28 \text{ kHz}$$

DELTA MODULATION

4.21. Consider a sinusoidal signal $m(t) = A \cos \omega_m t$ applied to a delta modulator with step size Δ. Show that slope overload distortion will occur if

$$A > \frac{\Delta}{\omega_m T_s} = \frac{\Delta}{2\pi}\left(\frac{f_s}{f_m}\right) \tag{4.52}$$

where $f_s = 1/T_s$ is the sampling frequency.

$$m(t) = A \cos \omega_m t \qquad \frac{dm(t)}{dt} = -A\omega_m \sin \omega_m t$$

From Eq. (*4.20*), to avoid the slope overload, we require that

$$\frac{\Delta}{T_s} \geq \left|\frac{dm(t)}{dt}\right|_{\max} = A\omega_m \qquad \text{or} \qquad A \leq \frac{\Delta}{\omega_m T_s}$$

Thus, if $A > \Delta/(\omega_m T_s)$, slope overload distortion will occur.

4.22. For a sinusoidal modulating signal

$$m(t) = A \cos \omega_m t \qquad \omega_m = 2\pi f_m$$

show that the maximum output signal-to-quantizing-noise ratio in a DM system under the assumption of no slope overload is given by

$$(\text{SNR})_o = \left(\frac{S}{N_q}\right)_o = \frac{3f_s^3}{8\pi^2 f_m^2 f_M} \tag{4.53}$$

where $f_s = 1/T_s$ is the sampling rate and f_M is the cutoff frequency of a low-pass filter at the output end of the receiver.

From Eq. (*4.52*), for no-slope-overload condition, we must have

$$A < \frac{\Delta}{\omega_m T_s} = \frac{\Delta}{2\pi}\left(\frac{f_s}{f_m}\right)$$

Thus, the maximum permissible value of the output signal power equals

$$P_{\max} = \frac{A^2}{2} = \frac{\Delta^2 f_s^2}{8\pi^2 f_m^2} \tag{4.54}$$

From Eq. (4.21), the mean-square quantizing error, or the quantizing noise power, is $\langle q_e^2 \rangle = \Delta^2/3$. Let the bandwidth of a postreconstruction low-pass filter at the output end of the receiver be $f_M \geq f_m$ and $f_M \ll f_s$. Then, assuming that the quantizing noise power P_q is uniformly distributed over the frequency band up to f_s, the output quantizing noise power within the bandwidth f_M is

$$N_q = \left(\frac{\Delta^2}{3}\right)\frac{f_M}{f_s} \tag{4.55}$$

Combining Eqs. (4.54) and (4.55), we see that the maximum output signal-to-quantizing-noise ratio is

$$\left(\frac{S}{N_q}\right)_o = \frac{P_{\max}}{N_q} = \frac{3f_s^3}{8\pi^2 f_m^2 f_M}$$

4.23. Determine the output SNR in a DM system for a 1-kHz sinusoid, sampled at 32 kHz, without slope overload, and followed by a 4-kHz postreconstruction filter.

From Eq. (4.53), we obtain

$$(\text{SNR})_o = \frac{3\left[(32)(10^3)\right]^3}{8\pi^2(10^3)^2(4)(10^3)} = 311.3 = 24.9 \text{ dB}$$

4.24. The data rate for Prob. 4.23 is 32 kb/s, which is the same bit rate obtained by sampling at 8 kHz with 4 b per sample in a PCM system. Find the average output SNR of a 4-b PCM quantizer for the sampling of a full-scale sinusoid with $f_s = 8$ kHz, and compare it with the result of Prob. 4.23.

From Eq. (4.48), we have

$$(\text{SNR})_{0\text{ dB}} = 1.76 + 6.02(4) = 25.84 \text{ dB}$$

Comparing this result with that of Prob. 4.23, we conclude that for all the simplicity of DM, it does not perform as well as even a 4-b PCM.

4.25. A DM system is designed to operate at 3 times the Nyquist rate for a signal with a 3-kHz bandwidth. The quantizing step size is 250 mV.

(a) Determine the maximum amplitude of a 1-kHz input sinusoid for which the delta modulator does not show slope overload.

(b) Determine the postfiltered output signal-to-quantizing-noise ratio for the signal of part (a).

(a)

$$m(t) = A\cos\omega_m t = A\cos 2\pi(10^3)t$$

$$\left|\frac{dm(t)}{dt}\right|_{\max} = A(2\pi)(10^3)$$

By Eq. (4.52), the maximum allowable amplitude of the input sinusoid is

$$A_{\max} = \frac{\Delta}{\omega_m T_s} = \frac{\Delta}{\omega_m} f_s = \frac{250}{2\pi(10^3)} 3(2)(3)(10^3) = 716.2 \text{ mV}$$

(b) From Eq. (4.53), and assuming that the cutoff frequency of the low-pass filter is f_m, we have

$$(\text{SNR})_o = \left(\frac{S}{N_q}\right)_o = \frac{3\left[(3)(6)(10^3)\right]^3}{8\pi^2(10^3)^3} = 221.6 = 23.5 \text{ dB}$$

SIGNALING FORMATS

4.26. Consider the binary sequence 0100101. Draw the waveforms for the following signaling formats.

(*a*) Unipolar NRZ signaling format

(*b*) Bipolar RZ signaling format

(*c*) AMI (alternate mark inversion) RZ signaling format

See Fig. 4-26.

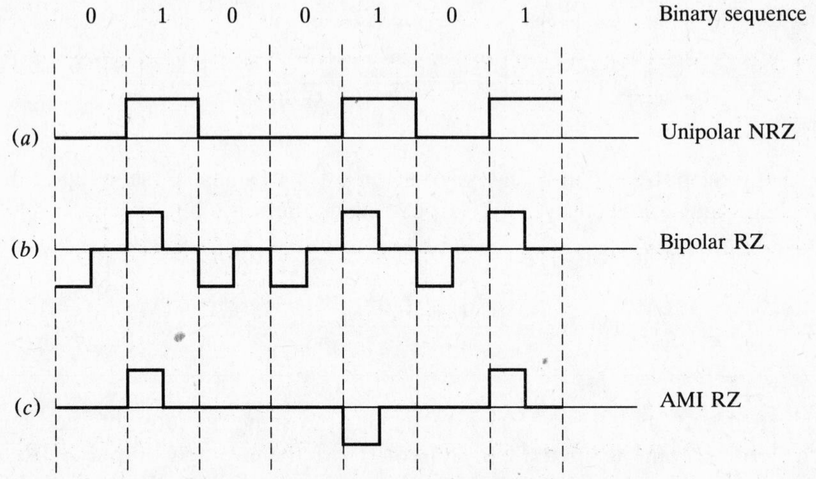

Fig. 4-26 Signaling formats for Prob. 4.26

4.27. Discuss the advantages and disadvantages of the three signaling formats illustrated in Fig. 4-26 of Prob. 4.26.

The unipolar NRZ signaling format, although conceptually simple, has disadvantages: There are no pulse transitions for long sequences of 0s or 1s, which are necessary if one wishes to extract timing or synchronizing information; and there is no way to detect when and if an error has occurred from the received pulse sequence.

The bipolar RZ signaling format guarantees the availability of timing information, but there is no error detection capability.

The AMI RZ signaling format has an error detection property; if two sequential pulses (ignoring intervening 0s) are detected with the same polarity, it is evident that an error has occurred. However, to guarantee the availability of timing information, it is necessary to restrict the allowable number of consecutive 0s.

4.28. Consider a binary sequence with a long sequence of 1s followed by a single 0 and then a long sequence of 1s. Draw the waveforms for this sequence, using the following signaling formats:

(*a*) Unipolar NRZ signaling

(*b*) Bipolar NRZ signaling

(*c*) AMI RZ signaling

(*d*) Split-Phase (Manchester) signaling

See Fig. 4-27.

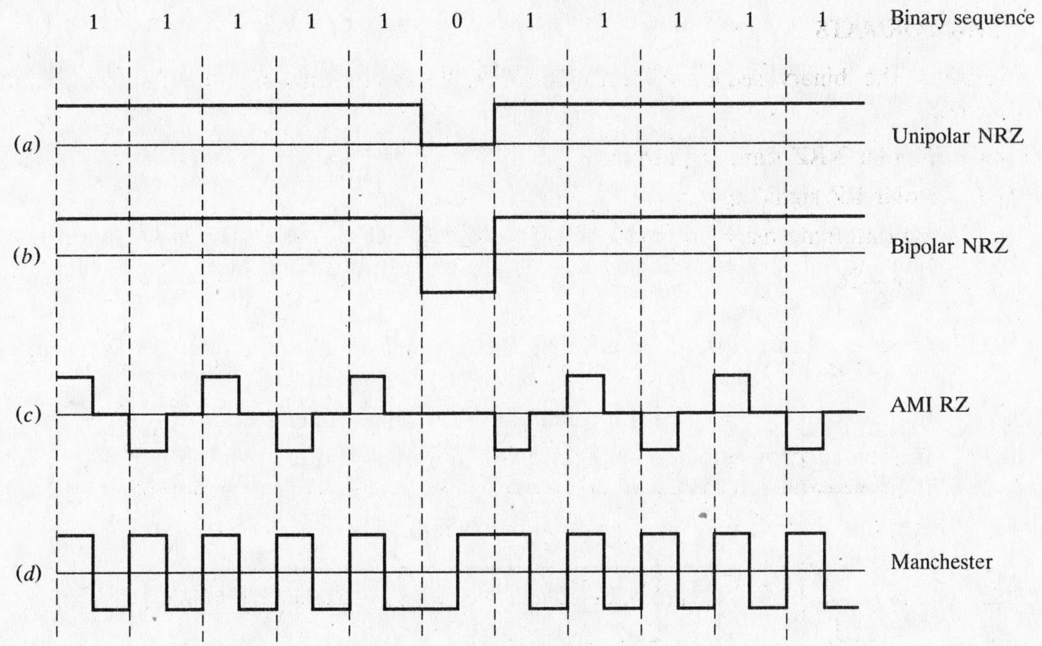

Fig. 4-27

4.29. The AMI RZ signaling waveform representing the binary sequence 0100101011 is transmitted over a noisy channel. The received waveform is shown in Fig. 4-28 which contains a single error. Locate the position of this error, and justify your answer.

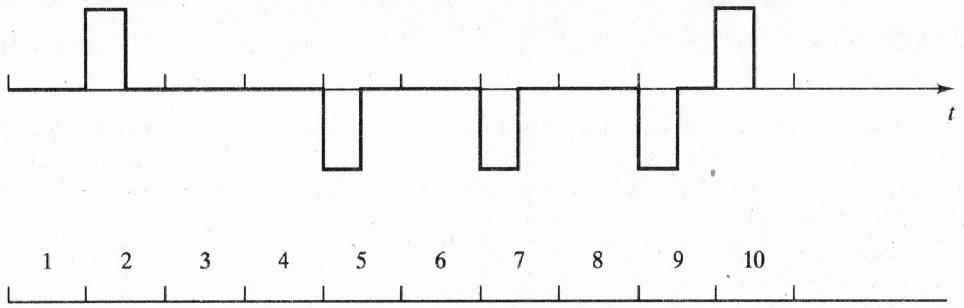

Fig. 4-28

The error is located at the bit position 7 (as indicated in Fig.4-28), where we have a negative pulse. This bit is in error, because with the AMI signaling format, positive and negative pulses (of equal amplitude) are used alternatively for symbol 1, and no pulse is used for symbol 0. The pulse in position 7 representing the third digit 1 in the data stream should have had positive polarity.

TIME-DIVISION MULTIPLEXING

4.30. Two analog signals $m_1(t)$ and $m_2(t)$ are to be transmitted over a common channel by means of time-division multiplexing. The highest frequency of $m_1(t)$ is 3 kHz, and that of $m_2(t)$ is 3.5

kHz. What is the minimum value of the permissible sampling rate?

The highest-frequency component of the composite signal $m_1(t)+m_2(t)$ is 3.5 kHz. Hence the minimum sampling rate is

$$2(3500) = 7000 \text{ samples/s}$$

4.31. A signal $m_1(t)$ is band-limited to 3.6 kHz, and three other signals $m_2(t)$, $m_3(t)$, and $m_4(t)$ are band-limited to 1.2 kHz each. These signals are to be transmitted by means of time-division multiplexing.

(*a*) Set up a scheme for accomplishing this multiplexing requirement, with each signal sampled at its Nyquist rate.

(*b*) What must be the speed of the commutator (in samples per second)?

(*c*) If the commutator output is quantized with $L = 1024$ and the result is binary-coded, what is the output bit rate?

(*d*) Determine the minimum transmission bandwidth of the channel.

(*a*)

Message	Bandwidth	Nyquist rate
$m_1(t)$	3.6 kHz	7.2 kHz
$m_2(t)$	1.2 kHz	2.4 kHz
$m_3(t)$	1.2 kHz	2.4 kHz
$m_4(t)$	1.2 kHz	2.4 kHz

If the sampling commutator rotates at the rate of 2400 rotations per second, then in one rotation we obtain one sample from each of $m_2(t)$, $m_3(t)$, and $m_4(t)$ and three samples from $m_1(t)$. This means that the commutator must have at least six poles connected to the signals, as shown in Fig. 4-29.

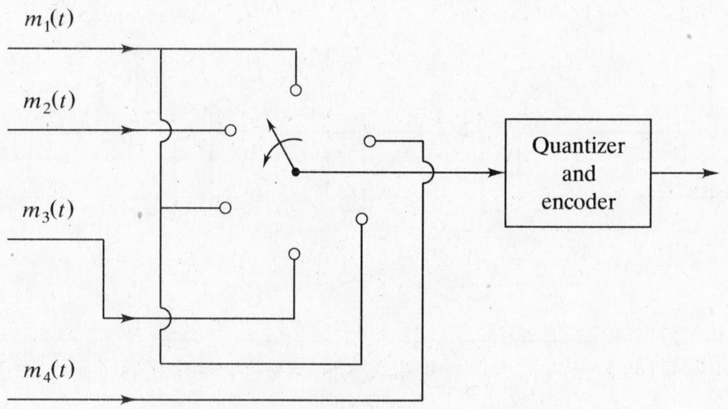

Fig. 4-29 Time-division multiplexing scheme

(*b*) $m_1(t)$ has 7200 samples/s. And $m_2(t)$, $m_3(t)$, and $m_4(t)$ each have 2400 samples/s. Hence, there are a total of 14 400 samples/s.

(*c*) $L = 1024 = 2^{10} = 2^n$

Thus, the output bit rate is 10(14 400) = 144 kb/s.

(*d*) The minimum channel bandwidth is

$$f_B = \tfrac{1}{2}(7.2 + 2.4 + 2.4 + 2.4) = 7.2 \text{ kHz}$$

4.32. The T1 carrier system used in digital telephony multiplexes 24 voice channels based on 8-b PCM. Each voice signal is usually put through a low-pass-filter with the cutoff frequency of about 3.4 kHz. The filtered voice signal is sampled at 8 kHz. In addition, a single bit is added at the end of the frame for the purpose of synchronization. Calculate (*a*) the duration of each bit, (*b*) the resultant transmission rate, and (*c*) the minimum required transmission bandwidth (Nyquist bandwidth).

(*a*) With a sampling rate of 8 kHz, each frame of the multiplexed signal occupies a period of

$$\frac{1}{8000} = 0.000125 \text{ s} = 125 \text{ microseconds } (\mu\text{s})$$

Since each frame consists of twenty-four 8-b words, plus a single synchronizing bit, it contains a total of

$$24(8) + 1 = 193 \text{ b}$$

Thus the duration of each bit is

$$T_b = \frac{125}{193}\ \mu\text{s} = 0.647\ \mu\text{s}$$

(*b*) The resultant transmission rate is

$$R_b = \frac{1}{T_b} = 1.544 \text{ Mb/s}$$

(*c*) From Eq. (*4.22*), the minimum required transmission bandwidth is

$$f_{T1} = \frac{1}{2T_b} = 772 \text{ kHz}$$

PULSE SHAPING AND INTERSYMBOL INTERFERENCE

4.33. Show that (a pulse-shape function) $h(t)$, with Fourier transform given by $H(\omega)$, that satisfies the criterion

$$\sum_{k=-\infty}^{\infty} H\left(\omega + \frac{2\pi k}{T}\right) = 1 \qquad \text{for } |\omega| \le \frac{\pi}{T} \tag{4.56}$$

has $h(nT)$ given by

$$h(nT) = \begin{cases} \dfrac{1}{T} & n = 0 \\ 0 & n \ne 0 \end{cases} \tag{4.57}$$

The criterion (*4.56*) is known as *Nyquist's pulse-shaping criterion.*

Taking the inverse Fourier transform of $H(\omega)$, we have

$$h(t) = \frac{1}{2\pi}\int_{-\infty}^{\infty} H(\omega)e^{j\omega t}\, d\omega$$

The range of integration in the above equation can be divided into segments of length $2\pi/T$ as

$$h(t) = \frac{1}{2\pi}\sum_{k=-\infty}^{\infty}\int_{(2k-1)\pi/T}^{(2k+1)\pi/T} H(\omega)e^{j\omega t}\, d\omega$$

and we can write $h(nT)$ as

$$h(nT) = \frac{1}{2\pi}\sum_{k=-\infty}^{\infty}\int_{(2k-1)\pi/T}^{(2k+1)\pi/T} H(\omega)e^{j\omega nT}\, d\omega \tag{4.58}$$

By the change of variable $u = \omega - 2\pi(k/T)$, Eq. (*4.58*) becomes

$$h(nT) = \frac{1}{2\pi} \sum_{k=-\infty}^{\infty} \int_{-\pi/T}^{\pi/T} H\left(u + \frac{2\pi k}{T}\right) e^{j(u+2\pi k/T)nT}\, du$$

Assuming that the integration and summation can be interchanged, we have

$$h(nT) = \frac{1}{2\pi} \int_{-\pi/T}^{\pi/T} \sum_{k=-\infty}^{\infty} H\left(u + \frac{2\pi k}{T}\right) e^{junT}\, du$$

Finally, if Eq. (*4.56*) is satisfied, then

$$h(nT) = \frac{1}{2\pi} \int_{-\pi/T}^{\pi/T} e^{junT}\, du$$

$$= \frac{1}{T}\frac{\sin n\pi}{n\pi} = \begin{cases} \dfrac{1}{T} & n = 0 \\ 0 & n \neq 0 \end{cases}$$

which verifies that $h(t)$ with a Fourier transform $H(\omega)$ satisfying criterion (*4.56*) produces zero ISI.

4.34. A certain telephone line bandwidth is 3.5 kHz. Calculate the data rate (in b/s) that can be transmitted if we use binary signaling with the raised-cosine pulses and a roll-off factor $\alpha = 0.25$.

Using Eq. (*4.28*), we see that the data rate is

$$R_b = \frac{1}{T} = \frac{2}{1 + 0.25}(3500) = 5600 \text{ b/s}$$

4.35. A communication channel of bandwidth 75 kHz is required to transmit binary data at a rate of 0.1 Mb/s using raised-cosine pulses. Determine the roll-off factor α.

$$T_b = \frac{1}{0.1(10^6)} = 10^{-5} \text{ s}$$

$$f_B = 75 \text{ kHz} = 75(10^3) \text{ Hz}$$

Using Eq. (*4.27*), we have

$$1 + \alpha = 2f_B T_b = 2(75)(10^3)(10^{-5}) = 1.5$$

Hence, we obtain

$$\alpha = 0.5$$

4.36. In a certain telemetry system, eight message signals having 2-kHz bandwidth each are time-division multiplexed using a binary PCM. The error in sampling amplitude cannot be greater than 1 percent of the peak amplitude. Determine the minimum transmission bandwidth required if raised-cosine pulses with roll-off factor $\alpha = 0.2$ are used. The sampling rate must be at least 25 percent above the Nyquist rate.

From Eq. (*4.9*), the maximum quantizing error must satisfy

$$(q_e)_{\max} = \frac{\Delta}{2} = \frac{m_p}{L} \leq 0.01 m_p$$

Hence $L \geq 100$, and we choose $L = 128 = 2^7$. The number of bits per sample required is 7.

Since the Nyquist sampling rate is $2f_M = 4000$ samples/s, the sampling rate for each signal is

$$f_s = 1.25(4000) = 5000 \text{ samples/s}$$

There are eight time-division multiplexed signals, requiring a total of

$$8(5000) = 40\,000 \text{ samples/s}$$

Since each sample is encoded by 7 bits, the resultant bit rate is

$$\frac{1}{T_b} = 7(40\,000) = 280 \text{ kb/s}$$

From Eq. (*4.27*), the minimum transmission bandwidth required is

$$f_B = \frac{1+0.2}{2}(280) = 168 \text{ kHz}$$

Supplementary Problems

4.37. If $m(t)$ is a band-limited signal, show that

$$\int_{-\infty}^{\infty} m(t)\phi_n(t)\,dt = T_s m(nT_s)$$

where $\phi_n(t)$ is the function defined in Eq. (*4.39*) of Prob. 4.4.

Hint: Use the orthogonality property (*4.40*) of $\phi_n(t)$.

4.38. The signals

$$m_1(t) = 10\cos 1000\pi t$$

and

$$m_2(t) = 10\cos 50\pi t$$

are both sampled with $f_s = 75$ Hz. Show that the two sequences of samples so obtained are identical.

Hint: Take the Fourier transforms of ideally sampled signals $m_{1s}(t)$ and $m_{2s}(t)$. *Note*: This problem indicates that by undersampling $m_1(t)$ and oversampling $m_2(t)$ appropriately, their sampled versions can be identical.

4.39. A signal

$$m(t) = \cos 200\pi t + 2\cos 320\pi t$$

is ideally sampled at $f_s = 300$ Hz. If the sampled signal is passed through an ideal low-pass filter with a cutoff frequency of 250 Hz, what frequency components will appear in the output?

Ans. 100-, 140-, 160-, and 200-Hz components

4.40. A duration-limited signal is a time function $m(t)$ for which

$$m(t) = 0 \qquad \text{for } |t| > T$$

Let $M(\omega) = \mathscr{F}[m(t)]$. Show that $M(\omega)$ can be uniquely determined from its values $M(n\pi/T)$ at a series of equidistant points spaced π/T apart. In fact, $M(\omega)$ is given by

$$M(\omega) = \sum_{n=-\infty}^{\infty} M\left(\frac{n\pi}{T}\right)\frac{\sin(\omega T - n\pi)}{\omega T - n\pi}$$

This is known as the *sampling theorem in the frequency domain.*

Hint: Interchange the roles of t and ω in the sampling theorem proof (Prob. 4.2).

4.41. American Standard Code for Information Interchange (ASCII) has 128 characters which are binary-coded. If a computer generates 100 000 characters per second, determine

(*a*) The number of bits required per character

(*b*) The data rate (or bit rate) R_b required to transmit the computer output

Ans. (*a*) 7 b per character, (*b*) $R_b = 0.7$ Mb/s

4.42. A PCM system uses a uniform quantizer followed by a 7-b binary encoder. The bit rate of the system is 50 Mb/s. What is the maximum message bandwidth for which system operation is satisfactory?

Ans. 3.57 MHz

4.43. Consider binary PCM transmission of a video signal with $f_s = 10$ MHz. Calculate the signaling rate needed to achieve $(\text{SNR})_o \geq 45$ dB.

Ans. 80 Mb/s

4.44. Show that in a PCM system, the output signal-to-quantizing-noise ratio can be expressed as

$$\left(\frac{S}{N_q}\right)_o = \frac{3}{2}\left(4^{f_B/f_m}\right)$$

where f_B is the channel bandwidth and f_m is the message bandwidth.

Hint: Use Eqs. (*4.14*), (*4.15*), and (*4.46*).

4.45. The bandwidth of a TV radio plus audio signal is 4.5 MHz. If this signal is converted to PCM with 1024 quantizing levels, determine the bit rate of the resulting PCM signal. Assume that the signal is sampled at a rate 20 percent above the Nyquist rate.

Ans. 108 Mb/s

4.46. A commonly used value A for the A law compander is $A = 87.6$. If $m_p = 20$ V and 256 quantizing levels are employed, what is the smallest and what is the largest effective separation between levels?

Ans. $\Delta_{\text{min}} = 9.8$ mV, $\Delta_{\text{max}} = 0.84$ V

4.47. Given the binary sequence 1101110, draw the transmitted pulse waveform for (*a*) AMI RZ signaling format and (*b*) split-phase (Manchester) signaling format.

Hint: See Fig. 4-11.

4.48. A given DM system operates with a sampling rate f_s and a fixed size Δ. If the input to the system is

$$m(t) = \alpha t \qquad \text{for } t > 0$$

determine the value of α for which slope overload occurs.

Ans. Δf_s

4.49. Consider a DM system whose receiver does not include a low-pass filter, as in Prob. 4.22. Show that under the assumption of no slope overload distortion, the maximum output signal-to-quantizing-noise ratio increases by 6 dB when the sampling rate is doubled. What is the improvement that results from the use of a low-pass filter at the receiver output?

Ans. 9-dB improvement

4.50. Twenty-four voice signals are sampled uniformly and then time-division-multiplexed. The sampling operation uses flat-top samples with 1-μs duration. The multiplexing operation includes provision for synchronization by adding an extra pulse of appropriate amplitude and 1-μs duration. The highest frequency component of each voice signal is 3.4 kHz.

(*a*) Assuming a sampling rate of 8 kHz, calculate the spacing between successive pulses of the multiplexed signal.

(*b*) Repeat (*a*), assuming the use of Nyquist rate sampling.

Ans. (*a*) 4 μs, (*b*) 5.68 μs

4.51. Five telemetry signals, each of bandwidth 1 kHz, are to be transmitted by binary PCM with TDM. The maximum tolerable error in sampling amplitude is 0.5 percent of the peak signal amplitude. The signals are sampled at least 20 percent above the Nyquist rate. Framing and synchronization require an additional 0.5 percent extra bits. Determine the minimum transmission data rate and the minimum required bandwidth for the TDM transmission.

Ans. $R_b = 964.8$ kb/s, $f_{\text{TDM}} = 482.9$ kHz

4.52. In a certain telemetry system, there are four analog signals $m_1(t)$, $m_2(t)$, $m_3(t)$, and $m_4(t)$. The bandwidth of $m_1(t)$ is 3.6 kHz, but the bandwidths of the remaining signals are 1.5 kHz each. Set up a suitable scheme for accomplishing the time-division multiplexing of these signals.

Ans. Use the same scheme as the one depicted in Fig. 4-29 of Prob. 4.31 with the commutator speed raised to 3000 rotations per second.

Chapter 5

Probability and Random Variables

5.1 INTRODUCTION

Thus far we have discussed the transmission of deterministic signals over a channel, and we have not emphasized the central role played by the concept of "randomness" in communication. The word *random* means unpredictable. If the receiver at the end of a channel knew in advance the message output from the originating source, there would be no need for communication. So there is a randomness in the message source. Moreover, transmitted signals are always accompanied by noise introduced in the system. These noise waveforms are also unpredictable. The objective of this chapter is to present the mathematical background essential for further study of communication.

5.2 PROBABILITY

A. Random Experiments:

An experiment is called a *random experiment* if its outcome cannot be predicted. Typical examples of a random experiment are the roll of a die, the toss of a coin, drawing a card from a deck, or selecting a message signal for transmission from several messages.

B. Fundamental Definitions:

The set of all possible outcomes of a random experiment is called the *sample space* S. An element in S is called a *sample point*. Each outcome of a random experiment corresponds to a sample point.

A set of sample points is called an *event*.

An assignment of real numbers to the events defined on S is known as the *probability measure*.

C. Auxiliary Definitions:

1. The *complement* of event A, denoted $\bar{A}$, is the event containing all sample points in S but not in A.
2. The *union* of events A and B, denoted $A \cup B$, is the event containing all sample points in either A or B or both.
3. The *intersection* of events A and B, denoted $A \cap B$, is the event containing all sample points in both A and B.
4. The event containing no sample point is called the *null event*, denoted $\varnothing$. Thus $\varnothing$ corresponds to an impossible event.
5. Two events A and B are called *mutually exclusive* or *disjoint* if they contain no common sample point, that is, if $A \cap B = \varnothing$.
6. If every sample point of A is a sample point of B, then A is a subset of B, denoted by $A \subset B$.

D. Algebra of Events:

The relations between the operations of complementation, union, and intersection are geometrically illustrated by Venn diagrams shown in Fig. 5-1. From Fig. 5-1 we see that

$$A \cup \bar{A} = S \tag{5.1}$$

$$A \cap \bar{A} = \varnothing = \bar{S} \tag{5.2}$$

$$A \cap S = A \tag{5.3}$$

$$\overline{(A \cup B)} = \bar{A} \cap \bar{B} \tag{5.4}$$

$$\overline{(A \cap B)} = \bar{A} \cup \bar{B} \tag{5.5}$$

$$A \cup B = (A \cap \bar{B}) \cup (A \cap B) \cup (\bar{A} \cap B) \tag{5.6}$$

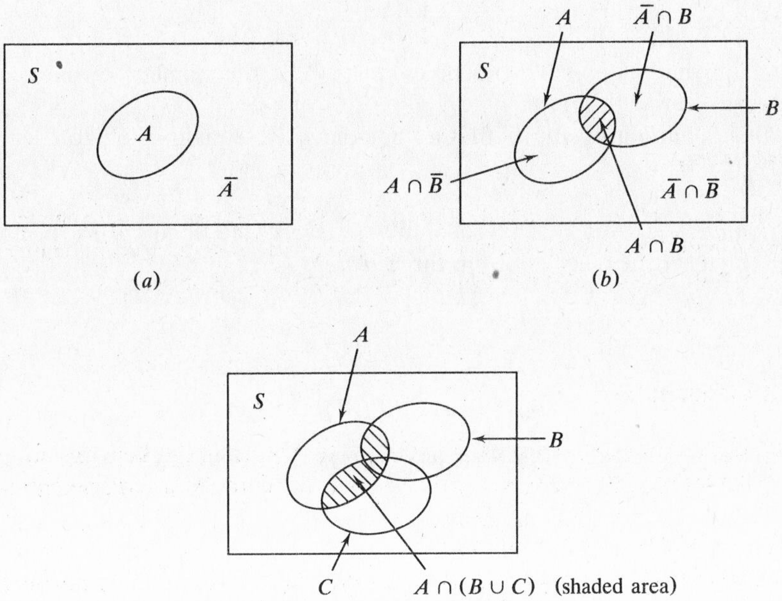

(c)

Fig. 5-1 Venn diagram

Equations (5.4) and (5.5) are known as *De Morgan's laws*. The union and intersection operations also satisfy the following laws:

Commutative laws:

$$A \cup B = B \cup A \tag{5.7}$$

$$A \cap B = B \cap A \tag{5.8}$$

Associative laws:

$$A \cup (B \cup C) = (A \cup B) \cup C \tag{5.9}$$

$$A \cap (B \cap C) = (A \cap B) \cap C \tag{5.10}$$

Distributive laws:

$$A \cap (B \cup C) = (A \cap B) \cup (A \cap C) \tag{5.11}$$

$$A \cup (B \cap C) = (A \cup B) \cap (A \cup C) \tag{5.12}$$

E. Probabilities of Events:

1. *Relative-Frequency Definition:*

Suppose that a random experiment is repeated n times. If an event A occurs n_A times, then its probability $P(A)$ is defined as

$$P(A) = \lim_{n \to \infty} \frac{n_A}{n} \tag{5.13}$$

Note that this limit may not exist.

2. *Classical Definition:*

In this definition, the probability $P(A)$ of event A is defined without experimentation. It is given by

$$P(A) = \frac{N_A}{N} \tag{5.14}$$

where N is the number of possible outcomes and N_A is the number of outcomes belonging to event A.

Note that this definition presumes that all outcomes are equally likely to occur.

3. *Axiomatic Definition:*

In this *axiomatic* definition, the probability of event A, denoted $P(A)$, is a real number assigned to A which satisfies the following three *axioms*:

Axiom 1: $$P(A) \geq 0 \tag{5.15}$$

Axiom 2: $$P(S) = 1 \tag{5.16}$$

Axiom 3: $$P(A \cup B) = P(A) + P(B) \qquad \text{if } A \cap B = \varnothing \tag{5.17}$$

By using the above axioms, the following useful laws of probability can be obtained (Probs. 5.2 to 5.4).

1. $$P(A) \leq 1 \tag{5.18}$$

2. $$P(\bar{A}) = 1 - P(A) \tag{5.19}$$

3. $$P(\varnothing) = 0 \tag{5.20}$$

4. $$P(A) \leq P(B) \qquad \text{if } A \subset B \tag{5.21}$$

5. $$P(A \cup B) = P(A) + P(B) - P(A \cap B) \tag{5.22}$$

F. Conditional Probability:

The *conditional probability* of an event A given event B, denoted by $P(A|B)$, is defined as

$$P(A|B) = \frac{P(A \cap B)}{P(B)} \tag{5.23}$$

where $P(B) \neq 0$ and $P(A \cap B)$ is the joint probability of A and B. From the definition (*5.23*) we can obtain the following *Bayes' rule* (Prob. 5.5):

$$P(B|A) = \frac{P(A|B)P(B)}{P(A)} \tag{5.24}$$

where $P(B|A)$ is the conditional probability of B given A.

G. Independent Events:

Two events A and B are called (*statistically*) *independent* if and only if

$$P(A \cap B) = P(A)P(B) \tag{5.25}$$

Thus, if A and B are independent, then by Eq. (5.23)

$$P(A|B) = P(A) \tag{5.26}$$

Let $A_1, A_2, \ldots, A_n$ be events defined in S. Then n events are (*mutually*) *independent* events if and only if the relations

$$\begin{aligned} P(A_i \cap A_j) &= P(A_i)P(A_j) \\ P(A_i \cap A_j \cap A_k) &= P(A_i)P(A_j)P(A_k) \\ &\vdots \\ P(A_i \cap A_j \cap \cdots \cap A_n) &= P(A_i)P(A_j) \cdots P(A_n) \end{aligned} \tag{5.27}$$

hold for all combinations of the indices such that

$$1 \le i < j < k < \cdots \le n$$

H. Total Probability:

Let N events $A_1, A_2, \ldots, A_N$ be mutually *exclusive* and *exhaustive*, that is,

$$A_i \cap A_j = \varnothing \qquad i \ne j = 1, 2, \ldots, N \tag{5.28}$$

and

$$A_1 \cup A_2 \cup \cdots \cup A_N = S \tag{5.29}$$

Let B be any event defined over S. Then

$$P(B) = \sum_{k=1}^{N} P(B \cap A_k) = \sum_{k=1}^{N} P(B|A_k)P(A_k) \tag{5.30}$$

which is known as the *total probability* of event B (Prob. 5.6).

5.3 RANDOM VARIABLES, DISTRIBUTION FUNCTIONS, AND DENSITY FUNCTIONS

A. Random Variables:

A *random variable* $X(\lambda)$ is a single-valued real function that assigns a real number, called the *value* of $X(\lambda)$, to each sample point λ of S such that

1. The set $\{\lambda\colon X(\lambda) \le x\}$ is an event for every real number x.
2. $P\{\lambda\colon X(\lambda) = -\infty\} = 0$ and $P\{\lambda\colon X(\lambda) \le \infty\} = 1$.

We use *uppercase* letters to denote *random variables* and *lowercase* letters to denote *fixed values* of random variables (that is, real numbers).

Thus, the random variable X induces a probability measure on the real line as follows:

$$\begin{aligned} P(X = x) &= P\{\lambda\colon X(\lambda) = x\} \\ P(X \le x) &= P\{\lambda\colon X(\lambda) \le x\} \\ P(x_1 < X \le x_2) &= P\{\lambda\colon x_1 < X(\lambda) \le x_2\} \end{aligned}$$

If X can take on only a *countable* number of distinct values, then X is called a *discrete* random variable. If X can assume any values within one or more intervals on the real line, then X is called a *continuous* random variable. The number of telephone calls arriving at an office in a finite time is an example of a discrete random variable, and the exact time of arrival of a telephone call is an example of a continuous random variable.

B. Distribution Function:

The *probability distribution function* of X is the function

$$F_X(x) = P(X \le x) \tag{5.31}$$

defined for every x from $-\infty$ to ∞.

Properties of $F_X(x)$:

1. $$F_X(-\infty) = 0 \tag{5.32a}$$
2. $$F_X(\infty) = 1 \tag{5.32b}$$
3. $$0 \le F_X(x) \le 1 \tag{5.32c}$$
4. $$F_X(x_1) \le F_X(x_2) \qquad \text{if } x_1 < x_2 \tag{5.32d}$$
5. $$P\{x_1 < X \le x_2\} = F_X(x_2) - F_X(x_1) \tag{5.32e}$$
6. $$F_X(x^+) = F_X(x) \qquad \text{where } F_X(x^+) = \lim_{0<\varepsilon\to 0} F(x+\varepsilon) \tag{5.32f}$$

C. Density Function:

The *probability density function* (pdf) of X is defined as

$$f_X(x) = \frac{dF_X(x)}{dx} \tag{5.33}$$

Properties of $f_X(x)$:

1. $$f_X(x) \ge 0 \tag{5.34a}$$
2. $$\int_{-\infty}^{\infty} f_X(x)\,dx = 1 \tag{5.34b}$$
3. $$F_X(x) = \int_{-\infty}^{x} f_X(\xi)\,d\xi \tag{5.34c}$$
4. $$P\{x_1 \le X \le x_2\} = \int_{x_1}^{x_2} f_X(x)\,dx \tag{5.34d}$$

If X is discrete, then

$$f_X(x) = \sum_i P(x_i)\delta(x - x_i) \tag{5.35}$$

where $P(x_i) = P(X = x_i)$ and $\delta(x)$ is the impulse function [see Eq. (*1.16*)].

D. Joint Distribution:

Let two random variables X and Y be defined on S. The event $\{X \le x, Y \le y\} = \{X \le x\} \cap \{Y \le y\}$ consists of all outcomes λ on S such that $X(\lambda) \le x$ and $Y(\lambda) \le y$.

The *joint probability distribution function* of X and Y is the function

$$F_{XY}(x, y) = P(X \le x, Y \le y) \tag{5.36}$$

defined for every x and y from $-\infty$ to ∞. Since $\{X \le \infty\}$ and $\{Y \le \infty\}$ are certain events, we obtain

$$\{X \le x, Y \le \infty\} = \{X \le x\} \qquad \{X \le \infty, Y \le y\} = \{Y \le y\}$$

so that

$$F_{XY}(x, \infty) = F_X(x) \tag{5.37a}$$

$$F_{XY}(\infty, y) = F_Y(y) \tag{5.37b}$$

The *joint probability density function* (joint pdf) of X and Y is given by

$$f_{XY}(x, y) = \frac{\partial^2}{\partial x\, \partial y} F_{XY}(x, y) \tag{5.38}$$

and
$$f_{XY}(x, y) \ge 0 \tag{5.39}$$

By twice integrating Eq. (5.38), we obtain

$$F_{XY}(x, y) = \int_{-\infty}^{y} \int_{-\infty}^{x} f_{XY}(\xi, \eta)\, d\xi\, d\eta \tag{5.40}$$

and
$$F_{XY}(\infty, \infty) = \int_{-\infty}^{\infty} \int_{-\infty}^{\infty} f_{XY}(\xi, \eta)\, d\xi\, d\eta = 1 \tag{5.41}$$

E. Marginal Distribution:

The functions $F_X(x)$ and $F_Y(y)$ are called *marginal probability distributions* if they are derived by

$$F_X(x) = F_{XY}(x, \infty) = \int_{-\infty}^{x} d\xi \int_{-\infty}^{\infty} f_{XY}(\xi, y)\, dy \tag{5.42a}$$

$$F_Y(y) = F_{XY}(\infty, y) = \int_{-\infty}^{y} d\eta \int_{-\infty}^{\infty} f_{XY}(x, \eta)\, dx \tag{5.42b}$$

The *marginal probability densities functions* (marginal pdf's) $f_X(x)$ and $f_Y(y)$ are given by

$$f_X(x) = F'_X(x) \qquad f_Y(y) = F'_Y(y)$$

Thus, differentiating Eqs. (5.42a) and (5.42b), we obtain

$$f_X(x) = \int_{-\infty}^{\infty} f_{XY}(x, y)\, dy \tag{5.43a}$$

$$f_Y(y) = \int_{-\infty}^{\infty} f_{XY}(x, y)\, dx \tag{5.43b}$$

F. Conditional Distribution:

The *conditional probability distribution function* of X given event B is defined as

$$F_X(x|B) = P(X \le x|B) = \frac{P(X \le x, B)}{P(B)} \qquad P(B) \ne 0 \tag{5.44}$$

where $P(X \le x, B)$ is the probability of the joint event $\{X \le x\} \cap B$.

The *conditional probability density function* (conditional pdf) of X given B is simply

$$f_X(x|B) = \frac{dF_X(x|B)}{dx} \tag{5.45}$$

Let X and Y be two random variables defined on S. The conditional pdf of X given the event

$\{Y=y\}$ is

$$f_{X|Y}(x|y)=\frac{f_{XY}(x,y)}{f_Y(y)} \qquad f_Y(y)\neq 0 \tag{5.46}$$

where $f_Y(y)$ is the marginal pdf of Y.

G. Independent Random Variables:

Random variables X and Y are called *independent* if

$$F_{XY}(x,y)=F_X(x)F_Y(y) \tag{5.47}$$

or

$$f_{XY}(x,y)=f_X(x)f_Y(y) \tag{5.48}$$

Thus, if X and Y are independent, then from Eqs. (*5.46*) and (*5.48*)

$$f_{X|Y}(x|y)=f_X(x) \tag{5.49}$$

5.4 FUNCTIONS OF RANDOM VARIABLES

A. Random Variable $g(X)$:

Given a random variable X and a function $g(x)$, the expression

$$Y=g(X) \tag{5.50}$$

is a new random variable. For a given λ in S, $X(\lambda)$ is a number and $g[X(\lambda)]$ is another number, which is the value $Y(\lambda)=g[X(\lambda)]$ assigned to the random variable Y. Thus

$$F_Y(y)=P(Y\leq y)=P(g(X)\leq y) \tag{5.51}$$

B. Determination of $f_Y(y)$:

To find the pdf $f_Y(y)$, we solve the equation $y=g(x)$. Denoting its real roots by x_k, we have

$$y=g(x_1)=\cdots=g(x_k)=\cdots \tag{5.52}$$

Then

$$f_Y(y)=\sum_k \frac{f_X(x_k)}{|g'(x_k)|} \tag{5.53}$$

where $g'(x)$ is the derivative of $g(x)$.

C. One Function of Two Random Variables:

Given two random variables X and Y and a function $g(x,y)$, the expression

$$Z=g(X,Y) \tag{5.54}$$

is a new random variable. With z a given number, we denote by D_z the region of the xy plane such that $g(x,y)\leq z$. Then

$$\{Z\leq z\}=\{g(X,Y)\leq z\}=\{(X,Y)\in D_z\}$$

where $\{(X,Y)\in D_z\}$ is the event consisting of all outcomes λ such that the point $(X(\lambda),Y(\lambda))$ is in

D_z. Hence

$$F_Z(z) = P(Z \le z) = P\{(X,Y) \in D_z\} = \iint_{D_z} f_{XY}(x,y)\,dx\,dy \tag{5.55}$$

The pdf of Z can be determined by

$$f_Z(z) = \frac{dF_Z(z)}{dz} \tag{5.56}$$

D. Two Functions of Two Random Variables:

Given X and Y with joint pdf $f_{XY}(x, y)$ and two functions $g(x, y)$ and $h(x, y)$, we form the new random variables

$$Z = g(X,Y) \qquad W = h(X,Y) \tag{5.57}$$

To find the joint pdf $f_{ZW}(z,w)$ we solve the system

$$g(x,y) = z \qquad h(x,y) = w \tag{5.58}$$

denoting by (x_k, y_k) its roots

$$g(x_k, y_k) = z \qquad h(x_k, y_k) = w$$

Then

$$f_{ZW}(z,w) = \sum_k \frac{f_{XY}(x_k, y_k)}{|J(x_k, y_k)|} = \sum_k f_{XY}(x_k, y_k)|\bar{J}(x_k, y_k)| \tag{5.59}$$

where

$$J(x,y) = \begin{vmatrix} \dfrac{\partial z}{\partial x} & \dfrac{\partial z}{\partial y} \\ \dfrac{\partial w}{\partial x} & \dfrac{\partial w}{\partial y} \end{vmatrix} \qquad \bar{J}(x,y) = \begin{vmatrix} \dfrac{\partial x}{\partial z} & \dfrac{\partial x}{\partial w} \\ \dfrac{\partial y}{\partial z} & \dfrac{\partial y}{\partial w} \end{vmatrix} \tag{5.60}$$

and $J(x, y)$ is the *jacobian* of the transformation (*5.58*).

5.5 STATISTICAL AVERAGES

A. Expected Value:

The *expected value* or *mean* of X is given by

$$\mu_X = E[X] = \int_{-\infty}^{\infty} x f_X(x)\,dx \tag{5.61}$$

If X is discrete, then using Eq. (*5.35*), we have

$$\mu_X = E[X] = \int_{-\infty}^{\infty} x\left[\sum_i P(x_i)\delta(x - x_i)\right] dx = \sum_i x_i P(x_i) \tag{5.62}$$

For an equally likely case, that is, when $P(x_i) = 1/N$ $(i = 1, \ldots, N)$, Eq. (*5.62*) becomes the arithmetic average of x_i:

$$\mu_X = \frac{1}{N}\sum_{i=1}^{N} x_i \tag{5.63}$$

The expected value of $Y = g(X)$ is given by

$$E[Y] = \int_{-\infty}^{\infty} y f_Y(y)\, dy \tag{5.64}$$

or

$$E[Y] = E[g(X)] = \int_{-\infty}^{\infty} g(x) f_X(x)\, dx \tag{5.65}$$

If X is discrete, then as in Eq. (5.62), Eq. (5.65) yields

$$E[g(X)] = \sum_i g(x_i) P(x_i) \tag{5.66}$$

The expected value of $Z = g(X, Y)$ is given by

$$E[Z] = \int_{-\infty}^{\infty} z f_Z(z)\, dz \tag{5.67}$$

or

$$E[Z] = E[g(X,Y)] = \int_{-\infty}^{\infty}\int_{-\infty}^{\infty} g(x,y) f_{XY}(x,y)\, dx\, dy \tag{5.68}$$

If X and Y are discrete, Eq. (5.68) yields

$$E[g(X,Y)] = \sum_i \sum_k g(x_i, y_k) P(x_i, y_k) \tag{5.69}$$

where $P(x_i, y_k) = P[X = x_i, Y = y_k]$.

Note that the *expectation operation* is linear, that is,

$$E[X+Y] = E[X] + E[Y] \tag{5.70}$$

$$E[cX] = cE[X] \tag{5.71}$$

where c is a constant (Prob. 5.40).

B. Moments and Variance:

The *nth moment* of X is defined by

$$m_n = E[X^n] = \int_{-\infty}^{\infty} x^n f_X(x)\, dx \tag{5.72}$$

The *nth central moment* of X is defined by

$$E\left[(X - \mu_X)^n\right] = \int_{-\infty}^{\infty} (x - \mu_x)^n f_X(x)\, dx \tag{5.73}$$

where $\mu_X = E[X]$.

The *second central moment* of X is called the *variance* of X, that is,

$$\mathrm{var}[X] = \sigma_X^2 = E\left[(X - \mu_X)^2\right] \tag{5.74}$$

The positive square root of the variance, or σ_X, is called the *standard deviation* of X. The variance or standard variation is a measure of the "spread" of the values of X from its mean μ_X. By using Eqs. (5.70) and (5.71), the expression in Eq. (5.74) can be simplified to

$$\sigma_X^2 = E[X^2] - \mu_X^2 = E[X^2] - (E[X])^2 \tag{5.75}$$

C. Joint Moments and Covariance:

Let X and Y be two random variables with joint pdf $f_{XY}(x, y)$. The *joint moment* of X and Y is defined by

$$m_{nk} = E[X^n Y^k] = \int_{-\infty}^{\infty} \int_{-\infty}^{\infty} x^n y^k f_{XY}(x, y)\, dx\, dy \tag{5.76}$$

where n and k are any positive integers.

The sum $n + k$ is called the *order* of the moment.

A joint moment for the special case $n = k = 1$ is called the *correlation* of X and Y, denoted by R_{XY}:

$$R_{XY} = m_{11} = E[XY] \tag{5.77}$$

The *covariance* of X and Y is defined by

$$C_{XY} = E[(X - \mu_X)(Y - \mu_Y)] \tag{5.78}$$

Expanding Eq. (*5.78*), we obtain

$$C_{XY} = R_{XY} - \mu_X \mu_Y = E[XY] - E[X]E[Y] \tag{5.79}$$

The *correlation coefficient* of X and Y is defined by

$$\rho_{XY} = \frac{C_{XY}}{\sigma_X \sigma_Y} \tag{5.80}$$

where σ_X and σ_Y are the standard deviations of X and Y, respectively.

It can be shown that (see Prob. 5.49)

$$|\rho_{XY}| \le 1 \qquad |C_{XY}| \le \sigma_X \sigma_Y \tag{5.81}$$

Two random variables X and Y are called *uncorrelated* if and only if their covariance is zero, that is,

$$C_{XY} = 0 \tag{5.82}$$

From Eqs. (*5.79*) and (*5.82*), we conclude that X and Y are uncorrelated if and only if

$$E[XY] = E[X]E[Y] \tag{5.83}$$

Two random variables X and Y are called *orthogonal* if

$$R_{XY} = E[XY] = 0 \tag{5.84}$$

From Eq. (*5.84*) it is obvious that if X and Y are uncorrelated random variables with nonzero mean values, then X and Y cannot be orthogonal.

D. Moment Generating Function:

The moment generating function of a random variable X is defined by

$$M_X(\lambda) = E[e^{\lambda X}] = \int_{-\infty}^{\infty} f_X(x) e^{\lambda x}\, dx \tag{5.85}$$

where λ is a real variable. Then

$$m_k = E[X^k] = M_X^{(k)}(0) \qquad k = 1, 2, \ldots \tag{5.86}$$

where

$$M_X^{(k)}(0) = \left. \frac{d^k M_X(\lambda)}{d\lambda^k} \right|_{\lambda=0}$$

E. Characteristic Functions:

The *characteristic function* of a random variable X is defined by

$$\Phi_X(\omega) = E[e^{j\omega X}] = \int_{-\infty}^{\infty} f_X(x)e^{j\omega x}\,dx \tag{5.87}$$

where ω is a real variable. This function is maximum at the origin, and

$$|\Phi_X(\omega)| \le \Phi_X(0) = 1 \tag{5.88}$$

As we see from Eq. (*5.87*), $\Phi_X(\omega)$ is the Fourier transform (with the sign of j reversed) of $f_X(x)$. Because of this fact, if $\Phi_X(\omega)$ is known, $f_X(x)$ can be found from the inverse Fourier transform

$$f_X(x) = \frac{1}{2\pi}\int_{-\infty}^{\infty} \Phi_X(\omega)e^{-j\omega x}\,d\omega \tag{5.89}$$

The *joint characteristic function* of X and Y is defined by

$$\Phi_{XY}(\omega_1,\omega_2) = E[e^{j(\omega_1 X+\omega_2 Y)}] = \int_{-\infty}^{\infty}\int_{-\infty}^{\infty} f_{XY}(x,y)e^{j(\omega_1 x+\omega_2 y)}\,dx\,dy \tag{5.90}$$

where ω_1 and ω_2 are real variables. This expression is recognized as the two-dimensional Fourier transform (with the sign of j reversed) of $f_{XY}(x,y)$. From the inverse Fourier transform we have

$$f_{XY}(x,y) = \frac{1}{(2\pi)^2}\int_{-\infty}^{\infty}\int_{-\infty}^{\infty} \Phi_{XY}(\omega_1,\omega_2)e^{-j(\omega_1 x+\omega_2 y)}\,d\omega_1\,d\omega_2 \tag{5.91}$$

From Eqs. (*5.87*) and (*5.90*), we have

$$\Phi_X(\omega) = \Phi_{XY}(\omega,0) \tag{5.92a}$$

$$\Phi_Y(\omega) = \Phi_{XY}(0,\omega) \tag{5.92b}$$

which are called *marginal characteristic functions*.

5.6 SPECIAL DISTRIBUTIONS

There are several distributions which arise very often in communication problems.

A. Binomial Distribution:

A random variable X is said to have a *binomial distribution* of order n if it takes the values $0, 1, \ldots, n$ with

$$P(X=k) = \binom{n}{k} p^k q^{n-k} \tag{5.93}$$

where $0 < p < 1$, $p+q=1$, $k = 0, 1, \ldots, n$, and

$$\binom{n}{k} = \frac{n!}{k!(n-k)!}$$

The pdf of X is given by

$$f_X(x) = \sum_{k=0}^{n} \binom{n}{k} p^k q^{n-k}\delta(x-k) \tag{5.94}$$

The corresponding distribution function is

$$F_X(x) = \sum_{k=0}^{n} \binom{n}{k} p^k q^{n-k}u(x-k) \tag{5.95}$$

Figure 5-2 illustrates the binomial density and distribution functions for $n = 6$ and $p = 0.6$.

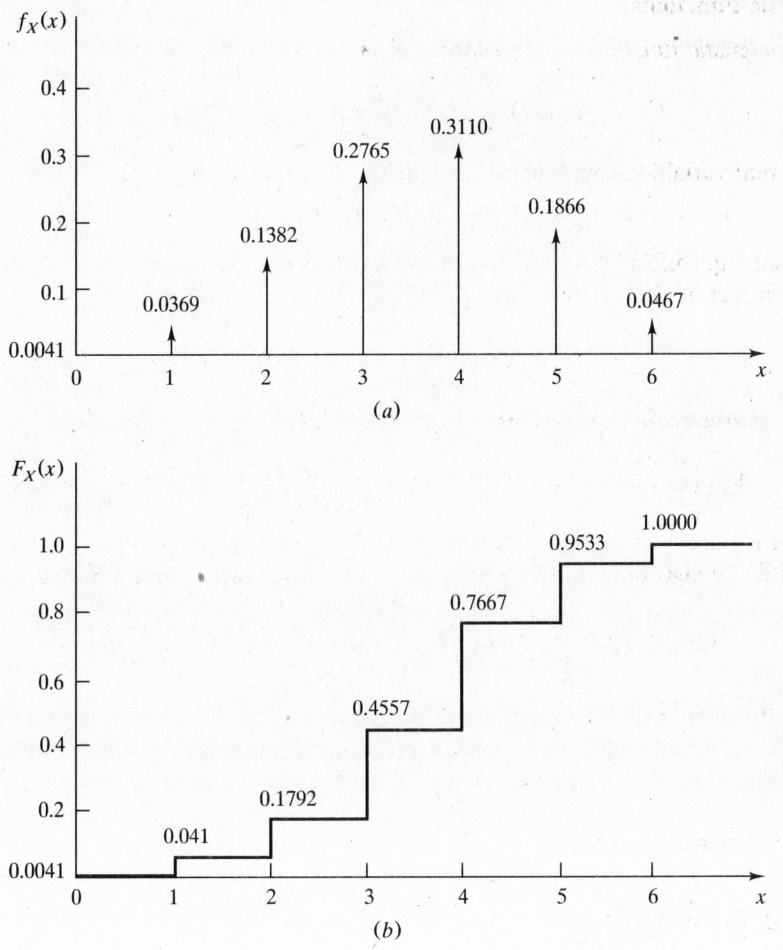

Fig. 5-2 Binomial density and distribution functions for $n = 6$ and $p = 0.6$

The mean and variance of X are

$$\mu_X = E[X] = np \tag{5.96}$$

$$\sigma_X^2 = npq = np(1-p) \tag{5.97}$$

The binomial random variable X is an integer-valued discrete random variable associated with repeated trials of an experiment. Consider performing some experiment and observing only whether event A occurs. If A occurs, we call the experiment a success; and if it does not occur ($\bar{A}$ occurs), a failure. Suppose that the probability that A occurs is $P(A) = p$, hence $P(\bar{A}) = q = 1 - p$. We repeat this experiment n times (trials) under the following assumptions:

1. $P(A)$ is constant on each trial.
2. The n trials are independent.

A point in the sample space is a sequence of n A's and $\bar{A}$'s. A point with k A's and $n-k$ $\bar{A}$'s will be assigned a probability of $p^k q^{n-k}$. Thus, if X is the random variable associated with the number of times that A occurs in n trials, then the values of X are the integers $k = 0, 1, \dots, n$.

In the study of communications, the binomial distribution applies to digital transmission when X stands for the number of errors in a message of n digits. (See Probs. 5.12 and 5.36.)

B. Poisson Distribution:

A random variable X is said to have a *Poisson distribution* with parameter α if it takes the values $0, 1, \ldots, n, \ldots$ with

$$P(X=k) = e^{-\alpha}\frac{\alpha^k}{k!} \qquad k = 0, 1, \ldots \tag{5.98}$$

The pdf of X is given by

$$f_X(x) = e^{-\alpha}\sum_{k=0}^{\infty}\frac{\alpha^k}{k!}\delta(x-k) \tag{5.99}$$

The corresponding distribution function is

$$F_X(x) = e^{-\alpha}\sum_{k=0}^{\infty}\frac{\alpha^k}{k!}u(x-k) \tag{5.100}$$

When plotted, these functions appear quite similar to those for the binomial random variables (Fig. 5-2).

The mean and variance of X are

$$\mu_X = E[X] = \alpha \tag{5.101}$$

$$\sigma_X^2 = \alpha \tag{5.102}$$

The Poisson distribution arises in some problems involving counting, for example, monitoring the number of telephone calls arriving at a switching center during various intervals of time. In digital communication, the Poisson distribution is pertinent to the problem of the transmission of many data bits when the error rates are low. The binomial distribution becomes awkward to handle in such cases. However, if the mean value of the error rate remains finite and equal to α, we can approximate the binomial distribution by the Poisson distribution. (See Probs. 5.13 and 5.14.)

C. Normal (or Gaussian) Distribution:

A random variable X is called *normal* (or *gaussian*) if its pdf is of the form

$$f_X(x) = \frac{1}{\sqrt{2\pi}\,\sigma}e^{-(x-\mu)^2/(2\sigma^2)} \tag{5.103}$$

The distribution function of X is

$$F_X(x) = \frac{1}{\sqrt{2\pi}\,\sigma}\int_{-\infty}^{x} e^{-(\xi-\mu)^2/(2\sigma^2)}\,d\xi$$

$$= \frac{1}{\sqrt{2\pi}}\int_{-\infty}^{(x-\mu)/\sigma} e^{-\lambda^2/2}\,d\lambda \tag{5.104}$$

This integral cannot be evaluated in a closed form and must be evaluated numerically. It is convenient to use the function $Q(z)$, defined as

$$Q(z) = \frac{1}{\sqrt{2\pi}}\int_{z}^{\infty} e^{-\lambda^2/2}\,d\lambda \tag{5.105}$$

Then Eq. (*5.104*) can be written as

$$F_X(x) = 1 - Q\left(\frac{x-\mu}{\sigma}\right) \tag{5.106}$$

The function $Q(z)$ is shown as the *complementary error function*, or simply the Q function. The function $Q(z)$ is tabulated in Table C-1 (App. C).

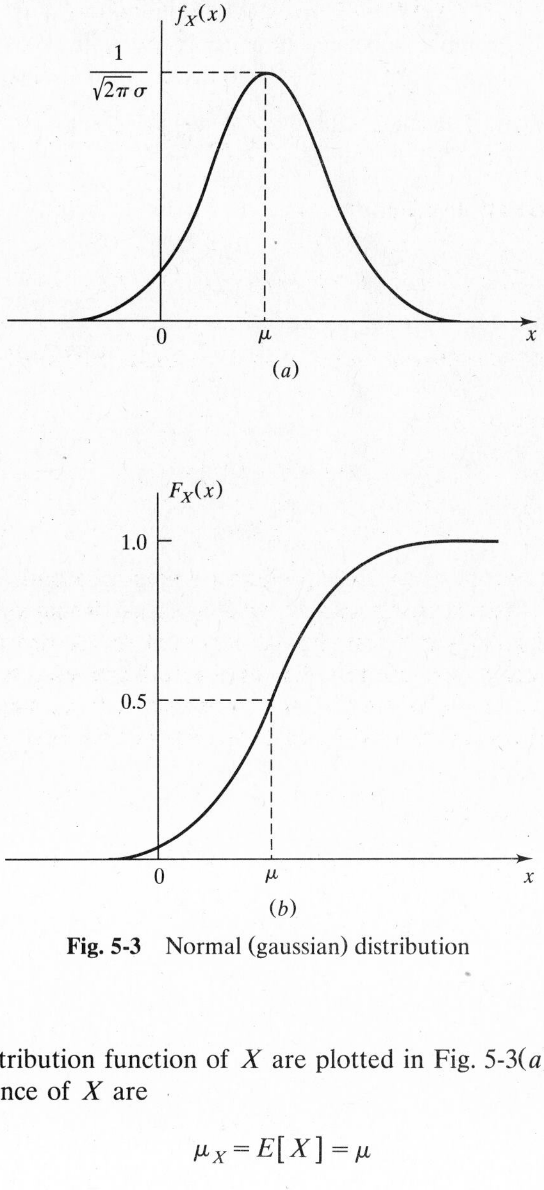

Fig. 5-3 Normal (gaussian) distribution

The pdf and the distribution function of X are plotted in Fig. 5-3(a) and (b), respectively. The mean and variance of X are

$$\mu_X = E[X] = \mu \tag{5.107}$$

$$\sigma_X^2 = E\left[(X - \mu_X)^2\right] = \sigma^2 \tag{5.108}$$

We shall use the notation $N(\mu; \sigma^2)$ to denote that X is normal (or gaussian) with mean μ and variance σ^2. In particular, $X = N(0; 1)$, that is, X with zero mean and unit variance, is defined as a *normalized gaussian* random variable.

The normal (or gaussian) distribution has played a significant role in the study of random phenomena in nature. Many naturally occurring random phenomena are approximately normal. Another reason for the importance of the normal distribution is a remarkable theorem called the *central-limit theorem*. This theorem states that the sum of a large number of independent random variables, under certain conditions, can be approximated by a normal distribution.

Solved Problems

PROBABILITY

5.1. An experiment consists of observing the sum of the number of dots showing when two dice are thrown. Describe the sample space, and assuming equally likely outcomes, find (*a*) the probability that the sum is 7 and (*b*) the probability that sum is greater than 10.

For this random experiment, the sample space contains 36 sample points, as shown in Fig. 5-4. Let λ_{ij} denote the elementary event (sampling point) consisting of the following outcome (Fig. 5-4): $\lambda_{ij}=(i,j)$, where i represents the number appearing on the one die and j represents the number appearing on the other die. Assume that the dice are fair and that all the outcomes are equally likely. So $P(\lambda_{ij})=\frac{1}{36}$.

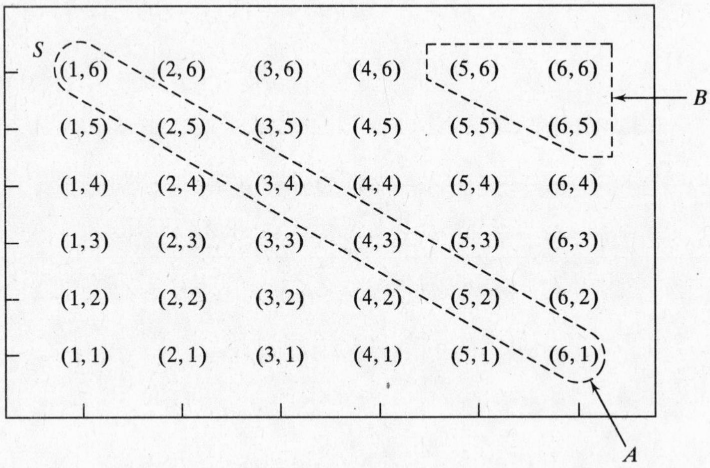

Fig. 5-4 Sample space for a throw of two dice

(*a*) Let A denote the event that the sum is 7. Since the events λ_{ij} are mutually exclusive, and from Fig. 5-4, we have

$$\begin{aligned}P(A)&=P(\lambda_{16}\cup\lambda_{25}\cup\lambda_{34}\cup\lambda_{43}\cup\lambda_{52}\cup\lambda_{61})\\&=P(\lambda_{16})+P(\lambda_{25})+P(\lambda_{34})+P(\lambda_{43})+P(\lambda_{52})+P(\lambda_{61})\\&=6\left(\tfrac{1}{36}\right)=\tfrac{1}{6}\end{aligned}$$

(*b*) Let B denote the event that the sum is greater than 10. Then from Fig. 5-4 we obtain

$$\begin{aligned}P(B)&=P(\lambda_{56}\cup\lambda_{65}\cup\lambda_{66})=P(\lambda_{56})+P(\lambda_{65})+P(\lambda_{66})\\&=3\left(\tfrac{1}{36}\right)=\tfrac{1}{12}\end{aligned}$$

5.2. Using the axioms of probability, prove (*a*) Eq. (*5.19*), that is,

$$P(\bar{A})=1-P(A)$$

and (*b*) Eq. (*5.18*), that is,

$$P(A)\le 1$$

(*a*) $$S=A\cup\bar{A} \text{ and } A\cap\bar{A}=\varnothing$$

Then the use of axioms 1 and 3 yields

$$P(S) = 1 = P(A) + P(\bar{A})$$

Thus

$$P(\bar{A}) = 1 - P(A)$$

(*b*) From Eq. (*5.19*), we have

$$P(A) = 1 - P(\bar{A})$$

Now, by axiom 1, $P(\bar{A}) \geq 0$. Thus we conclude that

$$P(A) \leq 1$$

5.3. Using the axioms of probability, verify Eq. (*5.20*), that is,

$$P(\varnothing) = 0$$

$$A = A \cup \varnothing \qquad \text{and} \qquad A \cap \varnothing = \varnothing$$

Therefore, by axiom 3,

$$P(A) = P(A \cup \varnothing) = P(A) + P(\varnothing)$$

and we conclude that

$$P(\varnothing) = 0$$

5.4. Using the axioms of probability, verify Eq. (*5.22*), that is,

$$P(A \cup B) = P(A) + P(B) - P(A \cap B)$$

First we decompose $A \cup B$, A, and B as unions of disjoint events. From the Venn diagram of Fig. 5-5,

$$A \cup B = (A \cap \bar{B}) \cup (A \cap B) \cup (\bar{A} \cap B)$$

$$A = (A \cap B) \cup (A \cap \bar{B})$$

$$A = (A \cap B) \cup (\bar{A} \cap B)$$

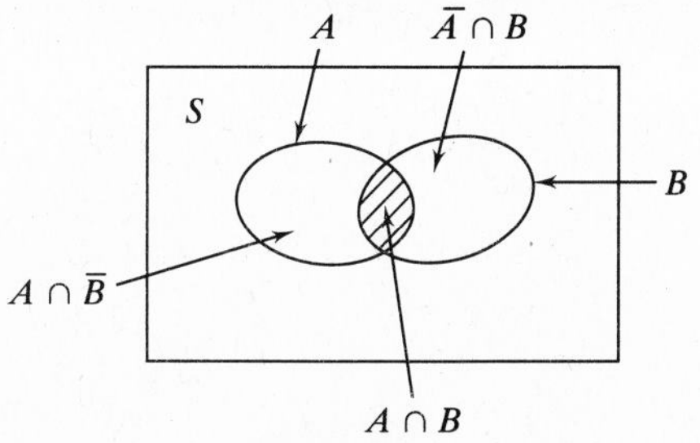

Fig. 5-5

Thus, by axiom 3,

$$P(A \cup B) = P(A \cap \bar{B}) + P(A \cap B) + P(\bar{A} \cap B) \tag{5.109}$$

$$P(A) = P(A \cap B) + P(A \cap \bar{B}) \tag{5.110}$$

$$P(B) = P(A \cap B) + P(\bar{A} \cap B) \tag{5.111}$$

From Eqs. (*5.110*) and (*5.111*),

$$P(A \cap \bar{B}) = P(A) - P(A \cap B)$$

$$P(\bar{A} \cap B) = P(B) - P(A \cap B)$$

Substituting these equations into Eq. (*5.109*), we obtain

$$P(A \cup B) = P(A) + P(B) - P(A \cap B)$$

5.5. Verify Bayes' rule (*5.24*), that is,

$$P(B|A) = \frac{P(A|B)P(B)}{P(A)}$$

From the conditional probability definition (*5.23*) and Eq. (*5.8*), we have

$$P(A|B) = \frac{P(A \cap B)}{P(B)} \tag{5.112}$$

$$P(B|A) = \frac{P(B \cap A)}{P(A)} = \frac{P(A \cap B)}{P(A)} \tag{5.113}$$

Combining Eqs. (*5.112*) and (*5.113*) yields

$$P(A \cap B) = P(A|B)P(B) = P(B|A)P(A) \tag{5.114}$$

from which we obtain

$$P(B|A) = \frac{P(A|B)P(B)}{P(A)}$$

5.6. Verify Eq. (*5.30*), that is,

$$P(B) = \sum_{k=1}^{N} P(B \cap A_k) = \sum_{k=1}^{N} P(B|A_k)P(A_k)$$

Since $B \cap S = B$ [and using Eq. (*5.29*)], we have

$$B = B \cap S = B \cap (A_1 \cup A_2 \cup \cdots \cup A_N)$$

$$= (B \cap A_1) \cup (B \cap A_2) \cup \cdots \cup (B \cap A_N) \tag{5.115}$$

Now the events $B \cap A_k$ $(k = 1, 2, \ldots, N)$ are mutually exclusive, as seen from the Venn diagram of Fig. 5-6. Then by axiom 3 of the probability definition and Eq. (*5.114*), we obtain

$$P(B) = P(B \cap S) = \sum_{k=1}^{N} P(B \cap A_k) = \sum_{k=1}^{N} P(B|A_k)P(A_k)$$

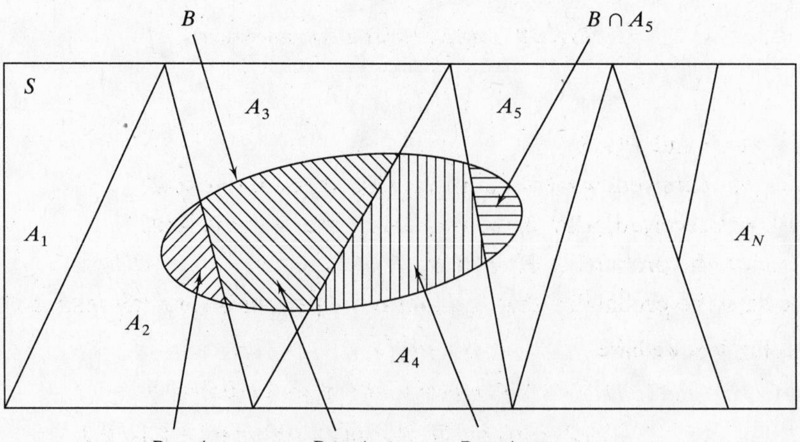

Fig. 5-6

5.7. Consider the switch circuit shown in Fig. 5-7. Let A, B, and C denote the events that switches a, b, and c are closed, respectively. Each of switches may fail to close with probability q. Assuming that the separate switch actions are independent, find the probability that a closed path exists between the terminals in the circuit if $q = 0.5$.

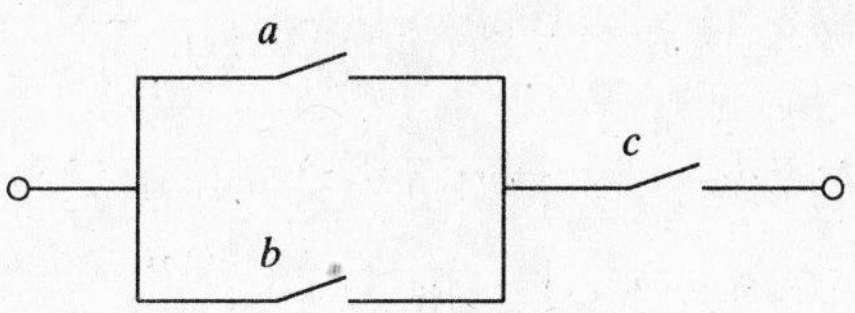

Fig. 5-7 A switch circuit

From Fig. 5-7 it is clear that a closed path exists between the terminals if either a or b is closed and if c is closed. Since $P(A) = P(B) = P(C) = p = 1 - q = 0.5$, by using Eqs. (5.22) and (5.25) we have

$$
\begin{aligned}
P(\text{closed path exists}) &= P[(A \cup B) \cap C] \\
&= P(A \cup B)P(C) \\
&= [P(A) + P(B) - P(A)P(B)]P(C) \\
&= (p + p - p^2)p = (2 - p)p^2 = 0.375
\end{aligned}
$$

5.8. In a binary communication system (Fig. 5-8), a 0 or 1 is transmitted. Because of channel noise, a 0 can be received as a 1 and vice versa. Let m_0 and m_1 denote the events of transmitting 0 and 1, respectively. Let r_0 and r_1 denote the events of receiving 0 and 1, respectively. Let $P(m_0) = 0.5$, $P(r_1|m_0) = p = 0.1$, and $P(r_0|m_1) = q = 0.2$.

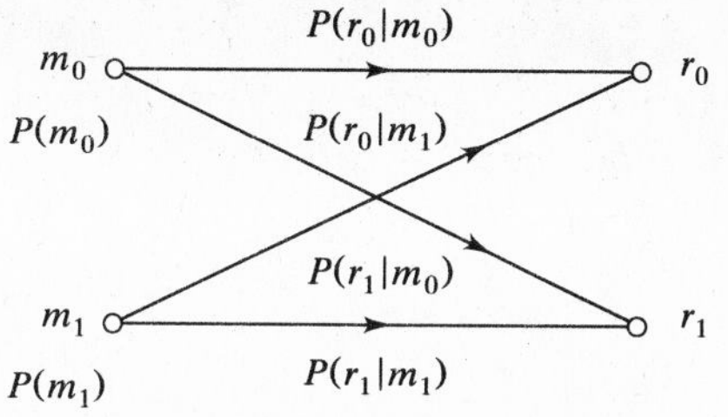

Fig. 5-8 Binary communication system

(*a*) Find $P(r_0)$ and $P(r_1)$.

(*b*) If a 0 was received, what is the probability that a 0 was sent?

(*c*) If a 1 was received, what is the probability that a 1 was sent?

(*d*) Calculate the probability of error P_e.

(*e*) Calculate the probability that the transmitted signal is correctly read at the receiver.

(*a*) From Fig. 5-8, we have

$$
\begin{aligned}
P(m_1) &= 1 - P(m_0) = 1 - 0.5 = 0.5 \\
P(r_0|m_0) &= 1 - P(r_1|m_0) = 1 - p = 1 - 0.1 = 0.9 \\
P(r_1|m_1) &= 1 - P(r_0|m_1) = 1 - q = 1 - 0.2 = 0.8
\end{aligned}
$$

Using Eq. (5.30), we obtain

$$P(r_0) = P(r_0|m_0)P(m_0) + P(r_0|m_1)P(m_1) = 0.9(0.5) + 0.2(0.5) = 0.55$$

$$P(r_1) = P(r_1|m_0)P(m_0) + P(r_1|m_1)P(m_1) = 0.1(0.5) + 0.8(0.5) = 0.45$$

(*b*) Using Bayes' rule (5.24), we have

$$P(m_0|r_0) = \frac{P(m_0)P(r_0|m_0)}{P(r_0)} = \frac{(0.5)(0.9)}{0.55} = 0.818$$

(*c*) Similarly,

$$P(m_1|r_1) = \frac{P(m_1)P(r_1|m_1)}{P(r_1)} = \frac{(0.5)(0.8)}{0.45} = 0.889$$

(*d*)
$$P_e = P(r_1|m_0)P(m_0) + P(r_0|m_1)P(m_1) = 0.1(0.5) + 0.2(0.5) = 0.15$$

(*e*) The probability that the transmitted signal is correctly read at the receiver is

$$P_c = P(r_0|m_0)P(m_0) + P(r_1|m_1)P(m_1) = 0.9(0.5) + 0.8(0.5) = 0.85$$

Note that the probability of error P_e is

$$P_e = 1 - P_c = 1 - 0.85 = 0.15$$

5.9. Consider an experiment consisting of the observation of six successive pulse positions on a communication link. Suppose that at each of the six possible pulse positions there can be a positive pulse, a negative pulse, or no pulse. Suppose also that the individual experiments which determine the kind of pulse at each possible position are independent. Let us denote the event that the ith pulse is positive by $\{x_i = +1\}$, that it is negative by $\{x_i = -1\}$, and that it is zero by $\{x_i = 0\}$. Assume that

$$P(x_i = +1) = p = 0.4 \qquad P(x_i = -1) = q = 0.3 \qquad \text{for } i = 1, 2, \dots, 6$$

(*a*) Find the probability that all pulses are positive.

(*b*) Find the probability that the first three pulses are positive, the next two are zero, and the last is negative.

(*a*) Since the individual experiments are independent, by Eq. (5.27) the probability that all pulses are positive is given by

$$P[(x_1 = +1) \cap (x_2 = +1) \cap \cdots \cap (x_6 = +1)] = P(x_1 = +1)P(x_2 = +1) \cdots P(x_6 = +1)$$
$$= p^6 = (0.4)^6 = 0.0041$$

(*b*) From the given assumptions, we have

$$P(x_i = 0) = 1 - p - q = 0.3$$

Thus, the probability that the first three pulses are positive, the next two are zero, and the last is negative is given by

$$P[(x_1 = +1) \cap (x_2 = +1) \cap (x_3 = +1) \cap (x_4 = 0) \cap (x_5 = 0) \cap (x_6 = -1)]$$
$$= P(x_1 = +1)P(x_2 = +1)P(x_3 = +1)P(x_4 = 0)P(x_5 = 0)P(x_6 = -1)$$
$$= p^3(1 - p - q)^2 q = (0.4)^3(0.3)^2(0.3) = 0.0017$$

5.10. Let A and B be events defined in a sample space S. Show that if both $P(A)$ and $P(B)$ are nonzero, then events A and B cannot be both mutually exclusive and independent.

Let A and B be mutually exclusive events and $P(A) \neq 0$, $P(B) \neq 0$. Then $P(A \cap B) = P(\varnothing) = 0$ and $P(A)P(B) \neq 0$. Therefore

$$P(A \cap B) \neq P(A)P(B)$$

that is, A and B are not independent.

5.11. Show that if events A and B are independent, then

$$P(A \cap \bar{B}) = P(A)P(\bar{B}) \tag{5.116}$$

From the Venn diagram of Fig. 5-1, or using Eqs. (*5.1*), (*5.3*), and (*5.11*), we have

$$A = (A \cap B) \cup (A \cap \bar{B})$$

and

$$(A \cap B) \cap (A \cap \bar{B}) = \varnothing$$

Then by the probability axiom 3 [Eq. (*5.17*)]

$$P(A) = P(A \cap B) + P(A \cap \bar{B})$$

Since A and B are independent, using Eqs. (*5.25*) and (*5.19*), we obtain

$$\begin{aligned} P(A \cap \bar{B}) &= P(A) - P(A \cap B) \\ &= P(A) - P(A)P(B) \\ &= P(A)[1 - P(B)] = P(A)P(\bar{B}) \end{aligned}$$

RANDOM VARIABLES, DISTRIBUTION FUNCTIONS, AND DENSITY FUNCTIONS

5.12. A binary source generates digits 1 and 0 randomly with probabilities 0.6 and 0.4, respectively.

(*a*) What is the probability that two 1s and three 0s will occur in a five-digit sequence?

(*b*) What is the probability that at least three 1s will occur in a five-digit sequence?

(*a*) Let X be the random variable denoting the number of 1s generated in a five-digit sequence. Since there are only two possible outcomes (1 or 0) and the probability of generating 1 is constant and there are five digits, it is clear that X has a binomial distribution described by Eq. (*5.93*) with $n = 5$ and $k = 2$. Hence, the probability that two 1s and three 0s will occur in a five-digit sequence is

$$P(X = 2) = \binom{5}{2}(0.6)^2(0.4)^3 = 0.23$$

(*b*) The probability that at least three 1s will occur in a five-digit sequence is

$$P(X \geq 3) = 1 - P(X \leq 2)$$

where

$$P(X \leq 2) = \sum_{k=0}^{2} \binom{5}{k}(0.6)^k(0.4)^{5-k} = 0.317$$

Hence

$$P(X \geq 3) = 1 - 0.317 = 0.683$$

5.13. Show that when n is very large ($n \gg k$) and p very small ($p \ll 1$), the binomial distribution [Eq. (*5.93*)] can be approximated by the following Poisson distribution [Eq. (*5.98*)]:

$$P(X = k) \approx e^{-np}\frac{(np)^k}{k!} \tag{5.117}$$

From Eq. (*5.93*)

$$P(X=k)=\binom{n}{k}p^k q^{n-k}=\frac{n!}{k!(n-k)!}p^k q^{n-k}$$
$$=\frac{n(n-1)\cdots(n-k+1)}{k!}p^k q^{n-k} \tag{5.118}$$

When $n \gg k$ and $p \ll 1$, then

$$n(n-1)\cdots(n-k+1)\approx n\cdot n\cdot\cdots\cdot n=n^k$$
$$q=1-p\approx e^{-p} \qquad q^{n-k}\approx e^{-(n-k)p}\approx e^{-np}$$

Substituting these relations into Eq. (*5.118*), we obtain

$$P(X=k)\approx e^{-np}\frac{(np)^k}{k!}$$

5.14. A noisy transmission channel has a per-digit error probability $p_e = 0.01$.

(*a*) Calculate the probability of more than one error in 10 received digits.

(*b*) Repeat (*a*), using the Poisson approximation, Eq. (*5.117*).

(*a*) Let X be a binomial random variable denoting the number of errors in 10 received digits. Then using Eq. (*5.93*), we obtain

$$P(X>1)=1-P(X=0)-P(X=1)$$
$$=1-\binom{10}{0}(0.01)^0(0.99)^{10}-\binom{10}{1}(0.01)^1(0.99)^9$$
$$=0.0042$$

(*b*) Using Eq. (*5.117*) with $np_e = 10(0.01) = 0.1$, we have

$$P(X>1)\approx 1-e^{-0.1}\frac{(0.1)^0}{0!}-e^{-0.1}\frac{(0.1)^1}{1!}=0.0047$$

5.15. Find the values of constants a and b such that

$$F_X(x)=[1-ae^{-x/b}]u(x)$$

is a valid distribution function. [Note that $u(x)$ is the unit step function.]

Since $u(-\infty)=0$, property 1 of $F_X(x)$ [Eq. (*5.32a*), $F_X(-\infty)=0$] is satisfied. Property 2 of $F_X(x)$ [Eq. (*5.32b*), $F_X(\infty)=1$] is satisfied if $b>0$. To satisfy property 3 of $F_X(x)$ $[0\le F_X(x)\le 1]$, we must have $0\le a\le 1$.

For $0\le a\le 1$ and $b>0$, $F_X(x)$ is sketched in Fig. 5-9. From Fig. 5-9 we see that all other properties of $F_X(x)$ [Eqs. (*5.32e*) and (*5.32f*)] are satisfied.

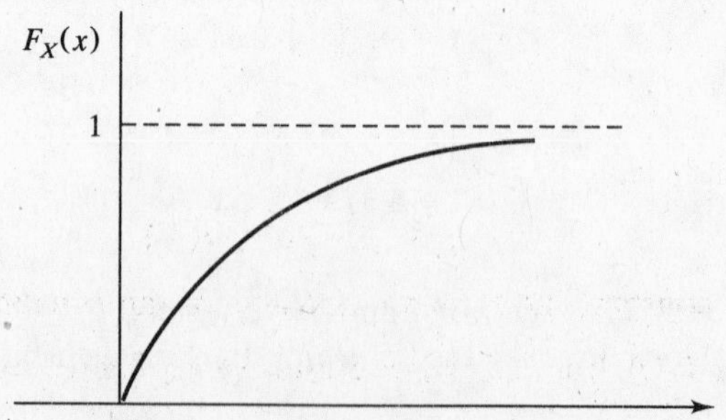

Fig. 5-9

5.16. The pdf of a random variable X is given by

$$f_X(x) = \begin{cases} k & a \le x \le b \\ 0 & \text{otherwise} \end{cases}$$

where k is a constant.

(a) Determine the value of k.

(b) Let $a = -1$ and $b = 2$. Calculate $P(|X| \le c)$ for $c = \frac{1}{2}$.

(a) From property 1 of $f_X(x)$ [Eq. (5.34a)] k must be a positive constant. From property 2 of $f_X(x)$ [Eq. (5.34b)]

$$\int_{-\infty}^{\infty} f_X(x)\,dx = \int_a^b k\,dx = k(b-a) = 1$$

from which we obtain $k = 1/(b-a)$. Thus

$$f_X(x) = \begin{cases} \dfrac{1}{b-a} & a \le x \le b \\ 0 & \text{otherwise} \end{cases} \tag{5.119}$$

A random variable X having the above pdf is called a *uniform* random variable.

(b) With $a = -1$ and $b = 2$ we have

$$f_X(x) = \begin{cases} \frac{1}{3} & -1 \le x \le 2 \\ 0 & \text{otherwise} \end{cases}$$

which is plotted in Fig. 5-10. From Eq. (5.34d)

$$P(|X| \le \tfrac{1}{2}) = P(-\tfrac{1}{2} \le X \le \tfrac{1}{2}) = \int_{-1/2}^{1/2} f_X(x)\,dx = \int_{-1/2}^{1/2} \tfrac{1}{3}\,dx = \tfrac{1}{3}$$

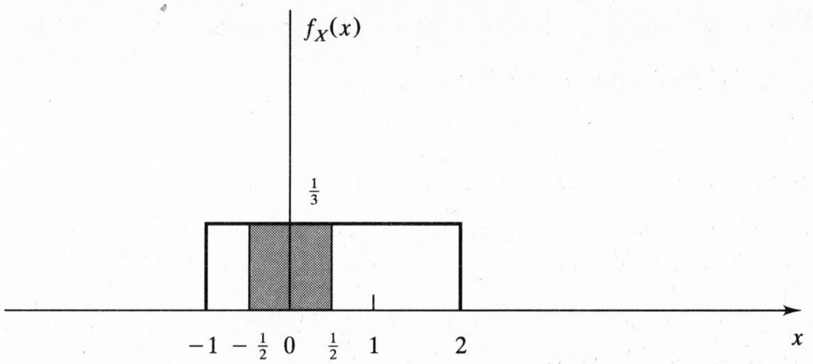

Fig. 5-10 Uniform distribution

5.17. The pdf of X is given by

$$f_X(x) = ke^{-ax}u(x)$$

where a is a positive constant. Determine the value of the constant k, and sketch $f_X(x)$.

From property 1 of $f_X(x)$ [Eq. (5.34a)] we must have $k \ge 0$. From property 2 of $f_X(x)$ [Eq. (5.34b)]

$$\int_{-\infty}^{\infty} f_X(x)\,dx = k\int_0^{\infty} e^{-ax}\,dx = \frac{k}{a} = 1$$

from which we obtain $k = a$. Thus

$$f_X(x) = ae^{-ax}u(x) \qquad a > 0 \tag{5.120}$$

which is sketched in Fig. 5-11.

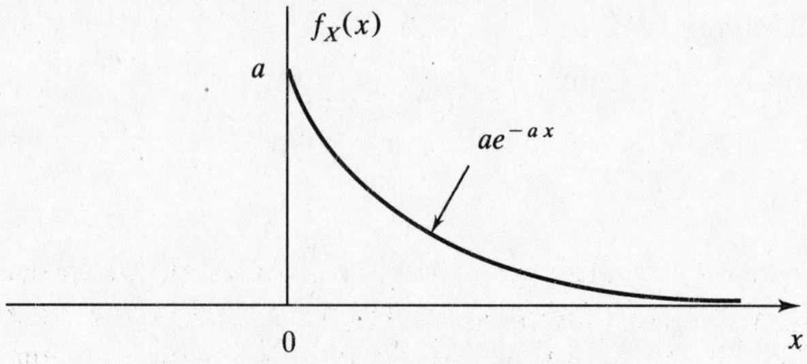

Fig. 5-11 Exponential distribution

A random variable X with the pdf given by Eq. (5.120) is called an *exponential* random variable with parameter a.

5.18. All manufactured devices and machines fail to work sooner or later. If the failure rate is constant, the time to failure T is modeled as an exponential random variable. Suppose that a particular class of computer memory chips has been found to have the exponential failure law of Eq. (5.120) in hours.

(a) Measurements show that the probability that the time to failure exceeds 10^4 hours (h) for chips in the given class is e^{-1} (≈ 0.368). Calculate the value of parameter a for this case.

(b) Using the value of parameter a determined in part (a), calculate the time t_0 such that the probability is 0.05 that the time to failure is less than t_0.

(a) Using Eqs. (5.34c) and (5.120), we see that the distribution function of T is given by

$$F_T(t) = \int_{-\infty}^{t} f_T(\tau)\,d\tau = (1 - e^{-at})u(t)$$

Now

$$\begin{aligned} P(T > 10^4) &= 1 - P(T \le 10^4) \\ &= 1 - F_T(10^4) = 1 - (1 - e^{-a(10^4)}) = e^{-a(10^4)} = e^{-1} \end{aligned}$$

from which we obtain $a = 10^{-4}$.

(b) We want

$$F_T(t_0) = P(T \le t_0) = 0.05$$

Hence,

$$1 - e^{-at_0} = 1 - e^{-(10^{-4})t_0} = 0.05$$

or

$$e^{-(10^{-4})t_0} = 0.95$$

from which we obtain

$$t_0 = -10^4 \ln 0.95 = 513 \text{ h}$$

5.19. The joint pdf of X and Y is given by

$$f_{XY}(x, y) = ke^{-(ax+by)}u(x)u(y)$$

where a and b are positive constants. Determine the value of constant k.

The value of k is determined by Eq. (*5.41*), that is,

$$F_{XY}(\infty,\infty)=\int_{-\infty}^{\infty}\int_{-\infty}^{\infty} f_{XY}(\xi,\eta)\,d\xi\,d\eta$$
$$=k\int_0^{\infty}\int_0^{\infty} e^{-(a\xi+b\eta)}\,d\xi\,d\eta$$
$$=k\int_0^{\infty} e^{-a\xi}\,d\xi\int_0^{\infty} e^{-b\eta}\,d\eta=\frac{k}{ab}=1$$

Hence, $k=ab$.

5.20. A manufacturer has been using two different manufacturing processes to make computer memory chips. Let X be the random variable characterizing the time to failure of chips made by one process and Y be the random variable characterizing the time to failure of chips made by the other process. Assume that X and Y have the joint pdf

$$f_{XY}(x,y)=abe^{-(ax+by)}u(x)u(y) \qquad (5.121)$$

where $a=10^{-4}$ and $b=1.2(10^{-4})$.

Determine the probability that the time to failure is greater for chips characterized by X than it is for chips characterized by Y.

The region D for which $x>y$ is shaded in Fig. 5-12. Then the probability that the time to failure is greater for chips characterized by X than it is for chips characterized by Y is

$$P(x>y)=\iint_D f_{XY}(x,y)\,dx\,dy=\int_0^{\infty}\int_0^{x} abe^{-(ax+by)}\,dy\,dx$$
$$=\int_0^{\infty} ae^{-ax}\int_0^{x} be^{-by}\,dy\,dx=\int_0^{\infty} ae^{-ax}(1-e^{-bx})\,dx$$
$$=\frac{b}{a+b}=\frac{1.2(10^{-4})}{10^{-4}+1.2(10^{-4})}=0.545$$

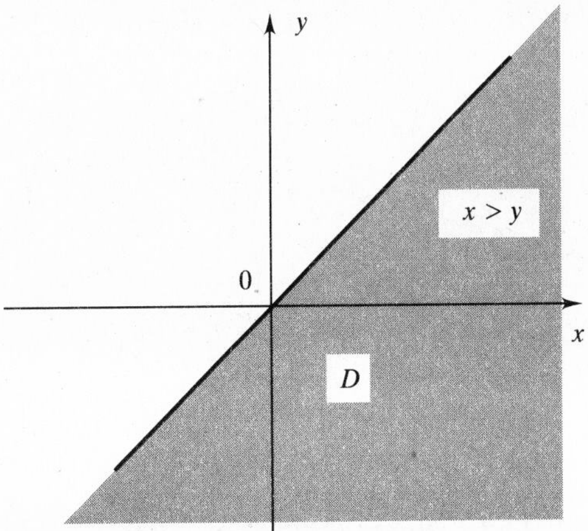

Fig. 5-12

5.21. The joint pdf of X and Y is given by

$$f_{XY}(x,y)=xye^{-(x^2+y^2)/2}u(x)u(y)$$

(*a*) Find the marginal pdf's $f_X(x)$ and $f_Y(y)$.

(*b*) Are X and Y independent?

(*a*) By Eqs. (*5.43a*) and (*5.43b*), we have

$$f_X(x) = \int_{-\infty}^{\infty} f_{XY}(x, y)\,dy = \int_0^{\infty} xye^{-(x^2+y^2)/2}u(x)\,dy$$

$$= xe^{-x^2/2}u(x)\int_0^{\infty} ye^{-y^2/2}\,dy = xe^{-x^2/2}u(x)$$

Since $f_{XY}(x, y)$ is symmetric with respect to x and y, interchanging x and y, we obtain

$$f_Y(y) = ye^{-y^2/2}u(y)$$

(*b*) Since $f_{XY}(x, y) = f_X(x)f_Y(y)$, we conclude that X and Y are independent.

5.22. The random variables X and Y are said to be *jointly normal* random variables if their joint pdf is given by

$$f_{XY}(x, y) = \frac{1}{2\pi\sigma_X\sigma_Y(1-\rho^2)^{1/2}} \exp\left\{-\frac{1}{2(1-\rho^2)}\right.$$

$$\left.\times\left[\left(\frac{x-\mu_X}{\sigma_X}\right)^2 - 2\rho\left(\frac{x-\mu_X}{\sigma_X}\right)\left(\frac{y-\mu_Y}{\sigma_Y}\right) + \left(\frac{y-\mu_Y}{\sigma_Y}\right)^2\right]\right\} \tag{5.122}$$

(*a*) Find the marginal pdf's of X and Y.

(*b*) Show that X and Y are independent when $\rho = 0$.

(*a*) By Eq. (*5.43a*) the marginal pdf of X is

$$f_X(x) = \int_{-\infty}^{\infty} f_{XY}(x, y)\,dy$$

By completing the square in the exponent of Eq. (*5.122*), we obtain

$$f_X(x) = \frac{\exp\left[-\frac{1}{2}\left(\frac{x-\mu_X}{\sigma_X}\right)^2\right]}{\sqrt{2\pi}\,\sigma_X} \int_{-\infty}^{\infty} \frac{1}{\sqrt{2\pi}\,\sigma_Y(1-\rho^2)^{1/2}}$$

$$\times \exp\left\{-\frac{1}{2(1-\rho^2)}\left[\frac{y-\mu_Y}{\sigma_Y} - \rho\left(\frac{x-\mu_X}{\sigma_X}\right)\right]^2\right\} dy$$

$$= \frac{\exp\left[-\frac{1}{2}\left(\frac{x-\mu_X}{\sigma_X}\right)^2\right]}{\sqrt{2\pi}\,\sigma_X} \int_{-\infty}^{\infty} \frac{1}{\sqrt{2\pi}\,\sigma_Y(1-\rho^2)^{1/2}}$$

$$\times \exp\left\{-\frac{1}{2\sigma_Y^2(1-\rho^2)}\left[y-\mu_Y-\rho\frac{\sigma_Y}{\sigma_X}(x-\mu_X)\right]^2\right\} dy$$

Comparing the integrand with Eq. (*5.103*), we see that the integrand is a normal pdf with mean

$$\mu_Y + \rho\frac{\sigma_Y}{\sigma_X}(x-\mu_X)$$

and variance

$$\sigma_Y^2(1-\rho^2)$$

Thus, the integral must be unity, and we obtain

$$f_X(x) = \frac{1}{\sqrt{2\pi}\,\sigma_X} \exp\left[-(x-\mu_X)^2/2\sigma_X^2\right] \tag{5.123}$$

In a similar manner, the marginal pdf of Y is

$$f_Y(y) = \int_{-\infty}^{\infty} f_{XY}(x, y)\, dx = \frac{1}{\sqrt{2\pi}\,\sigma_Y} \exp\left[-(y-\mu_Y)^2/2\sigma_Y^2\right] \tag{5.124}$$

(*b*) When $\rho = 0$, Eq. (5.122) reduces to

$$\begin{aligned} f_{XY}(x, y) &= \frac{1}{2\pi\sigma_X\sigma_Y} \exp\left\{-\frac{1}{2}\left[\left(\frac{x-\mu_X}{\sigma_X}\right)^2 + \left(\frac{y-\mu_Y}{\sigma_Y}\right)^2\right]\right\} \\ &= \frac{1}{\sqrt{2\pi}\,\sigma_X} \exp\left[-\frac{1}{2}\left(\frac{x-\mu_X}{\sigma_X}\right)^2\right] \frac{1}{\sqrt{2\pi}\,\sigma_Y} \exp\left[-\frac{1}{2}\left(\frac{y-\mu_Y}{\sigma_Y}\right)^2\right] \\ &= f_X(x) f_Y(y) \end{aligned}$$

Hence, X and Y are independent.

FUNCTIONS OF RANDOM VARIABLES

5.23. Let $Y = 2X + 3$. If a random variable X is uniformly distributed over $[-1, 2]$, find $f_Y(y)$ and sketch $f_X(x)$ and $f_Y(y)$.

From Eq. (5.119) (Prob. 5.16), we have

$$f_X(x) = \begin{cases} \frac{1}{3} & -1 \le x \le 2 \\ 0 & \text{otherwise} \end{cases}$$

The equation $y = g(x) = 2x + 3$ has a single solution $x_1 = (y - 3)/2$, the range of y is $[1, 7]$, and $g'(x) = 2$. Thus $-1 \le x_1 \le 2$ and by Eq. (5.53)

$$f_Y(y) = \tfrac{1}{2} f_X(x_1) = \begin{cases} \frac{1}{6} & 1 \le y \le 7 \\ 0 & \text{otherwise} \end{cases}$$

The pdf's $f_X(x)$ and $f_Y(y)$ are sketched in Fig. 5-13.

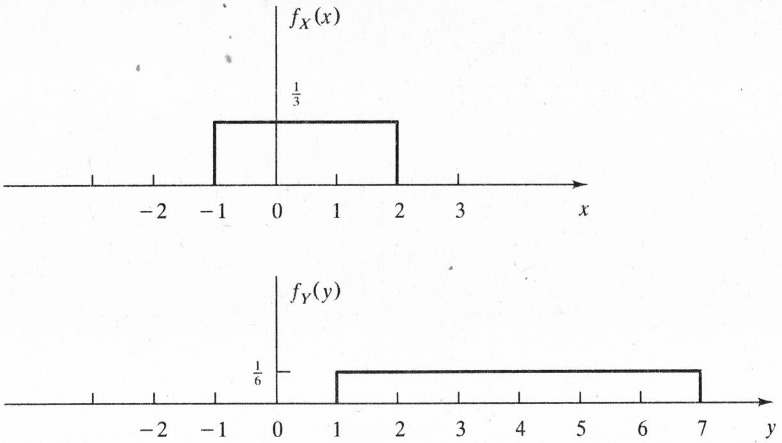

Fig. 5-13

5.24. Let $Y = aX + b$. Show that if $X = N(\mu; \sigma^2)$, then $Y = N(a\mu + b; a^2\sigma^2)$.

The equation $y = g(x) = ax + b$ has a single solution $x_1 = (y - b)/a$, and $g'(x) = a$. The range of y is $(-\infty, \infty)$. Hence, by Eq. (5.53)

$$f_Y(y) = \frac{1}{|a|} f_X\left(\frac{y-b}{a}\right) \tag{5.125}$$

Since $X = N(\mu; \sigma^2)$, by Eq. (*5.103*)

$$f_X(x) = \frac{1}{\sqrt{2\pi}\,\sigma} \exp\left[-\frac{1}{2\sigma^2}(x-\mu)^2\right] \tag{5.126}$$

Hence, by Eq. (*5.125*)

$$f_Y(y) = \frac{1}{\sqrt{2\pi}\,|a|\sigma} \exp\left[-\frac{1}{2\sigma^2}\left(\frac{y-b}{a}-\mu\right)^2\right]$$
$$= \frac{1}{\sqrt{2\pi}\,|a|\sigma} \exp\left[-\frac{1}{2a^2\sigma^2}(y-a\mu-b)^2\right] \tag{5.127}$$

which is the pdf of $N(a\mu + b; a^2\sigma^2)$. Thus, if $X = N(\mu; \sigma^2)$, then $Y = N(a\mu + b; a^2\sigma^2)$.

5.25. Let $Y = X^2$. Find and sketch $f_Y(y)$ if $X = N(0; 1)$.

If $y < 0$, then the equation $y = x^2$ has no real solutions; hence $f_Y(y) = 0$.
If $y > 0$, then $y = x^2$ has two solutions

$$x_1 = \sqrt{y} \qquad x_2 = -\sqrt{y}$$

Now, $y = g(x) = x^2$ and $g'(x) = 2x$. Hence, by Eq. (*5.53*)

$$f_Y(y) = \frac{1}{2\sqrt{y}}\left[f_X(\sqrt{y}) + f_X(-\sqrt{y})\right]u(y) \tag{5.128}$$

Since $X = N(0; 1)$ from Eq. (*5.103*), we have

$$f_X(x) = \frac{1}{\sqrt{2\pi}} e^{-x^2/2} \tag{5.129}$$

Since $f_X(x)$ is an even function from Eq. (*5.128*), we have

$$f_Y(y) = \frac{1}{\sqrt{y}} f_X(\sqrt{y})u(y) = \frac{1}{\sqrt{2\pi y}} e^{-y/2}u(y) \tag{5.130}$$

which is sketched in Fig. 5-14.

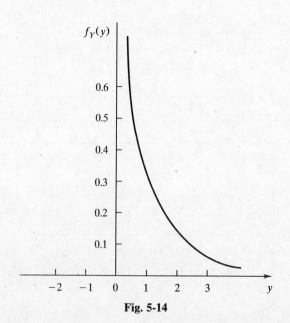

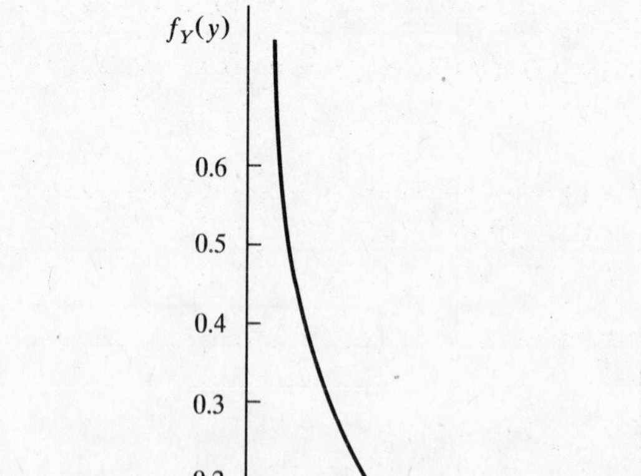

Fig. 5-14

5.26. Consider the half-wave linear rectifier transformation

$$Y = \begin{cases} X & X \geq 0 \\ 0 & X < 0 \end{cases}$$

Find $f_Y(y)$ if $X = N(0; 1)$.

Clearly $f_Y(y) = 0$ for $y < 0$. For $y > 0$, $y = g(x) = x$ has a single solution $x_1 = y$, and $g'(x) = 1$. Hence, by Eq. (*5.53*)

$$f_Y(y) = f_X(y) \qquad \text{and} \qquad F_Y(y) = F_X(y) \qquad y > 0$$

Thus $F_Y(y)$ is discontinuous at $y = 0$ with discontinuity

$$F_Y(0^+) - F_Y(0^-) = F_X(0)$$

Hence

$$f_Y(y) = F_X(0)\delta(y) + f_X(y)u(y) \tag{5.131}$$

Because $f_X(x)$ is an even function [Eq. (*5.129*)], we have

$$F_X(0) = \int_{-\infty}^{0} f_X(x)\, dx = \frac{1}{2}$$

Hence, Eq. (*5.131*) becomes

$$f_Y(y) = \frac{1}{2}\delta(y) + \frac{1}{\sqrt{2\pi}} e^{-y^2/2} u(y) \tag{5.132}$$

5.27. Let $Y = \sin X$, where X is uniformly distributed with

$$f_X(x) = \begin{cases} \dfrac{1}{2\pi} & 0 \leq x \leq 2\pi \\ 0 & \text{otherwise} \end{cases}$$

Find $f_Y(y)$ and $F_Y(y)$, and sketch these functions.

If $|y| > 1$, then the equation $y = \sin x$ has no solutions (Fig. 5-15); hence $f_Y(y) = 0$. If $|y| < 1$, then $y = \sin x$ has two solutions in the interval $0 \leq x < 2\pi$ (Fig. 5-15).

$$x_1 = \sin^{-1} y \qquad \text{and} \qquad x_2 = \pi - x_1$$

Since

$$y'(x_1) = \cos x_1 = \cos\left(\sin^{-1} y\right) = \sqrt{1 - y^2}$$

$$y'(x_2) = \cos x_2 = \cos\left(\pi - x_1\right) = -\cos x_1 = -\sqrt{1 - y^2}$$

by Eq. (*5.53*), we have

$$f_Y(y) = \frac{1}{2\pi\sqrt{1 - y^2}} + \frac{1}{2\pi\sqrt{1 - y^2}} = \frac{1}{\pi\sqrt{1 - y^2}} \qquad |y| < 1 \tag{5.133}$$

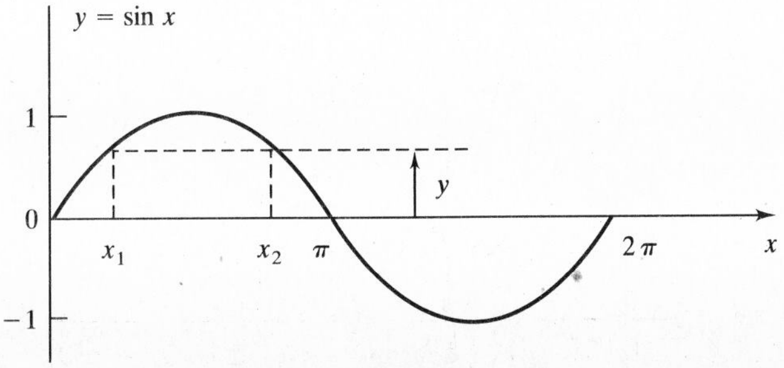

Fig. 5-15

Integrating the above density [Eq. (5.34c)], we obtain

$$F_Y(y) = \begin{cases} 0 & y < -1 \\ \dfrac{1}{2} + \dfrac{\sin^{-1} y}{\pi} & -1 \le y < 1 \\ 1 & 1 \le y \end{cases} \tag{5.134}$$

These functions are sketched in Fig. 5-16.

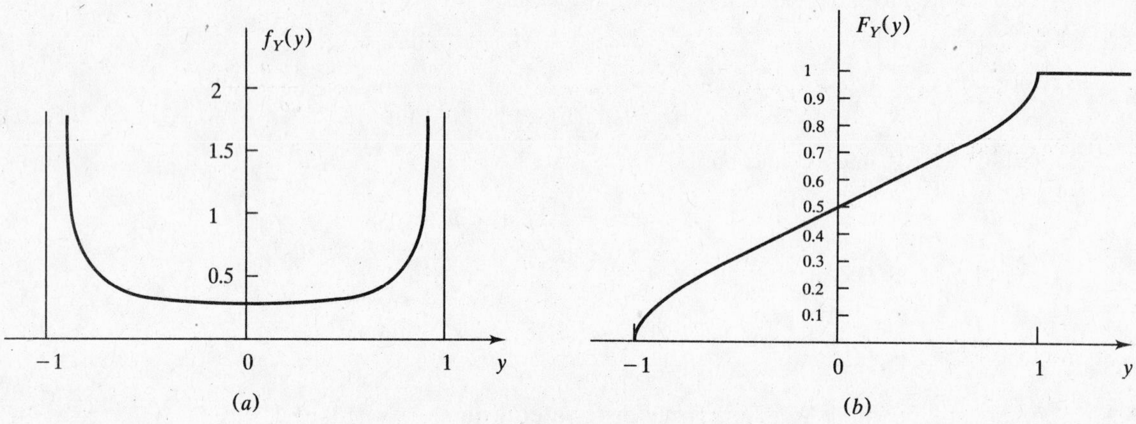

Fig. 5-16

5.28. The input to a noisy communication channel is a binary random variable X with $P(X=0) = P(X=1) = \frac{1}{2}$. The output of channel Z is given by $X+Y$, where Y is the additive noise introduced by the channel. Assuming that X and Y are independent and $Y = N(0;1)$, find the density function of Z.

Using Eqs. (5.30) and (5.31), we have

$$F_Z(z) = P(Z \le z) = P(Z \le z|X=0)P(X=0) + P(Z \le z|X=1)P(X=1)$$

Since

$$Z = X + Y$$
$$P(Z \le z|X=0) = P(X+Y \le z|X=0) = P(Y \le z) = F_Y(z)$$

Similarly,

$$P(Z \le z|X=1) = P(X+Y \le z|X=1) = P(Y \le z-1) = F_Y(z-1)$$

Hence,

$$F_Z(z) = \frac{1}{2}F_Y(z) + \frac{1}{2}F_Y(z-1)$$

Since $Y = N(0;1)$,

$$f_Y(y) = \frac{1}{\sqrt{2\pi}} e^{-y^2/2}$$

and

$$f_Z(z) = \frac{dF_Z(z)}{dz} = \frac{1}{2}f_Y(z) + \frac{1}{2}f_Y(z-1)$$
$$= \frac{1}{2}\left[\frac{1}{\sqrt{2\pi}}e^{-z^2/2} + \frac{1}{\sqrt{2\pi}}e^{-(z-1)^2/2}\right] \tag{5.135}$$

5.29. Consider the transformation

$$Z = aX + bY \qquad W = cX + dY \tag{5.136}$$

Find the joint density function $f_{ZW}(z, w)$ in terms of $f_{XY}(x, y)$.

If $ad - bc \neq 0$, then the system

$$ax + by = z \qquad cx + dy = w$$

has one and only one solution:

$$x = \alpha z + \beta w \qquad y = \gamma z + \eta w$$

where

$$\alpha = \frac{d}{ad - bc} \qquad \beta = \frac{-b}{ad - bc} \qquad \gamma = \frac{-c}{ad - bc} \qquad \eta = \frac{a}{ad - bc}$$

Since [Eq. (5.60)]

$$J(x, y) = \begin{vmatrix} \dfrac{\partial z}{\partial x} & \dfrac{\partial z}{\partial y} \\ \dfrac{\partial w}{\partial x} & \dfrac{\partial w}{\partial y} \end{vmatrix} = \begin{vmatrix} a & b \\ c & d \end{vmatrix} = ad - bc$$

Eq. (5.59) yields

$$f_{ZW}(z, w) = \frac{1}{|ad - bc|} f_{XY}(\alpha z + \beta w, \gamma z + \eta w) \tag{5.137}$$

5.30. Let $Z = X + Y$. Find the pdf of Z if X and Y are independent random variables.

We introduce an auxiliary random variable W, defined by

$$W = Y$$

The system $z = x + y$, $w = y$ has a single solution:

$$x = z - w \qquad y = w$$

Since

$$J(x, y) = \begin{vmatrix} \dfrac{\partial z}{\partial x} & \dfrac{\partial z}{\partial y} \\ \dfrac{\partial w}{\partial x} & \dfrac{\partial w}{\partial y} \end{vmatrix} = \begin{vmatrix} 1 & 1 \\ 0 & 1 \end{vmatrix} = 1$$

Eq. (5.59) yields [or by setting $a = b = d = 1$ and $c = 0$ in Eq. (5.137)]

$$f_{ZW}(z, w) = f_{XY}(z - w, w) \tag{5.138}$$

Hence, by Eq. (5.43a), we obtain

$$f_Z(z) = \int_{-\infty}^{\infty} f_{ZW}(z, w)\, dw = \int_{-\infty}^{\infty} f_{XY}(z - w, w)\, dw \tag{5.139}$$

If X and Y are independent, then

$$f_Z(z) = \int_{-\infty}^{\infty} f_X(z - w) f_Y(w)\, dw \tag{5.140}$$

which is the convolution of functions $f_X(x)$ and $f_Y(y)$.

5.31. Suppose that X and Y are independent normalized normal random variables. Find the pdf of $Z = X + Y$.

The pdf's of X and Y are

$$f_X(x) = \frac{1}{\sqrt{2\pi}} e^{-x^2/2} \qquad f_Y(y) = \frac{1}{\sqrt{2\pi}} e^{-y^2/2}$$

Then, by Eq. (*5.140*), we have

$$\begin{aligned} f_Z(z) &= \int_{-\infty}^{\infty} f_X(z-w) f_Y(w)\,dw \\ &= \int_{-\infty}^{\infty} \frac{1}{\sqrt{2\pi}} e^{-(z-w)^2/2} \frac{1}{\sqrt{2\pi}} e^{-w^2/2}\,dw \\ &= \frac{1}{2\pi}\int_{-\infty}^{\infty} \exp\left[-\frac{1}{2}(z^2 - 2zw + 2w^2)\right] dw \\ &= \frac{1}{2\pi}\int_{-\infty}^{\infty} \exp\left\{-\frac{1}{2}\left[\frac{z^2}{2} + \left(\sqrt{2}\,w - \frac{z}{\sqrt{2}}\right)^2\right]\right\} dw \\ &= \frac{1}{\sqrt{2\pi}} e^{-z^2/4} \frac{1}{\sqrt{2\pi}} \int_{-\infty}^{\infty} \exp\left[-\frac{1}{2}\left(\sqrt{2}\,w - \frac{z}{\sqrt{2}}\right)^2\right] dw \end{aligned}$$

Let $u = \sqrt{2}\,w - z/\sqrt{2}$. Then

$$f_Z(z) = \frac{1}{\sqrt{2\pi}} e^{-z^2/4} \frac{1}{\sqrt{2}} \int_{-\infty}^{\infty} \frac{1}{\sqrt{2\pi}} e^{-u^2/2}\,du$$

Since the integrand is the pdf of $N(0; 1)$, the integral is equal to unity, and we obtain

$$f_Z(z) = \frac{1}{\sqrt{2\pi}\sqrt{2}} e^{-z^2/4} = \frac{1}{\sqrt{2\pi}\sqrt{2}} e^{-z^2/[2(\sqrt{2})^2]} \tag{5.141}$$

which is the pdf of $N(0; \sqrt{2})$.

Thus, Z is a normal random variable with zero mean and variance $\sqrt{2}$.

5.32. Consider the transformation

$$R = \sqrt{X^2 + Y^2} \qquad \Theta = \tan^{-1}\frac{Y}{X} \tag{5.142}$$

Find $f_{R\Theta}(r, \theta)$ in terms of $f_{XY}(x, y)$.

We assume that $r \geq 0$ and $0 \leq \theta \leq 2\pi$. With this assumption, the system

$$\sqrt{x^2 + y^2} = r \qquad \tan^{-1}\frac{y}{x} = \theta$$

has a single solution:

$$x = r\cos\theta \qquad y = r\sin\theta$$

Since [Eq. (*5.60*)]

$$\bar{J}(x, y) = \begin{vmatrix} \dfrac{\partial x}{\partial r} & \dfrac{\partial x}{\partial \theta} \\ \dfrac{\partial y}{\partial r} & \dfrac{\partial y}{\partial \theta} \end{vmatrix} = \begin{vmatrix} \cos\theta & -r\sin\theta \\ \sin\theta & r\cos\theta \end{vmatrix} = r$$

Eq. (*5.59*) yields

$$f_{R\Theta}(r, \theta) = r f_{XY}(r\cos\theta, r\sin\theta) \tag{5.143}$$

5.33. A voltage V is a function of time t and is given by

$$V(t) = X\cos\omega t + Y\sin\omega t \qquad (5.144)$$

in which ω is a constant angular frequency and $X = Y = N(0;\sigma^2)$ and they are independent.

(*a*) Show that $V(t)$ may be written as

$$V(t) = R\cos(\omega t - \Theta) \qquad (5.145)$$

(*b*) Find the density functions of R and Θ, and show that R and Θ are independent.

(*a*)

$$\begin{aligned} V(t) &= X\cos\omega t + Y\sin\omega t \\ &= \sqrt{X^2+Y^2}\left(\frac{X}{\sqrt{X^2+Y^2}}\cos\omega t + \frac{Y}{\sqrt{X^2+Y^2}}\sin\omega t\right) \\ &= \sqrt{X^2+Y^2}(\cos\Theta\cos\omega t + \sin\Theta\sin\omega t) \\ &= R\cos(\omega t - \Theta) \end{aligned}$$

where

$$R = \sqrt{X^2+Y^2} \qquad \text{and} \qquad \Theta = \tan^{-1}\frac{Y}{X}$$

(*b*) Since $X = Y = N(0;\sigma^2)$ and are independent, from Eqs. (*5.48*) and (*5.103*)

$$\begin{aligned} f_{XY}(x,y) &= \frac{1}{\sqrt{2\pi}\,\sigma}e^{-x^2/(2\sigma^2)}\frac{1}{\sqrt{2\pi}\,\sigma}e^{-y^2/(2\sigma^2)} \\ &= \frac{1}{2\pi\sigma^2}e^{-(x^2+y^2)/(2\sigma^2)} \end{aligned} \qquad (5.146)$$

Thus, using the result of Prob. 5.32 [Eq. (*5.143*)], we have

$$\begin{aligned} f_{R\Theta}(r,\theta) &= rf_{XY}(r\cos\theta, r\sin\theta) \\ &= \frac{r}{2\pi\sigma^2}e^{-r^2/(2\sigma^2)} \end{aligned} \qquad (5.147)$$

Using Eqs. (*5.43a*) and (*5.43b*), we obtain

$$f_R(r) = \int_0^{2\pi} f_{R\Theta}(r,\theta)\,d\theta = \frac{r}{2\pi\sigma^2}e^{-r^2/(2\sigma^2)}\int_0^{2\pi} d\theta = \frac{r}{\sigma^2}e^{-r^2/(2\sigma^2)} \qquad (5.148)$$

$$f_\Theta(\theta) = \int_0^\infty f_{R\Theta}(r,\theta)\,dr = \frac{1}{2\pi\sigma^2}\int_0^\infty re^{-r^2/(2\sigma^2)}\,dr = \frac{1}{2\pi} \qquad (5.149)$$

and

$$f_{R\Theta}(r,\theta) = f_R(r)f_\Theta(\theta) \qquad (5.150)$$

Hence, R and Θ are independent.

Note that Θ is a uniform random variable, and R is called a *Rayleigh* random variable.

STATISTICAL AVERAGES

5.34. The random variable X takes the values 0 and 1 with probabilities α and $\beta = 1 - \alpha$, respectively. Find the mean and the variance of X.

Using Eqs. (*5.62*) and (*5.66*), we have

$$\mu_X = E[X] = 0(\alpha) + 1(\beta) = \beta$$

$$E[X^2] = 0^2(\alpha) + 1^2(\beta) = \beta$$

From Eq. (5.75)

$$\sigma_X^2 = E[X^2] - (E[X])^2 = \beta - \beta^2 = \beta(1-\beta) = \alpha\beta$$

5.35. Let $Y = \sin X$, where X is uniformly distributed in the interval $(0, 2\pi)$. Find the mean and the second moment of X by (a) Eq. (5.64) and (b) Eq. (5.65).

(a) From the result of Prob. 5.27, we have

$$f_Y(y) = \begin{cases} \dfrac{1}{\pi\sqrt{1-y^2}} & |y| < 1 \\ 0 & |y| > 1 \end{cases}$$

Using Eq. (5.64), we obtain

$$E[Y] = \int_{-1}^{1} \frac{y}{\pi\sqrt{1-y^2}}\, dy = 0$$

because the integrand is an odd function, and

$$E[Y^2] = \int_{-1}^{1} \frac{y^2}{\pi\sqrt{1-y^2}}\, dy = \frac{1}{2}$$

from a table of integrals.

(b) Since X is uniformly distributed in the interval $(0, 2\pi)$, we have

$$f_X(x) = \begin{cases} \dfrac{1}{2\pi} & 0 \le x \le 2\pi \\ 0 & \text{otherwise} \end{cases}$$

Thus, using Eq. (5.65) gives

$$E[Y] = E[\sin X] = \frac{1}{2\pi}\int_0^{2\pi} \sin x\, dx = 0$$

and

$$E[Y^2] = E[\sin^2 X] = \frac{1}{2\pi}\int_0^{2\pi} \sin^2 x\, dx = \frac{1}{2\pi}\int_0^{2\pi} \frac{1}{2}(1 - \cos 2x)\, dx = \frac{1}{2}$$

5.36. Binary data are transmitted over a noisy communication channel in a block of 16 binary digits. The probability that a received digit is in error due to channel noise is 0.01. Assume that the errors occurring in various digit positions within a block are independent.

(a) Find the mean (average number of) errors per block.

(b) Find the variance of the number of errors per block.

(c) Find the probability that the number of errors per block is greater than or equal to 4.

(a) Let X be the random variable representing the number of errors per block. Then X has a binomial distribution with $n = 16$ and $p = 0.01$. By Eq. (5.96) the average number of errors per block is

$$E[X] = np = (16)(0.01) = 0.16$$

(b) By Eq. (5.97)

$$\sigma_X^2 = np(1-p) = (16)(0.01)(0.99) = 0.158$$

(c)

$$P(X \ge 4) = 1 - P(X \le 3)$$

Using Eq. (*5.95*), we have

$$P(X \leq 3) = \sum_{k=0}^{3} \binom{16}{k} (0.01)^k (0.99)^{16-k} = 0.986$$

Hence

$$P(X \geq 4) = 1 - 0.986 = 0.014$$

5.37. Let X be a Poisson random variable defined by Eq. (*5.98*).

(*a*) Show that

$$\sum_{k=0}^{\infty} P(X=k) = 1 \tag{5.151}$$

(*b*) Verify Eqs. (*5.101*) and (*5.102*), that is,

$$\mu_X = E[X] = \alpha \quad \text{and} \quad \sigma_X^2 = \alpha$$

(*a*) From Eq. (*5.98*)

$$P(X=k) = e^{-\alpha} \frac{\alpha^k}{k!}$$

Thus,

$$\sum_{k=0}^{\infty} P(X=k) = \sum_{k=0}^{\infty} e^{-\alpha} \frac{\alpha^k}{k!} = e^{-\alpha} \sum_{k=0}^{\infty} \frac{\alpha^k}{k!} = e^{-\alpha} e^{\alpha} = 1$$

(*b*) By Eq. (*5.62*)

$$\mu_X = E[X] = \sum_{k=0}^{\infty} kP(X=k) = 0 + \sum_{k=1}^{\infty} e^{-\alpha} \frac{\alpha^k}{(k-1)!}$$

$$= \alpha e^{-\alpha} \sum_{k=1}^{\infty} \frac{\alpha^{k-1}}{(k-1)!} = \alpha e^{-\alpha} \sum_{m=0}^{\infty} \frac{\alpha^m}{m!} = \alpha e^{-\alpha} e^{\alpha} = \alpha$$

Similarly,

$$E[X(X-1)] = \sum_{k=0}^{\infty} k(k-1)P(X=k) = 0 + 0 + \sum_{k=2}^{\infty} e^{-\alpha} \frac{\alpha^k}{(k-2)!}$$

$$= \alpha^2 e^{-\alpha} \sum_{k=2}^{\infty} \frac{\alpha^{k-2}}{(k-2)!} = \alpha^2 e^{-\alpha} \sum_{m=0}^{\infty} \frac{\alpha^m}{m!} = \alpha^2 e^{-\alpha} e^{\alpha} = \alpha^2$$

or

$$E[X^2 - X] = E[X^2] - E[X] = E[X^2] - \alpha = \alpha^2$$

Thus

$$E[X^2] = \alpha^2 + \alpha \tag{5.152}$$

Then using Eq. (*5.75*) gives

$$\sigma_X^2 = E[X^2] - (E[X])^2 = (\alpha^2 + \alpha) - \alpha^2 = \alpha$$

5.38. Let $X = N(\mu; \sigma^2)$. Verify Eqs. (*5.107*) and (*5.108*), that is,

$$\mu_X = E[X] = \mu \qquad \sigma_X^2 = E\left[(X-\mu)^2\right] = \sigma^2$$

Substituting Eq. (*5.103*) into Eq. (*5.61*), we have

$$\mu_X = E[X] = \frac{1}{\sqrt{2\pi}\,\sigma} \int_{-\infty}^{\infty} x e^{-(x-\mu)^2/(2\sigma^2)}\, dx$$

Changing the variable of integration to $y=(x-\mu)/\sigma$, we have

$$E[X]=\frac{1}{\sqrt{2\pi}}\int_{-\infty}^{\infty}(\sigma y+\mu)e^{-y^2/2}\,dy$$

$$=\frac{\sigma}{\sqrt{2\pi}}\int_{-\infty}^{\infty}ye^{-y^2/2}\,dy+\mu\int_{-\infty}^{\infty}\frac{1}{\sqrt{2\pi}}e^{-y^2/2}\,dy$$

The first integral is zero since its integrand is an odd function. The second integral is unity since its integrand is the pdf of $N(0;1)$. Thus

$$\mu_X=E[X]=\mu$$

From property 2 of $f_X(x)$ [Eq. (5.34b)], we have

$$\int_{-\infty}^{\infty}e^{-(x-\mu)^2/(2\sigma^2)}\,dx=\sigma\sqrt{2\pi} \tag{5.153}$$

Differentiating with respect to σ, we obtain

$$\int_{-\infty}^{\infty}\frac{(x-\mu)^2}{\sigma^3}e^{-(x-\mu)^2/(2\sigma^2)}\,dx=\sqrt{2\pi}$$

Multiplying both sides by $\sigma^2/\sqrt{2\pi}$, we have

$$\frac{1}{\sqrt{2\pi}\,\sigma}\int_{-\infty}^{\infty}(x-\mu)^2e^{-(x-\mu)^2/(2\sigma^2)}\,dx=E\left[(X-\mu)^2\right]=\sigma_X^2=\sigma^2$$

5.39. Let $X=N(0;\sigma^2)$. Show that

$$m_n=E[X^n]=\begin{cases}0 & n=2k+1\\ 1\cdot 3\cdot\cdots\cdot(n-1)\sigma^n & n=2k\end{cases} \tag{5.154}$$

$$X=N(0;\sigma^2)\longrightarrow f_X(x)=\frac{1}{\sqrt{2\pi}\,\sigma}e^{-x^2/(2\sigma^2)}$$

The odd moments m_{2k+1} of X are 0 because $f_X(-x)=f_X(x)$. Differentiating the identity

$$\int_{-\infty}^{\infty}e^{-\alpha x^2}\,dx=\sqrt{\frac{\pi}{\alpha}} \tag{5.155}$$

k times with respect to α when $n=2k$, we obtain

$$\int_{-\infty}^{\infty}x^{2k}e^{-\alpha x^2}\,dx=\frac{1\cdot 3\cdot\cdots\cdot(2k-1)}{2^k}\sqrt{\frac{\pi}{\alpha^{2k+1}}}$$

Setting $\alpha=1/(2\sigma^2)$, we have

$$m_{2k}=E[X^{2k}]=\frac{1}{\sqrt{2\pi}\,\sigma}\int_{-\infty}^{\infty}x^{2k}e^{-x^2/(2\sigma^2)}\,dx$$

$$=1\cdot 3\cdot\cdots\cdot(2k-1)\sigma^{2k}$$

5.40. Verify Eq. (5.70), that is,

$$E[X+Y]=E[X]+E[Y]$$

Let $f_{XY}(x,y)$ be the joint density function of X and Y. Then using Eq. (5.68), we have

$$E[X+Y]=\int_{-\infty}^{\infty}\int_{-\infty}^{\infty}(x+y)f_{XY}(x,y)\,dx\,dy$$

$$=\int_{-\infty}^{\infty}\int_{-\infty}^{\infty}xf_{XY}(x,y)\,dx\,dy+\int_{-\infty}^{\infty}\int_{-\infty}^{\infty}yf_{XY}(x,y)\,dx\,dy$$

Using Eqs. (*5.43a*) and (*5.61*), we have

$$\int_{-\infty}^{\infty}\int_{-\infty}^{\infty} x f_{XY}(x,y)\,dx\,dy = \int_{-\infty}^{\infty} x\left[\int_{-\infty}^{\infty} f_{XY}(x,y)\,dy\right]dx$$
$$= \int_{-\infty}^{\infty} x f_X(x)\,dx = E[X]$$

In a similar manner, we have

$$\int_{-\infty}^{\infty}\int_{-\infty}^{\infty} y f_{XY}(x,y)\,dx\,dy = \int_{-\infty}^{\infty} y f_Y(y)\,dy = E[Y]$$

Thus

$$E[X+Y] = E[X] + E[Y]$$

5.41. If X and Y are independent, then show that

$$E[XY] = E[X]E[Y] \tag{5.156}$$

and

$$E[g_1(X)g_2(Y)] = E[g_1(X)]E[g_2(Y)] \tag{5.157}$$

If X and Y are independent, then by Eqs. (*5.48*) and (*5.68*) we have

$$E[XY] = \int_{-\infty}^{\infty}\int_{-\infty}^{\infty} xy f_X(x) f_Y(y)\,dx\,dy$$
$$= \int_{-\infty}^{\infty} x f_X(x)\,dx \int_{-\infty}^{\infty} y f_Y(y)\,dy$$
$$= E[X]E[Y]$$

Similarly,

$$E[g_1(X)g_2(Y)] = \int_{-\infty}^{\infty}\int_{-\infty}^{\infty} g_1(x)g_2(y) f_X(x) f_Y(y)\,dx\,dy$$
$$= \int_{-\infty}^{\infty} g_1(x) f_X(x)\,dx \int_{-\infty}^{\infty} g_2(y) f_Y(y)\,dy$$
$$= E[g_1(X)]E[g_2(Y)]$$

5.42. Find the covariance of X and Y if (*a*) they are independent and (*b*) Y is related to X by $Y = aX + b$.

(*a*) If X and Y are independent, then by Eqs. (*5.79*) and (*5.156*)

$$C_{XY} = E[XY] - E[X]E[Y]$$
$$= E[X]E[Y] - E[X]E[Y] = 0 \tag{5.158}$$

(*b*)

$$E[XY] = E[X(aX+b)] = aE[X^2] + bE[X] = aE[X^2] + b\mu_X$$
$$\mu_Y = E[Y] = E[aX+b] = aE[X] + b = a\mu_X + b$$

Thus

$$C_{XY} = E[XY] - E[X]E[Y]$$
$$= aE[X^2] + b\mu_X - \mu_X(a\mu_X + b)$$
$$= a\left(E[X^2] - \mu_X^2\right) = a\sigma_X^2 \tag{5.159}$$

Note that the result of (*a*) states that if X and Y are independent, then they are uncorrelated. But the converse is not necessarily true. (See Prob. 5.44.)

5.43. Let $Z = aX + bY$, where a and b are arbitrary constants. Show that if X and Y are independent, then

$$\sigma_Z^2 = a^2\sigma_X^2 + b^2\sigma_Y^2 \tag{5.160}$$

By Eqs. (5.70) and (5.71),

$$\mu_Z = E[Z] = E[aX + bY] = aE[X] + bE[Y] = a\mu_X + b\mu_Y$$

By Eq. (5.74)

$$\begin{aligned}\sigma_Z^2 &= E\left[(Z-\mu_Z)^2\right] = E\left\{[(aX+bY)-(a\mu_X+b\mu_Y)]^2\right\}\\ &= E\left\{[a(X-\mu_X)+b(Y-\mu_Y)]^2\right\}\\ &= a^2E\left[(X-\mu_X)^2\right] + 2abE[(X-\mu_X)(Y-\mu_Y)] + b^2E\left[(Y-\mu_Y)^2\right]\\ &= a^2\sigma_X^2 + 2abE[(X-\mu_X)(Y-\mu_Y)] + b^2\sigma_Y^2 \end{aligned} \tag{5.161}$$

Since X and Y are independent, by Eq. (5.157)

$$E[(X-\mu_X)(Y-\mu_Y)] = E[X-\mu_X]E[Y-\mu_Y] = 0$$

Hence

$$\sigma_Z^2 = a^2\sigma_X^2 + b^2\sigma_Y^2$$

5.44. Let X and Y be defined by

$$X = \cos\Theta \qquad \text{and} \qquad Y = \sin\Theta$$

where Θ is a random variable uniformly distributed over $[0, 2\pi]$.

(a) Show that X and Y are uncorrelated.

(b) Show that X and Y are not independent.

(a) From Eq. (5.119)

$$f_\Theta(\theta) = \begin{cases} \dfrac{1}{2\pi} & 0 \le \theta \le 2\pi \\ 0 & \text{otherwise} \end{cases}$$

Using Eqs. (5.61) and (5.65), we have

$$E[X] = \int_{-\infty}^{\infty} x f_X(x)\,dx = \int_{-\infty}^{\infty} (\cos\theta) f_\Theta(\theta)\,d\theta = \frac{1}{2\pi}\int_0^{2\pi} \cos\theta\,d\theta = 0$$

Similarly,

$$E[Y] = \frac{1}{2\pi}\int_0^{2\pi} \sin\theta\,d\theta = 0$$

and

$$\begin{aligned} E[XY] &= \frac{1}{2\pi}\int_0^{2\pi} \cos\theta\sin\theta\,d\theta \\ &= \frac{1}{4\pi}\int_0^{2\pi} \sin 2\theta\,d\theta = 0 = E[X]E[Y] \end{aligned}$$

Thus, from Eq. (5.83), X and Y are uncorrelated.

(b)

$$E[X^2] = \frac{1}{2\pi}\int_0^{2\pi} \cos^2\theta\,d\theta = \frac{1}{2\pi}\int_0^{2\pi} \frac{1}{2}(1+\cos 2\theta)\,d\theta = \frac{1}{2}$$

$$E[Y^2] = \frac{1}{2\pi}\int_0^{2\pi} \sin^2\theta\,d\theta = \frac{1}{2\pi}\int_0^{2\pi} \frac{1}{2}(1-\cos 2\theta)\,d\theta = \frac{1}{2}$$

$$E[X^2Y^2] = \frac{1}{2\pi}\int_0^{2\pi} \cos^2\theta\sin^2\theta\,d\theta = \frac{1}{8\pi}\int_0^{2\pi} \frac{1}{2}(1-\cos 4\theta)\,d\theta = \frac{1}{8}$$

Hence
$$E[X^2Y^2]=\frac{1}{8}\neq\frac{1}{4}=E[X^2]E[Y^2]$$

If X and Y were independent, then by Eq. (*5.157*) we would have $E[X^2Y^2]=E[X^2]E[Y^2]$. Therefore, X and Y are not independent.

5.45. Show that ρ in Eq. (*5.122*) (Prob. 5.22) is the correlation coefficient of X and Y.

By definition (*5.80*) and Eq. (*5.78*), the correlation coefficient of X and Y is

$$\rho_{XY}=E\left[\left(\frac{X-\mu_X}{\sigma_X}\right)\left(\frac{Y-\mu_Y}{\sigma_Y}\right)\right]$$
$$=\int_{-\infty}^{\infty}\int_{-\infty}^{\infty}\left[\left(\frac{X-\mu_X}{\sigma_X}\right)\left(\frac{Y-\mu_Y}{\sigma_Y}\right)\right]f_{XY}(x,y)\,dx\,dy \qquad (5.162)$$

where

$$f_{XY}(x,y)=\frac{1}{2\pi\sigma_X\sigma_Y(1-\rho^2)^{1/2}}$$
$$\times\exp\left\{-\frac{1}{2(1-\rho^2)}\left[\left(\frac{x-\mu_X}{\sigma_X}\right)^2-2\rho\left(\frac{x-\mu_X}{\sigma_X}\right)\left(\frac{y-\mu_Y}{\sigma_Y}\right)+\left(\frac{y-\mu_Y}{\sigma_Y}\right)^2\right]\right\}$$

By making a change in variables $v=(x-\mu_X)/\sigma_X$ and $w=(y-\mu_Y)/\sigma_Y$, Eq. (*5.162*) can be written as

$$\rho_{XY}=\int_{-\infty}^{\infty}\int_{-\infty}^{\infty}vw\frac{1}{2\pi(1-\rho^2)^{1/2}}\exp\left[-\frac{1}{2(1-\rho^2)}(v^2-2\rho vw+w^2)\right]dv\,dw$$
$$=\int_{-\infty}^{\infty}\frac{w}{\sqrt{2\pi}}\left\{\int_{-\infty}^{\infty}\frac{v}{\sqrt{2\pi}(1-\rho^2)^{1/2}}\exp\left[-\frac{(v-\rho w)^2}{2(1-\rho^2)}\right]dv\right\}e^{-w^2/2}\,dw$$

The term in the curly braces is identified as the mean of $V=N[\rho w;1-\rho^2]$, and so

$$\rho_{XY}=\int_{-\infty}^{\infty}\frac{w}{\sqrt{2\pi}}(\rho w)e^{-w^2/2}\,dw=\rho\int_{-\infty}^{\infty}w^2\frac{1}{\sqrt{2\pi}}e^{-w^2/2}\,dw$$

The last integral is the variance of $W=N(0;1)$, so it is equal to 1 and we obtain $\rho_{XY}=\rho$.

5.46. If $f_X(x)=0$ for $x<0$, then show that, for any $\alpha>0$,

$$P(X\geq\alpha)\leq\frac{\mu_X}{\alpha} \qquad (5.163)$$

where $\mu_X=E[X]$. This is known as the *Markov inequality*.

From Eq. (*5.34d*)

$$P(X\geq\alpha)=\int_{\alpha}^{\infty}f_X(x)\,dx$$

Since $f_X(x)=0$ for $x<0$,

$$\mu_X=E[X]=\int_0^{\infty}xf_X(x)\,dx\geq\int_{\alpha}^{\infty}xf_X(x)\,dx\geq\alpha\int_{\alpha}^{\infty}f_X(x)\,dx$$

Hence

$$\int_{\alpha}^{\infty}f_X(x)\,dx=P(X\geq\alpha)\leq\frac{\mu_X}{\alpha}$$

5.47. For any $\epsilon > 0$, show that

$$P(|X - \mu_X| \geq \epsilon) \leq \frac{\sigma_X^2}{\epsilon^2} \tag{5.164}$$

where $\mu_X = E[X]$ and σ_X^2 is the variance of X. This is known as the *Chebyshev inequality*.

From Eq. (*5.34d*)

$$P(|X - \mu_X| \geq \epsilon) = \int_{-\infty}^{\mu_X - \epsilon} f_X(x)\,dx + \int_{\mu_X+\epsilon}^{\infty} f_X(x)\,dx = \int_{|x-\mu_X|\geq\epsilon} f_X(x)\,dx$$

By Eq. (*5.74*)

$$\sigma_X^2 = \int_{-\infty}^{\infty} (x - \mu_X)^2 f_X(x)\,dx \geq \int_{|x-\mu_X|\geq\epsilon} (x-\mu_X)^2 f_X(x)\,dx \geq \epsilon^2 \int_{|x-\mu_X|\geq\epsilon} f_X(x)\,dx$$

Hence

$$\int_{|x-\mu_X|\geq\epsilon} f_X(x)\,dx \leq \frac{\sigma_X^2}{\epsilon^2}$$

or

$$P(|X - \mu_X| \geq \epsilon) \leq \frac{\sigma_X^2}{\epsilon^2}$$

5.48. Let X and Y be real random variables with finite second moments. Show that

$$(E[XY])^2 \leq E[X^2]E[Y^2] \tag{5.165}$$

This is known as the *Cauchy-Schwarz inequality*.

Because the mean-square value of a random variable can never be negative,

$$E\left[(X - \alpha Y)^2\right] \geq 0$$

for any value of α. Expanding this, we obtain

$$E[X^2] - 2\alpha E[XY] + \alpha^2 E[Y^2] \geq 0$$

Choose a value of α for which the left-hand side of this inequality is minimum

$$\alpha = \frac{E[XY]}{E[Y^2]}$$

which results in the inequality

$$E[X^2] - \frac{(E[XY])^2}{E[Y^2]} \geq 0$$

or

$$(E[XY])^2 \leq E[X^2]E[Y^2]$$

5.49. Verify Eq. (*5.81*).

From the Cauchy-Schwarz inequality Eq. (*5.165*) we have

$$\{E[(X-\mu_X)(Y-\mu_Y)]\}^2 \leq E\left[(X-\mu_X)^2\right]E\left[(Y-\mu_Y)^2\right]$$

or

$$C_{XY}^2 \leq \sigma_X^2 \sigma_Y^2$$

Then

$$\rho_{XY}^2 = \frac{C_{XY}^2}{\sigma_X^2 \sigma_Y^2} \le 1$$

from which it follows that

$$|\rho_{XY}| \le 1$$

5.50. If X and Y are random variables with joint pdf $f_{XY}(x, y)$, then we define the *conditional expectation* by

$$E[Y|X=x] = E[Y|x] = \int_{-\infty}^{\infty} y f_{Y|X}(y|x)\, dy \tag{5.166}$$

where [by Eq. (5.46)]

$$f_{Y|X}(y|x) = \frac{f_{XY}(x, y)}{f_X(x)} \tag{5.167}$$

Determine $E[Y|x]$ for the two jointly normal random variables of Prob. 5.22.

Substituting Eqs. (5.122) and (5.123) into Eq. (5.167) and after some cancellation and rearranging, we obtain

$$f_{Y|X}(y|x) = \frac{1}{\sqrt{2\pi}\,\sigma_Y (1-\rho^2)^{1/2}} \exp\left\{-\frac{1}{2(1-\rho^2)}\left[\frac{y-\mu_Y}{\sigma_Y} - \rho\left(\frac{x-\mu_X}{\sigma_X}\right)\right]^2\right\}$$

$$= \frac{1}{\sqrt{2\pi}\,\sigma_Y (1-\rho^2)^{1/2}} \exp\left\{-\frac{1}{2\sigma_Y^2(1-\rho^2)}\left[y - \rho\frac{\sigma_Y}{\sigma_X}(x-\mu_X) - \mu_Y\right]^2\right\}$$

which is equal to the pdf of

$$Y = N\left[\rho\frac{\sigma_Y}{\sigma_X}(x-\mu_X) + \mu_Y; \sigma_Y^2(1-\rho^2)\right]$$

Thus, we conclude that

$$E[Y|x] = \rho\frac{\sigma_Y}{\sigma_X}(x-\mu_X) + \mu_Y \tag{5.168}$$

Note that when X and Y are independent, then $\rho = 0$ and $E[Y|x] = \mu_Y = E[Y]$.

5.51. Show that

$$E[E[Y|x]] = E[Y] \tag{5.169}$$

Using Eqs. (5.166) and (5.167), we have

$$E[E[Y|X]] = \int_{-\infty}^{\infty} E[Y|x] f_X(x)\, dx$$

$$= \int_{-\infty}^{\infty}\int_{-\infty}^{\infty} y f_{Y|X}(y|x) f_X(x)\, dx\, dy$$

$$= \int_{-\infty}^{\infty}\int_{-\infty}^{\infty} y \frac{f_{XY}(x, y)}{f_X(x)} f_X(x)\, dx\, dy$$

$$= \int_{-\infty}^{\infty} y \left[\int_{-\infty}^{\infty} f_{XY}(x, y)\, dx\right] dy$$

$$= \int_{-\infty}^{\infty} y f_Y(y)\, dy = E[Y]$$

5.52. Let $X = N(\mu; \sigma^2)$.

(*a*) Find the moment generating function $M_X(\lambda)$ of X.

(*b*) Using $M_X(\lambda)$, find the mean and variance of X.

(*a*) From definition (*5.85*)

$$M_X(\lambda) = E[e^{\lambda X}] = \int_{-\infty}^{\infty} e^{\lambda x} f_X(x)\, dx$$

$$= \int_{-\infty}^{\infty} e^{\lambda x} \frac{1}{\sqrt{2\pi}\,\sigma} e^{-(x-\mu)^2/(2\sigma^2)}\, dx \tag{5.170}$$

Changing the variable $y = (x-\mu)/\sigma$, we have

$$M_X(\lambda) = \int_{-\infty}^{\infty} \frac{1}{\sqrt{2\pi}} e^{-y^2/2} e^{\lambda(\sigma y + \mu)}\, dy$$

$$= e^{\lambda\mu} \int_{-\infty}^{\infty} \frac{1}{\sqrt{2\pi}} e^{-y^2/2 + \lambda\sigma y}\, dy$$

$$= e^{\lambda\mu + \lambda^2\sigma^2/2} \int_{-\infty}^{\infty} \frac{1}{\sqrt{2\pi}} e^{-(y-\lambda\sigma)^2/2}\, dy$$

Since the integrand is the pdf of $N(\lambda\sigma; 1)$, the integral is unity and we obtain

$$M_X(\lambda) = e^{\lambda\mu + \lambda^2\sigma^2/2} \tag{5.171}$$

(*b*) Using Eq. (*5.86*)

$$E[X] = \frac{d}{d\lambda} e^{\lambda\mu + \lambda^2\sigma^2/2}\bigg|_{\lambda=0}$$

$$= (\mu + \lambda\sigma^2) e^{\lambda\mu + \lambda^2\sigma^2/2}\Big|_{\lambda=0} = \mu$$

$$E[X^2] = \frac{d^2}{d\lambda^2} e^{\lambda\mu + \lambda^2\sigma^2/2}\bigg|_{\lambda=0}$$

$$= \left[\sigma^2 + (\mu + \lambda\sigma^2)^2\right] e^{\lambda\mu + \lambda^2\sigma^2/2}\Big|_{\lambda=0} = \sigma^2 + \mu^2$$

By Eq. (*5.75*)

$$\sigma_X^2 = \sigma^2 + \mu^2 - \mu^2 = \sigma^2$$

5.53. The moment generating function of X and Y is defined as

$$M_{XY}(\lambda_1, \lambda_2) = E[e^{\lambda_1 X + \lambda_2 Y}] \tag{5.172}$$

where λ_1 and λ_2 are real variables. Let X and Y be jointly normal random variables with the pdf given by Eq. (*5.122*) of Prob. 5.22. Find the moment generating function of X and Y.

From Eq. (*5.122*)

$$f_{XY}(x, y) = \frac{1}{2\pi\sigma_X\sigma_Y(1-\rho^2)^{1/2}} e^{-Q(x,y)}$$

where

$$Q(x, y) = \frac{1}{2(1-\rho^2)}\left[\left(\frac{x-\mu_X}{\sigma_X}\right)^2 - 2\rho\left(\frac{x-\mu_X}{\sigma_X}\right)\left(\frac{y-\mu_Y}{\sigma_Y}\right) + \left(\frac{y-\mu_Y}{\sigma_Y}\right)^2\right]$$

Now

$$M_{XY}(\lambda_1,\lambda_2)=E[e^{\lambda_1 X+\lambda_2 Y}]$$

$$=\frac{1}{2\pi\sigma_X\sigma_Y(1-\rho^2)^{1/2}}\int_{-\infty}^{\infty}\int_{-\infty}^{\infty}\theta^{\lambda_1 x+\lambda_2 y}e^{-Q(x,y)}\,dx\,dy$$

$$=\frac{1}{2\pi\sigma_X\sigma_Y(1-\rho^2)^{1/2}}\int_{-\infty}^{\infty}\int_{-\infty}^{\infty}e^{\xi(x,y)}\,dx\,dy$$

where

$$\xi(x,y)=\lambda_1 x+\lambda_2 y-\frac{1}{2(1-\rho^2)}\left[\left(\frac{x-\mu_X}{\sigma_X}\right)^2-2\rho\left(\frac{x-\mu_X}{\sigma_X}\right)\left(\frac{y-\mu_Y}{\sigma_Y}\right)+\left(\frac{y-\mu_Y}{\sigma_Y}\right)^2\right]$$

Changing variables $t=(x-\mu_X)/\sigma_X$ and $s=(y-\mu_Y)/\sigma_Y$, we have

$$\xi(x,y)=\eta(t,s)=\lambda_1(\sigma_X t+\mu_X)+\lambda_2(\sigma_Y s+\mu_Y)-\frac{1}{2(1-\rho^2)}(t^2-2\rho ts+s^2)$$

$$=\lambda_1\mu_X+\lambda_2\mu_Y+\lambda_1\sigma_X t-\frac{1}{2}t^2+\lambda_2\sigma_Y s-\frac{1}{2(1-\rho^2)}(s-\rho t)^2$$

and

$$M_{XY}(\lambda_1,\lambda_2)=\frac{1}{2\pi(1-\rho^2)^{1/2}}\int_{-\infty}^{\infty}\int_{-\infty}^{\infty}e^{\eta(t,s)}\,dt\,ds$$

$$=\frac{e^{\lambda_1\mu_X+\lambda_2\mu_Y}}{2\pi(1-\rho^2)^{1/2}}\int_{-\infty}^{\infty}e^{\lambda_1\sigma_X t-t^2/2}$$

$$\times\left\{\int_{-\infty}^{\infty}e^{\lambda_2\sigma_Y s}\exp\left[-\frac{1}{2(1-\rho^2)}(s-\rho t)^2\right]ds\right\}dt$$

Next consider

$$\frac{1}{\sqrt{2\pi}\,(1-\rho^2)^{1/2}}\int_{-\infty}^{\infty}e^{\lambda_2\sigma_Y s}\exp\left[-\frac{1}{2(1-\rho^2)}(s-\rho t)^2\right]ds \tag{5.173}$$

A comparison of Eq. (*5.173*) with Eq. (*5.170*) shows that it is the moment generating function of $N(\rho t;1-\rho^2)$ with λ replaced by $\lambda_2\sigma_Y$. Thus, by using Eq. (*5.171*), Eq. (*5.173*) becomes

$$\exp\left[\lambda_2\sigma_Y\rho t+\frac{1}{2}\lambda_2^2\sigma_Y^2(1-\rho^2)\right]$$

Then

$$M_{XY}(\lambda_1,\lambda_2)=\exp\left[\lambda_1\mu_X+\lambda_2\mu_Y+\frac{1}{2}\lambda_2^2\sigma_Y^2(1-\rho^2)\right]\frac{1}{\sqrt{2\pi}}\int_{-\infty}^{\infty}e^{\lambda_1\sigma_X t-t^2/2+\lambda_2\sigma_Y\rho t}\,dt$$

Now

$$\frac{1}{\sqrt{2\pi}}\int_{-\infty}^{\infty}e^{\lambda_1\sigma_X t-t^2/2+\lambda_2\sigma_Y\rho t}\,dt=\frac{1}{\sqrt{2\pi}}\int_{-\infty}^{\infty}e^{(\lambda_1\sigma_X+\lambda_2\sigma_Y\rho)t}e^{-t^2/2}\,dt$$

is the moment generating function of $N(0;1)$ with λ replaced by $\lambda_1\sigma_X+\lambda_2\sigma_Y\rho$ [Eq. (*5.170*)]. Thus

$$\frac{1}{\sqrt{2\pi}}\int_{-\infty}^{\infty}e^{\lambda_1\sigma_X t+\lambda_2\sigma_Y\rho t}e^{-t^2/2}\,dt=\exp\left[\frac{1}{2}(\lambda_1\sigma_X+\lambda_2\sigma_Y\rho)^2\right]$$

Hence,

$$M_{XY}(\lambda_1, \lambda_2) = \exp\left[\lambda_1\mu_X + \lambda_2\mu_Y + \frac{1}{2}\lambda_2^2\sigma_Y^2(1-\rho^2) + \frac{1}{2}(\lambda_1\sigma_X + \lambda_2\sigma_Y\rho)^2\right]$$

$$= \exp\left[\lambda_1\mu_X + \lambda_2\mu_Y + \frac{1}{2}(\lambda_1^2\sigma_X^2 + 2\lambda_1\lambda_2\sigma_X\sigma_Y\rho + \lambda_2^2\sigma_Y^2)\right] \tag{5.174}$$

5.54. Show that if X and Y are zero-mean jointly normal random variables, then

$$E[X^2Y^2] = E[X^2]E[Y^2] + 2(E[XY])^2 \tag{5.175}$$

By Eq. (*5.172*) the moment generating function of X and Y is given by

$$M_{XY}(\lambda_1, \lambda_2) = E[e^{\lambda_1 X + \lambda_2 Y}]$$

Expanding the exponential and using the linearity of the operator E, we obtain

$$M_{XY}(\lambda_1, \lambda_2) = \sum_{n=0}^{\infty} \frac{1}{n!} \sum_{k=0}^{n} \binom{n}{k} E[X^k Y^{n-k}]\lambda_1^k \lambda_2^{n-k} \tag{5.176}$$

The coefficient of $\lambda_1^2\lambda_2^2$ in Eq. (*5.176*) is

$$\frac{1}{4!}\binom{4}{2}E[X^2Y^2] = \frac{1}{4}E[X^2Y^2] \tag{5.177}$$

Setting $\mu_X = \mu_Y = 0$ in Eq. (*5.174*), we see that the joint moment generating function of X and Y is

$$M_{XY}(\lambda_1, \lambda_2) = \exp\left[\frac{1}{2}(\lambda_1^2\sigma_X^2 + 2\lambda_1\lambda_2\sigma_X\sigma_Y\rho + \lambda_2^2\sigma_Y^2)\right]$$

$$= \exp\left[\frac{1}{2}(\lambda_1^2\sigma_X^2 + 2\lambda_1\lambda_2 C_{XY} + \lambda_2^2\sigma_Y^2)\right] \tag{5.178}$$

where $C_{XY} = \sigma_X\sigma_Y\rho = E[XY]$ is the covariance of X and Y.

Let $m = \frac{1}{2}(\lambda_1^2\sigma_X^2 + 2\lambda_1\lambda_2 C_{XY} + \lambda_2^2\sigma_Y^2)$. Then

$$e^m = \sum_{n=0}^{\infty} \frac{1}{n!} m^n = 1 + m + \frac{1}{2}m^2 + \frac{1}{6}m^3 + \cdots$$

The factors $\lambda_1^2\lambda_2^2$ in Eq. (*5.178*) appear only in the term

$$\frac{1}{2}m^2 = \frac{1}{8}(\lambda_1^2\sigma_X^2 + 2\lambda_1\lambda_2 C_{XY} + \lambda_2^2\sigma_Y^2)^2$$

and then the coefficient of $\lambda_1^2\lambda_2^2$ in Eq. (*5.178*) is

$$\frac{1}{8}(2\sigma_X^2\sigma_Y^2 + 4C_{XY}^2) = \frac{1}{4}\sigma_X^2\sigma_Y^2 + \frac{1}{2}C_{XY}^2 \tag{5.179}$$

Hence, equating Eqs. (*5.177*) and (*5.179*), we obtain

$$\frac{1}{4}E[X^2Y^2] = \frac{1}{4}\sigma_X^2\sigma_Y^2 + \frac{1}{2}C_{XY}^2$$

or

$$E[X^2Y^2] = \sigma_X^2\sigma_Y^2 + 2C_{XY}^2 = E[X^2]E[Y^2] + 2(E[XY])^2$$

5.55. Using the characteristic equation technique, derive Eq. (*5.140*) of Prob. 5.30.

Let $Z = X + Y$, where X and Y are independent. Let

$$\Phi_X(\omega) = E[e^{j\omega X}] \qquad \Phi_Y(\omega) = E[e^{j\omega Y}]$$

Then

$$\Phi_Z(\omega) = E[e^{j\omega Z}] = E[e^{j\omega(X+Y)}]$$
$$= E[e^{j\omega X}]E[e^{j\omega Y}] = \Phi_X(\omega)\Phi_Y(\omega) \tag{5.180}$$

Applying the convolution theorem [Eq. (*1.53*)], we have

$$f_Z(z) = \mathscr{F}^{-1}[\Phi_Z(\omega)] = \mathscr{F}^{-1}[\Phi_X(\omega)\Phi_Y(\omega)]$$
$$= f_X(x) * f_Y(y) = \int_{-\infty}^{\infty} f_X(x) f_Y(z-x)\,dx$$

Since the convolution operation is commutative, we also can write

$$f_Z(z) = f_Y(y) * f_X(x) = \int_{-\infty}^{\infty} f_Y(y) f_X(z-y)\,dy$$

which is equal to Eq. (*5.140*), obtained in Prob. 5.30.

Supplementary Problems

5.56. Show that if events A and B are independent, then

$$P(\bar{A} \cap \bar{B}) = P(\bar{A})P(\bar{B})$$

Hint: Use Eq. (*5.116*) and the relation

$$\bar{A} = (\bar{A} \cap B) \cup (\bar{A} \cap \bar{B})$$

5.57. A certain computer becomes inoperable if two components A and B both fail. The probability that A fails is 0.01, and the probability that B fails is 0.005. However, the probability that B fails increases by a factor of 3 if A has failed.

(a) Calculate the probability that the computer becomes inoperable.

(b) Find the probability that A will fail if B has failed.

Ans. (a) 0.00015, (b) 0.03

5.58. A certain binary PCM system transmits the two binary states $X = +1$, $X = -1$ with equal probability. However, because of channel noise, the receiver makes recognition errors. Also, as a result of path distortion, the receiver may lose necessary signal strength to make any decision. Thus, there are three possible receiver states: $Y = +1$, $Y = 0$, and $Y = -1$, where $Y = 0$ corresponds to "loss of signal." Assume that $P(Y = -1|X = +1) = 0.1$, $P(Y = +1|X = -1) = 0.2$, and $P(Y = 0|X = +1) = P(Y = 0|X = -1) = 0.05$.

(a) Find the probabilities $P(Y = +1)$, $P(Y = -1)$, and $P(Y = 0)$.

(b) Find the probabilities $P(X = +1|Y = +1)$ and $P(X = -1|Y = -1)$.

Ans. (a) $P(Y = +1) = 0.525$, $P(Y = -1) = 0.425$, $P(Y = 0) = 0.05$
(b) $P(X = +1|Y = +1) = 0.81$, $P(X = -1|Y = -1) = 0.88$

5.59. Suppose 10 000 digits are transmitted over a noisy channel having per-digit error probability $p = 5 \times 10^{-5}$. Find the probability that there will be no more than two digit errors.

Ans. 0.9856

5.60. Show that Eq. (*5.103*) does in fact define a true probability density; in particular, show that

$$\int_{-\infty}^{\infty} f_X(x)\,dx = 1$$

Hint: Make a change of variable $[y = (x - \mu)/\sigma]$ and show that

$$I = \int_{-\infty}^{\infty} e^{-y^2/2}\,dy = \sqrt{2\pi}$$

which can be proved by evaluating I^2 by using the polar coordinates.

5.61. A noisy resistor produces a voltage $V_n(t)$. At $t = t_1$, the noise level $X = V_n(t_1)$ is known to be a gaussian random variable with density

$$f_X(x) = \frac{1}{\sqrt{2\pi}\,\sigma} e^{-x^2/(2\sigma^2)}$$

Compute the probability that $|X| > k\sigma$ for $k = 1, 2, 3$.

Ans. $P(|X| > \sigma) = 0.3173$, $P(|X| > 2\sigma) = 0.0455$, $P(|X| > 3\sigma) = 0.0027$

5.62. Consider the transformation $Y = 1/X$.

(*a*) Find $f_Y(y)$ in terms of $f_X(x)$.

(*b*) If $f_X(x) = \dfrac{\alpha/\pi}{\alpha^2 + x^2}$, find $f_Y(y)$.

Ans. (*a*) $f_Y(y) = \dfrac{1}{y^2} f_X\left(\dfrac{1}{y}\right)$

(*b*) $f_Y(y) = \dfrac{1/(\alpha\pi)}{1/\alpha^2 + y^2}$

Note that X and Y are known as *Cauchy* random variables.

5.63. Let X and Y be two independent random variables with

$$f_X(x) = \alpha e^{-\alpha x} u(x) \qquad f_Y(y) = \beta e^{-\beta x} u(y)$$

Find the density function of $Z = X + Y$.

Ans. $f_z(z) = \begin{cases} \dfrac{\alpha\beta}{\beta - \alpha}(e^{-\alpha z} - e^{-\beta z})u(z) & \beta \neq \alpha \\ \alpha^2 z e^{-\alpha z} u(z) & \beta = \alpha \end{cases}$

5.64. Let X be a random variable uniformly distributed over $[a, b]$. Find the mean and the variance of X.

Ans. $\mu_X = \dfrac{b + a}{2}$, $\sigma_X^2 = \dfrac{(b - a)^2}{12}$

5.65. Given the random variable X with mean μ_X and variance σ_X^2, find the linear transformation $Y = aX + b$ such that $\mu_Y = 0$ and $\sigma_Y^2 = 1$.

Ans. $a = \dfrac{1}{\sigma_X}$, $b = -\dfrac{\mu_X}{\sigma_X}$

5.66. Define random variables Z and W by

$$Z = X + aY \qquad W = X - aY$$

where a is a real number. Determine a such that Z and W are orthogonal.

Ans. $a = \sqrt{\dfrac{E[X^2]}{E[Y^2]}}$

5.67. (*a*) Find the moment generating function of X uniformly distributed over (a, b).
(*b*) Using the result of (*a*), find $E[X]$, $E[X^2]$, and $E[X^3]$.

Ans. (*a*) $\dfrac{e^{\lambda b} - e^{\lambda a}}{\lambda(b-a)}$

(*b*) $E[X] = \frac{1}{2}(b+a)$, $E[X^2] = \frac{1}{3}(b^2 + ab + a^2)$, $E[X^3] = \frac{1}{4}(b^3 + b^2 a + ba^2 + a^3)$

5.68. Let $\Phi_X(\omega)$ be the characteristic function of X. Let $Y = aX + b$. Find the characteristic function of Y in terms of $\Phi_X(\omega)$.

Ans. $\Phi_Y(\omega) = e^{j\omega b}\Phi_X(a\omega)$

5.69. Let $X = N(\mu; \sigma^2)$. Find the characteristic function of X.

Ans. $\Phi_X(\omega) = e^{j\mu\omega - \sigma^2\omega^2/2}$

5.70. Show that if X and Y are independent, then

$$\Phi_{XY}(\omega_1, \omega_2) = \Phi_X(\omega_1)\Phi_Y(\omega_2)$$

Hint: Use Eqs. (*5.48*) and (*5.88*).

5.71. Let X_i, $i = 1, 2, \ldots, n$, be a set of n independent normal random variables with mean μ_i and variance σ_i^2. Let

$$Y = \sum_{i=1}^{n} X_i = X_1 + X_2 + \cdots + X_n$$

Show that Y is a normal random variable with

$$\mu_Y = \sum_{i=1}^{n} \mu_i \qquad \text{and} \qquad \sigma_Y^2 = \sum_{i=1}^{n} \sigma_i^2$$

Hint: Find the characteristic function of Y, and use the result of Prob. 5.69.

5.72. Show that the nth moment of X can be found from its characteristic function $\Phi_X(\omega)$ by

$$E[X^n] = j^{-n} \left.\frac{d^n \Phi_X(\omega)}{d\omega^n}\right|_{\omega=0}$$

Hint: Differentiate both sides of Eq. (*5.85*) n times, and show that

$$\frac{d^n}{d\omega^n} E[e^{j\omega X}] = j^n E[X^n e^{j\omega X}]$$

5.73. Let X_1 and X_2 be two independent normal random variables with means μ_1 and μ_2 and variances σ_1^2 and σ_2^2, respectively.

(*a*) Using the characteristic function technique, find the pdf of $Z = a_1 X_1 + a_2 X_2$.
(*b*) Using the characteristic function of $X = N(0; \sigma^2)$, find $E[X^4]$.

Ans. (*a*) $f_Z(z) = \dfrac{1}{\sqrt{2\pi}\,\sigma_Z} \exp\left[-\dfrac{1}{2\sigma_Z^2}(z - \mu_Z)^2\right]$ with $\mu_Z = a_1\mu_1 + a_2\mu_2$ and $\sigma_Z^2 = a_1^2\sigma_1^2 + a_2^2\sigma_2^2$

(*b*) $E[X^4] = 3\sigma^4$

Chapter 6

Random Signals and Noise

6.1 INTRODUCTION

Models for random message signals and noise encountered in communication systems are developed in this chapter. Random signals cannot be explicitly described prior to their occurrence, and noises cannot be described by deterministic functions of time. However, when observed over a long period, a random signal or noise may exhibit certain regularities that can be described in terms of probabilities and statistical averages. Such a model, in the form of a probabilistic description of a collection of functions of times, is called a random process.

6.2 DEFINITIONS AND NOTATIONS OF RANDOM PROCESSES

Consider a random experiment with outcomes λ and a sample space S. If to every outcome $\lambda \in S$ we assign a real-valued time function $X(t, \lambda)$, we create a *random* (or *stochastic*) *process*. A random process $X(t, \lambda)$ is therefore a function of two parameters, the time t and the outcome λ. For a specific λ, say, λ_i, we have a single time function $X(t, \lambda_i) = x_i(t)$. This time function is called a *sample function*. The totality of all sample functions is called an *ensemble*. For a specific time t_j, $X(t_j, \lambda) = X_j$ denotes a random variable. For fixed $t\ (= t_j)$ and fixed $\lambda\ (= \lambda_i)$, $X(t_j, \lambda_i) = x_i(t_j)$ is a number.

Thus, a random process is sometimes defined as a family of random variables indexed by the parameter $t \in T$, where T is called the *index set*.

Figure 6-1 illustrates the concepts of the sample space of the random experiment, outcomes of the experiment, associated sample functions, and random variables resulting from taking two measurements of the sample functions.

In the following we use the notation $X(t)$ to represent $X(t, \lambda)$.

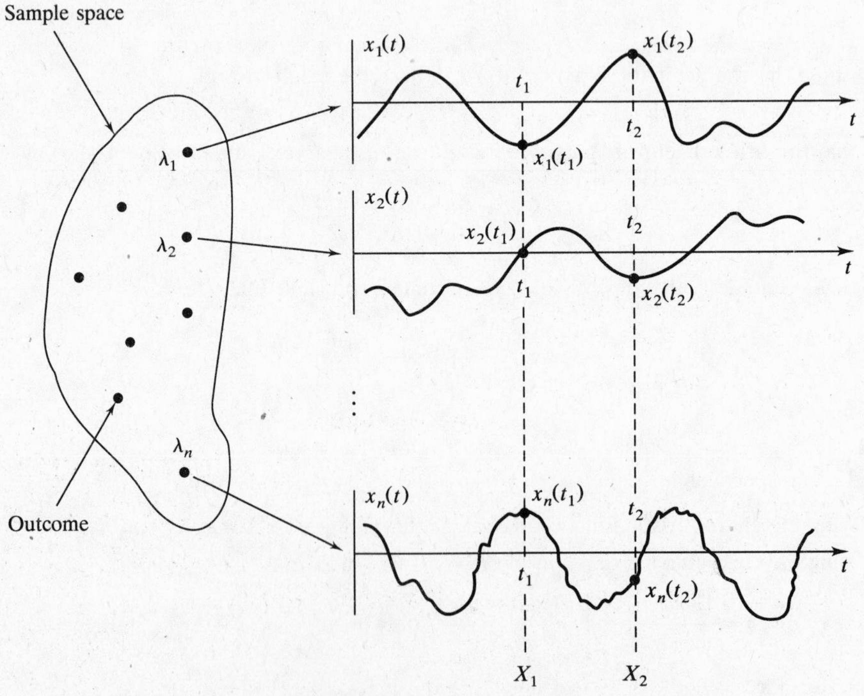

Fig. 6-1 Random process

6.3 STATISTICS OF RANDOM PROCESSES

A. Probabilistic Expressions:

Consider a random process $X(t)$. For a particular time t_1, $X(t_1) = X_1$ is a random variable, and its distribution function $F_X(x_1; t_1)$ is defined as

$$F_X(x_1; t_1) = P\{X(t_1) \le x_1\} \tag{6.1}$$

where x_1 is any real number.

And $F_X(x_1; t_1)$ is called the *first-order distribution* of $X(t)$. The corresponding first-order density function is obtained by

$$f_X(x_1; t_1) = \frac{\partial F_X(x_1; t_1)}{\partial x_1} \tag{6.2}$$

Similarly, given t_1 and t_2, $X(t_1) = X_1$ and $X(t_2) = X_2$ represent two random variables. Their joint distribution is called the *second-order distribution* and is given by

$$F_X(x_1, x_2; t_1, t_2) = P\{X(t_1) \le x_1, X(t_2) \le x_2\} \tag{6.3}$$

where x_1 and x_2 are any real numbers.

The corresponding second-order density function is obtained by

$$f_X(x_1, x_2; t_1, t_2) = \frac{\partial^2 F_X(x_1, x_2; t_1, t_2)}{\partial x_1 \partial x_2} \tag{6.4}$$

In a similar manner, for n random variables $X(t_i) = X_i$ $(i = 1, \ldots, n)$, the *nth-order distribution* is

$$F_X(x_1, \ldots, x_n; t_1, \ldots, t_n) = P\{X(t_1) \le x_1, \ldots, X(t_n) \le x_n\} \tag{6.5}$$

The corresponding nth-order density function is

$$f_X(x_1, \ldots, x_n; t_1, \ldots, t_n) = \frac{\partial^n F_X(x_1, \ldots, x_n; t_1, \ldots, t_n)}{\partial x_1 \cdots \partial x_n} \tag{6.6}$$

B. Statistical Averages:

As in the case of random variables, random processes are often described by using *statistical averages* (or *ensemble averages*).

The *mean* of $X(t)$ is defined by

$$\mu_X(t) = E[X(t)] = \int_{-\infty}^{\infty} x f_X(x; t)\, dx \tag{6.7}$$

where $X(t)$ is treated as a random variable for a fixed value of t.

The *autocorrelation* of $X(t)$ is defined by

$$\begin{aligned} R_{XX}(t_1, t_2) &= E[X(t_1)X(t_2)] \\ &= \int_{-\infty}^{\infty}\int_{-\infty}^{\infty} x_1 x_2 f_X(x_1, x_2; t_1, t_2)\, dx_1\, dx_2 \end{aligned} \tag{6.8}$$

The *autocovariance* of $X(t)$ is defined by

$$\begin{aligned} C_{XX}(t_1, t_2) &= E\{[X(t_1) - \mu_X(t_1)][X(t_2) - \mu_X(t_2)]\} \\ &= R_{XX}(t_1, t_2) - \mu_X(t_1)\mu_X(t_2) \end{aligned} \tag{6.9}$$

The *nth joint moment* of $X(t)$ is defined by

$$E[X(t_1) \cdots X(t_n)] = \int_{-\infty}^{\infty} \cdots \int_{-\infty}^{\infty} x_1 \cdots x_n f_X(x_1, \ldots, x_n; t_1, \ldots, t_n)\, dx_1 \cdots dx_n \tag{6.10}$$

C. Stationarity:

1. Strict-Sense Stationary:

A random process $X(t)$ is called *strict-sense stationary* (SSS) if its statistics are invariant to a shift of origin. In other words, the process $X(t)$ is SSS if

$$f_X(x_1,\ldots,x_n;t_1,\ldots,t_n)=f_X(x_1,\ldots,x_n;t_1+c,\ldots,t_n+c) \tag{6.11}$$

for any c.

From Eq. (*6.11*) it follows that $f_X(x_1;t_1)=f_X(x_1;t_1+c)$ for any c. Hence the first-order density of a stationary $X(t)$ is independent of t:

$$f_X(x_1;t)=f_X(x_1) \tag{6.12}$$

Similarly, $f_X(x_1,x_2;t_1,t_2)=f_X(x_1,x_2;t_1+c,t_2+c)$ for any c. Setting $c=-t_1$, we obtain

$$f_X(x_1,x_2;t_1,t_2)=f_X(x_1,x_2;t_2-t_1) \tag{6.13}$$

which indicates that if $X(t)$ is SSS, the joint density of the random variables $X(t)$ and $X(t+\tau)$ is independent of t and depends only on the time difference τ.

2. Wide-Sense Stationary:

A random process $X(t)$ is called *wide-sense stationary* (WSS) if its mean is constant

$$E[X(t)]=\mu_X \tag{6.14}$$

and its autocorrelation depends only on the time difference τ

$$E[X(t)X(t+\tau)]=R_{XX}(\tau) \tag{6.15}$$

From Eqs. (*6.9*) and (*6.15*) it follows that the autocovariance of a WSS process also depends only on the time difference τ:

$$C_{XX}(\tau)=R_{XX}(\tau)-\mu_X^2 \tag{6.16}$$

Setting $\tau=0$ in Eq. (*6.15*), we obtain

$$E[X^2(t)]=R_{XX}(0) \tag{6.17}$$

Thus the average power of a WSS process is independent of t and equals $R_{XX}(0)$.

Note that an SSS process is WSS but a WSS process is not necessarily SSS.

Two processes $X(t)$ and $Y(t)$ are called *jointly wide-sense stationary* (jointly WSS) if each is WSS and their cross-correlation depends only on the time difference τ:

$$R_{XY}(t,t+\tau)=E[X(t)Y(t+\tau)]=R_{XY}(\tau) \tag{6.18}$$

From Eq. (*6.18*) it follows that the cross-covariance of jointly WSS $X(t)$ and $Y(t)$ also depends only on the time difference τ:

$$C_{XY}(\tau)=R_{XY}(\tau)-\mu_X\mu_Y \tag{6.19}$$

D. Time Averages and Ergodicity:

The *time-averaged mean* of a sample function $x(t)$ of a random process $X(t)$ is defined as

$$\bar{x}=\langle x(t)\rangle=\lim_{T\to\infty}\frac{1}{T}\int_{-T/2}^{T/2}x(t)\,dt \tag{6.20}$$

where the symbol $\langle\cdot\rangle$ denotes *time-averaging*.

Similarly, the *time-averaged autocorrelation* of the sample function $x(t)$ is defined as

$$\bar{R}_{XX}(\tau)=\langle x(t)x(t+\tau)\rangle=\lim_{T\to\infty}\frac{1}{T}\int_{-T/2}^{T/2}x(t)x(t+\tau)\,dt \tag{6.21}$$

Note that $\bar{x}$ and $\bar{R}_{XX}(\tau)$ are random variables; their values depend on which sample function of $X(t)$ is used in the time-averaging evaluations.

If $X(t)$ is stationary, then by taking the expected value on both sides of Eqs. (6.20) and (6.21), we obtain

$$E[\bar{x}] = \lim_{T\to\infty} \frac{1}{T}\int_{-T/2}^{T/2} E[x(t)]\, dt = \mu_X \tag{6.22}$$

which indicates that the expected value of the time-averaged mean is equal to the ensemble mean, and

$$E[\bar{R}_{XX}(\tau)] = \lim_{T\to\infty} \frac{1}{T}\int_{-T/2}^{T/2} E[x(t)x(t+\tau)]\, dt = R_{XX}(\tau) \tag{6.23}$$

which also indicates that the expected value of the time-averaged autocorrelation is equal to the ensemble autocorrelation.

A random process $X(t)$ is said to be *ergodic* if time averages are the same for all sample functions and equal to the corresponding ensemble averages. Thus, in an ergodic process, all its statistics can be obtained by observing a single sample function $x(t) = X(t, \lambda)$ (λ fixed) of the process.

A stationary process $X(t)$ is called *ergodic in the mean* if

$$\bar{x} = \langle x(t)\rangle = E[X(t)] = \mu_X \tag{6.24}$$

Similarly, a stationary process $X(t)$ is called *ergodic in the autocorrelation* if

$$\bar{R}_{XX}(\tau) = \langle x(t)x(t+\tau)\rangle = E[X(t)X(t+\tau)] = R_{XX}(\tau) \tag{6.25}$$

Testing for the ergodicity of a random process is usually very difficult. A reasonable assumption in the analysis of most communication signals is that the random waveforms are ergodic in the mean and in the autocorrelation. Fundamental electrical engineering parameters, such as dc value, root-mean-square (rms) value, and average power can be related to the moments of an ergodic random process. They are summarized in the following:

1. $\bar{x} = \langle x(t)\rangle$ is equal to the dc level of the signal.
2. $[\bar{x}]^2 = \langle x(t)\rangle^2$ is equal to the normalized power in the dc component.
3. $\bar{R}_{XX}(0) = \langle x^2(t)\rangle$ is equal to the total average normalized power.
4. $\bar{\sigma}_X^2 = \langle x^2(t)\rangle - \langle x(t)\rangle^2$ is equal to the average normalized power in the time-varying or ac component of the signal.
5. $\bar{\sigma}_X$ is equal to the rms value of the ac component of the signal.

6.4 CORRELATIONS AND POWER SPECTRAL DENSITIES

In the following we assume that all random processes are WSS.

A. Autocorrelation $R_{XX}(\tau)$:

The autocorrelation of $X(t)$ is [Eq. (6.15)]

$$R_{XX}(\tau) = E[X(t)X(t+\tau)]$$

Properties of $R_{XX}(\tau)$:

1. $$R_{XX}(-\tau) = R_{XX}(\tau) \tag{6.26}$$

2. $$|R_{XX}(\tau)| \le R_{XX}(0) \tag{6.27}$$

3. $$R_{XX}(0) = E[X^2(t)] \tag{6.28}$$

B. Cross-Correlation $R_{XY}(\tau)$:

The cross-correlation of $X(t)$ and $Y(t)$ is [Eq. (*6.18*)]

$$R_{XY}(\tau) = E[X(t)Y(t+\tau)]$$

Properties of $R_{XY}(\tau)$:

1. $$R_{XY}(-\tau) = R_{YX}(\tau) \tag{6.29}$$

2. $$|R_{XY}(\tau)| \le \sqrt{R_{XX}(0)R_{YY}(0)} \tag{6.30}$$

3. $$|R_{XY}(\tau)| \le \tfrac{1}{2}[R_{XX}(0) + R_{YY}(0)] \tag{6.31}$$

C. Autocovariance $C_{XX}(\tau)$:

The autocovariance of $X(t)$ is [Eq. (*6.9*)]

$$\begin{aligned} C_{XX}(\tau) &= E[\{X(t) - E[X(t)]\}\{X(t+\tau) - E[X(t+\tau)]\}] \\ &= R_{XX}(\tau) - \mu_X^2 \end{aligned} \tag{6.32}$$

D. Cross-Covariance $C_{XY}(\tau)$:

The cross-covariance of $X(t)$ and $Y(t)$ is

$$\begin{aligned} C_{XY}(\tau) &= E[\{X(t) - E[X(t)]\}\{Y(t+\tau) - E[Y(t+\tau)]\}] \\ &= R_{XY}(\tau) - \mu_X\mu_Y \end{aligned} \tag{6.33}$$

Two processes $X(t)$ and $Y(t)$ are called (*mutually*) *orthogonal* if

$$R_{XY}(\tau) = 0 \tag{6.34}$$

They are called *uncorrelated* if

$$C_{XY}(\tau) = 0 \tag{6.35}$$

E. Power Spectrum Density:

Let $R_{XX}(\tau)$ be the autocorrelation of $X(t)$. Then the *power spectral density* (or *power spectrum*) of $X(t)$ is defined by the Fourier transform of $R_{XX}(\tau)$ (Sec. 1.6) as

$$S_{XX}(\omega) = \int_{-\infty}^{\infty} R_{XX}(\tau)e^{-j\omega\tau}\,d\tau \tag{6.36}$$

Thus

$$R_{XX}(\tau) = \frac{1}{2\pi}\int_{-\infty}^{\infty} S_{XX}(\omega)e^{j\omega\tau}\,d\omega \tag{6.37}$$

Equations (*6.36*) and (*6.37*) are known as the *Wiener-Khinchin relations*.

Properties of $S_{XX}(\omega)$:

1. $S_{XX}(\omega)$ is real and $S_{XX}(\omega) \ge 0$ (*6.38*)

2. $$S_{XX}(-\omega) = S_{XX}(\omega) \tag{6.39}$$

3. $$\frac{1}{2\pi}\int_{-\infty}^{\infty} S_{XX}(\omega)\,d\omega = R_{XX}(0) = E[X^2(t)] \tag{6.40}$$

F. Cross Spectral Densities:

The *cross spectral densities* (or *cross power spectra*) $S_{XY}(\omega)$ and $S_{YX}(\omega)$ of $X(t)$ and $Y(t)$ are defined by

$$S_{XY}(\omega)=\int_{-\infty}^{\infty} R_{XY}(\tau)e^{-j\omega\tau}\,d\tau \tag{6.41}$$

and

$$S_{YX}(\omega)=\int_{-\infty}^{\infty} R_{YX}(\tau)e^{-j\omega\tau}\,d\tau \tag{6.42}$$

The cross-correlations and cross spectral densities thus form Fourier transform pairs. Thus

$$R_{XY}(\tau)=\frac{1}{2\pi}\int_{-\infty}^{\infty} S_{XY}(\omega)e^{j\omega\tau}\,d\omega \tag{6.43}$$

$$R_{YX}(\tau)=\frac{1}{2\pi}\int_{-\infty}^{\infty} S_{YX}(\omega)e^{j\omega\tau}\,d\omega \tag{6.44}$$

Substituting the relation [Eq. (*6.29*)]

$$R_{XY}(\tau)=R_{YX}(-\tau)$$

in Eqs. (*6.41*) and (*6.42*), we obtain

$$S_{XY}(\omega)=S_{YX}(-\omega)=S^*_{YX}(\omega) \tag{6.45}$$

6.5 TRANSMISSION OF RANDOM PROCESSES THROUGH LINEAR SYSTEMS

A. System Response:

A *linear time-invariant* (LTI) system is represented by (Sec. 1.7)

$$Y(t)=L[X(t)] \tag{6.46}$$

where $Y(t)$ is the output random process of an LTI system, represented by a linear operator L with input random process $X(t)$.

Let $h(t)$ be the impulse response of an LTI system (Fig. 6-2). Then [Sec. 1.8, Eq. (*1.80*)]

$$Y(t)=h(t)*X(t)=\int_{-\infty}^{\infty} h(\alpha)X(t-\alpha)\,d\alpha \tag{6.47}$$

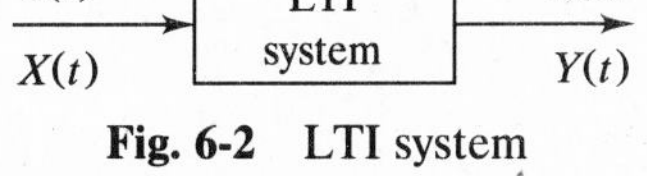

Fig. 6-2 LTI system

B. Mean and Autocorrelation of the Output:

$$\begin{aligned}\mu_Y(t)=E[Y(t)]&=E\left[\int_{-\infty}^{\infty} h(\alpha)X(t-\alpha)\,d\alpha\right]\\&=\int_{-\infty}^{\infty} h(\alpha)E[X(t-\alpha)]\,d\alpha\\&=\int_{-\infty}^{\infty} h(\alpha)\mu_X(t-\alpha)\,d\alpha=h(t)*\mu_X(t)\end{aligned} \tag{6.48}$$

where $*$ denotes convolution.

$$
\begin{aligned}
R_{YY}(t_1,t_2) &= E[Y(t_1)Y(t_2)] \\
&= E\left[\int_{-\infty}^{\infty}\int_{-\infty}^{\infty} h(\alpha)X(t_1-\alpha)h(\beta)X(t_2-\beta)\,d\alpha\,d\beta\right] \\
&= \int_{-\infty}^{\infty}\int_{-\infty}^{\infty} h(\alpha)h(\beta)E[X(t_1-\alpha)X(t_2-\beta)]\,d\alpha\,d\beta \\
&= \int_{-\infty}^{\infty}\int_{-\infty}^{\infty} h(\alpha)h(\beta)R_{XX}(t_1-\alpha,t_2-\beta)\,d\alpha\,d\beta
\end{aligned}
\tag{6.49}
$$

If the input $X(t)$ is WSS, then from Eq. (*6.48*) we have

$$
E[Y(t)] = \int_{-\infty}^{\infty} h(\alpha)\mu_X\,d\alpha = \mu_X\int_{-\infty}^{\infty} h(\alpha)\,d\alpha = \mu_X H(0) \tag{6.50}
$$

where $H(0)$ is the frequency response of the linear system at $\omega = 0$. Thus, the mean of the output is a constant.

The autocorrelation of the output given in Eq. (*6.49*) becomes

$$
R_{YY}(t_1,t_2) = \int_{-\infty}^{\infty}\int_{-\infty}^{\infty} h(\alpha)h(\beta)R_{XX}(t_2-t_1+\alpha-\beta)\,d\alpha\,d\beta \tag{6.51}
$$

which indicates that $R_{YY}(t_1,t_2)$ is a function of the time difference $\tau = t_2 - t_1$. Hence,

$$
R_{YY}(\tau) = \int_{-\infty}^{\infty}\int_{-\infty}^{\infty} h(\alpha)h(\beta)R_{XX}(\tau+\alpha-\beta)\,d\alpha\,d\beta \tag{6.52}
$$

Thus, we conclude that if the input $X(t)$ is WSS, the output $Y(t)$ is also WSS.

C. Power Spectral Density of the Output:

Taking the Fourier transform of both sides of Eq. (*6.52*), we obtain the power spectral density of the output

$$
\begin{aligned}
S_{YY}(\omega) &= \int_{-\infty}^{\infty} R_{YY}(\tau)e^{-j\omega\tau}\,d\tau \\
&= \int_{-\infty}^{\infty}\int_{-\infty}^{\infty}\int_{-\infty}^{\infty} h(\alpha)h(\beta)R_{XX}(\tau+\alpha-\beta)e^{-j\omega\tau}\,d\tau\,d\alpha\,d\beta \\
&= |H(\omega)|^2 S_{XX}(\omega)
\end{aligned}
\tag{6.53}
$$

Thus, we obtained the important result that the power spectral density of the output is the product of the power spectral density of the input and the magnitude squared of the frequency response of the system [Eq. (*1.101*)].

When the autocorrelation of the output $R_{YY}(\tau)$ is desired, it is easier to determine the power spectral density $S_{YY}(\omega)$ and then to evaluate the inverse Fourier transform (Prob. 6.21). Thus,

$$
\begin{aligned}
R_{YY}(\tau) &= \frac{1}{2\pi}\int_{-\infty}^{\infty} S_{YY}(\omega)e^{j\omega\tau}\,d\omega \\
&= \frac{1}{2\pi}\int_{-\infty}^{\infty} |H(\omega)|^2 S_{XX}(\omega)e^{j\omega\tau}\,d\omega
\end{aligned}
\tag{6.54}
$$

By Eq. (*6.17*), the average power in the output $Y(t)$ is

$$
E[Y^2(t)] = R_{YY}(0) = \frac{1}{2\pi}\int_{-\infty}^{\infty} |H(\omega)|^2 S_{XX}(\omega)\,d\omega \tag{6.55}
$$

6.6 SPECIAL CLASSES OF RANDOM PROCESSES

In this section, we consider some important random processes that are commonly encountered in the study of communication systems.

A. Gaussian Random Process:

Consider a random process $X(t)$, and define n random variables $X(t_1), \dots, X(t_n)$ corresponding to n time instants $t_1, \dots, t_n$. Let $\mathbf{X}$ be a *random vector* ($n \times 1$ matrix) defined by

$$\mathbf{X} = \begin{bmatrix} X(t_1) \\ \vdots \\ X(t_n) \end{bmatrix} \tag{6.56}$$

Let $\mathbf{x}$ be an n-dimensional vector ($n \times 1$ matrix) defined by

$$\mathbf{x} = \begin{bmatrix} x_1 \\ \vdots \\ x_n \end{bmatrix} \tag{6.57}$$

so that the event $\{X(t_1) \le x_1, \dots, X(t_n) \le x_n\}$ is written $\{\mathbf{X} \le \mathbf{x}\}$. Then $X(t)$ is called a *gaussian* (or *normal*) process if $\mathbf{X}$ has a jointly multivariate gaussian density function for every finite set of $\{t_i\}$ and every n.

The multivariate gaussian density function is given by

$$f_{\mathbf{X}}(\mathbf{x}) = \frac{1}{(2\pi)^{n/2}|\det \mathbf{C}|^{1/2}} \exp\left[-\frac{1}{2}(\mathbf{x}-\boldsymbol{\mu})^T \mathbf{C}^{-1}(\mathbf{x}-\boldsymbol{\mu})\right] \tag{6.58}$$

where T denotes the "transpose," $\boldsymbol{\mu}$ is the *vector means*, and $\mathbf{C}$ is the *covariance matrix*, given by

$$\boldsymbol{\mu} = E[\mathbf{X}] = \begin{bmatrix} \mu_1 \\ \vdots \\ \mu_n \end{bmatrix} = \begin{bmatrix} E[X(t_1)] \\ \vdots \\ E[X(t_n)] \end{bmatrix} \tag{6.59}$$

$$\mathbf{C} = \begin{bmatrix} C_{11} & \cdots & C_{1n} \\ \cdots & \cdots & \cdots \\ C_{n1} & \cdots & C_{nn} \end{bmatrix} \tag{6.60}$$

where

$$C_{ij} = C_{XX}(t_i, t_j) = R_{XX}(t_i, t_j) - \mu_i \mu_j \tag{6.61}$$

which is the covariance of $X(t_i)$ and $X(t_j)$, and $\det \mathbf{C}$ is the determinant of the matrix $\mathbf{C}$.

Some of the important properties of a gaussian process are as follows:

1. A gaussian process $X(t)$ is completely specified by the set of means
$$\mu_i = E[X(t_i)] \qquad i = 1, \dots, n$$
and the set of autocorrelations
$$R_{XX}(t_i, t_j) = E[X(t_i)X(t_j)] \qquad i, j = 1, \dots, n$$
2. If the set of random variables $X(t_i)$, $i = 1, \dots, n$, is uncorrelated, that is,
$$C_{ij} = 0 \qquad i \ne j$$
then $X(t_i)$ are independent.
3. If a gaussian process $X(t)$ is WSS, then $X(t)$ is SSS.

4. If the input process $X(t)$ of a linear system is gaussian, then the output process $Y(t)$ is also gaussian.

B. White Noise:

A random process $X(t)$ is called *white noise* if [Fig. 6-3(*a*)]

$$S_{XX}(\omega) = \frac{\eta}{2} \tag{6.62}$$

Taking the inverse Fourier transform of Eq. (6.62), we have

$$R_{XX}(\tau) = \frac{\eta}{2}\delta(\tau) \tag{6.63}$$

which is illustrated in Fig. 6-3(*b*). It is usually assumed that the mean of white noise is zero.

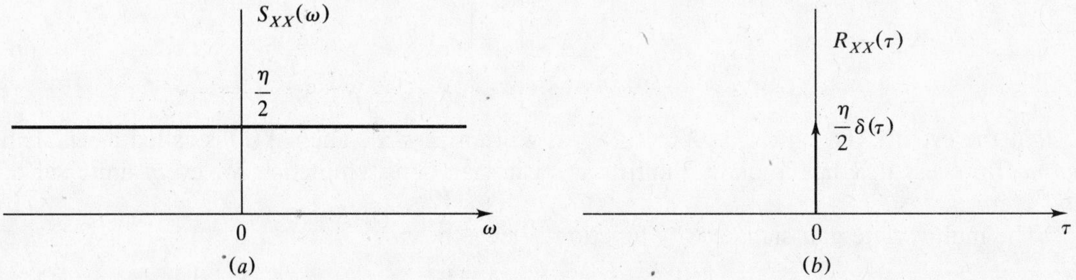

Fig. 6-3 White noise

C. Band-Limited White Noise:

A random process $X(t)$ is called *band-limited white noise* if

$$S_{XX}(\omega) = \begin{cases} \dfrac{\eta}{2} & |\omega| \le \omega_B \\ 0 & |\omega| > \omega_B \end{cases} \tag{6.64}$$

Then

$$R_{XX}(\tau) = \frac{1}{2\pi}\int_{-\omega_B}^{\omega_B} \frac{\eta}{2} e^{j\omega\tau}\, d\omega = \frac{\eta\omega_B}{2\pi}\frac{\sin \omega_B\tau}{\omega_B\tau} \tag{6.65}$$

And $S_{XX}(\omega)$ and $R_{XX}(\tau)$ of band-limited white noise are shown in Fig. 6-4.

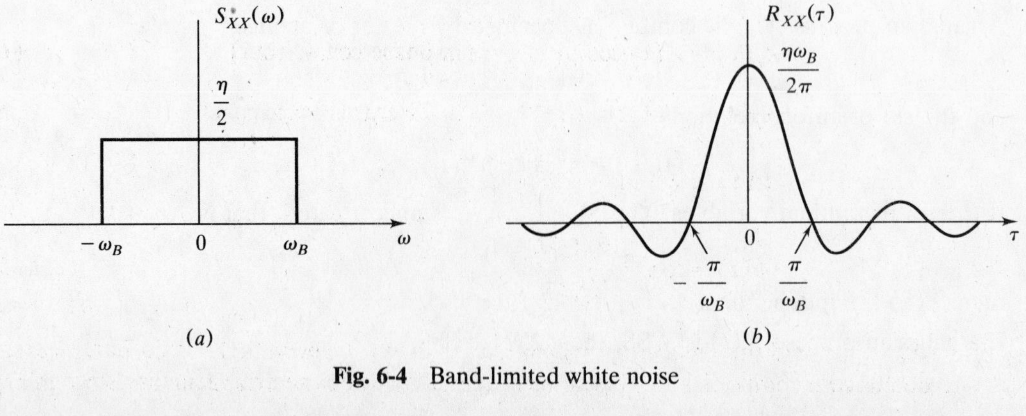

Fig. 6-4 Band-limited white noise

Note that the term *white* or *band-limited white* refers to the spectral shape of the process $X(t)$ only, and these terms do not imply that the distribution associated with $X(t)$ is gaussian.

D. Narrowband Random Process:

Suppose that $X(t)$ is a WSS process with zero mean and its power spectral density $S_{XX}(\omega)$ is nonzero only in some narrow frequency band of width $2W$ which is very small compared to a center frequency ω_c, as shown in Fig. 6-5. Then the process $X(t)$ is called a *narrowband random process.*

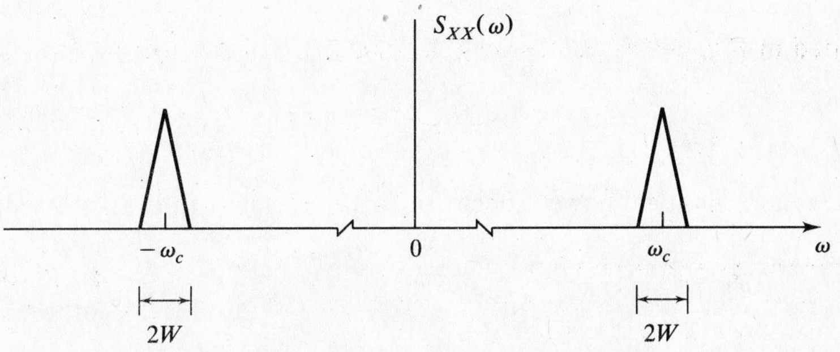

Fig. 6-5 Narrowband random process

In many communication systems, a narrowband process (or noise) is produced when white noise (or broadband noise) is passed through a narrowband linear filter. When a sample function of the narrowband process is viewed on an oscilloscope, the observed waveform appears as a sinusoid of random amplitude and phase. For this reason, the narrowband noise $X(t)$ is conveniently represented by the expression

$$X(t) = V(t)\cos\left[\omega_c t + \phi(t)\right] \tag{6.66}$$

where $V(t)$ and $\phi(t)$ are random processes which are called the *envelope function* and *phase function*, respectively.

Equation (6.66) can be rewritten as

$$\begin{aligned} X(t) &= V(t)\cos\phi(t)\cos\omega_c t - V(t)\sin\phi(t)\sin\omega_c t \\ &= X_c(t)\cos\omega_c t - X_s(t)\sin\omega_c t \end{aligned} \tag{6.67}$$

where

$$X_c(t) = V(t)\cos\phi(t) \qquad \text{(in-phase component)} \tag{6.68a}$$

$$X_s(t) = V(t)\sin\phi(t) \qquad \text{(quadrature component)} \tag{6.68b}$$

$$V(t) = \sqrt{X_c^2(t) + X_s^2(t)} \tag{6.69a}$$

$$\phi(t) = \tan^{-1}\frac{X_s(t)}{X_c(t)} \tag{6.69b}$$

Equation (6.67) is called the *quadrature representation* of $X(t)$. Given $X(t)$, its in-phase component $X_c(t)$ and quadrature component $X_s(t)$ can be obtained by using the arrangement shown in Fig. 6-6.

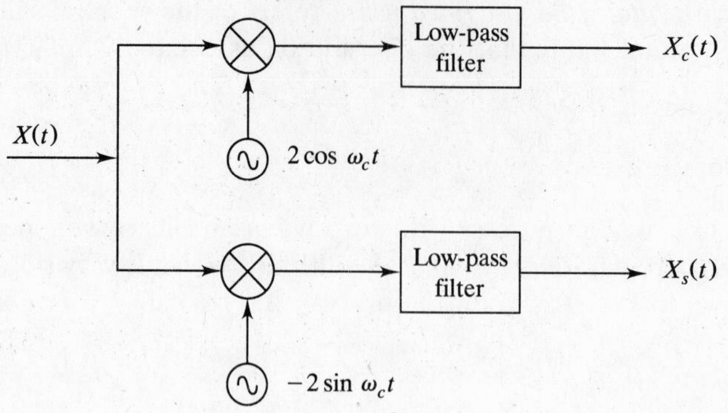

Fig. 6-6

Properties of $X_c(t)$ and $X_s(t)$:

1. $X_c(t)$ and $X_s(t)$ have identical power spectral densities related to that of $X(t)$ by

$$S_{X_c}(\omega) = S_{X_s}(\omega) = \begin{cases} S_{XX}(\omega-\omega_c) + S_{XX}(\omega+\omega_c) & |\omega| \le W \\ 0 & \text{otherwise} \end{cases} \tag{6.70}$$

2. $X_c(t)$ and $X_s(t)$ have the same means and variances as $X(t)$:

$$\mu_{X_c} = \mu_{X_s} = \mu_X = 0 \tag{6.71}$$

$$\sigma_{X_c}^2 = \sigma_{X_s}^2 = \sigma_X^2 \tag{6.72}$$

3. $X_c(t)$ and $X_s(t)$ are uncorrelated with each other:

$$E[X_c(t)X_s(t)] = 0 \tag{6.73}$$

4. If $X(t)$ is gaussian, then $X_c(t)$ and $X_s(t)$ are also gaussian.
5. If $X(t)$ is gaussian, then for fixed but arbitrary t, $V(t)$ is a random variable with Rayleigh distribution and $\phi(t)$ is a random variable uniformly distributed over $[0, 2\pi]$. (See Prob. 5.33.)

Solved Problems

RANDOM PROCESSES AND STATISTICS OF RANDOM PROCESSES

6.1. Consider a random process $X(t)$ given by

$$X(t) = A\cos(\omega t + \Theta) \tag{6.74}$$

where A and ω are constants and Θ is a uniform random variable over $[-\pi, \pi]$. Show that $X(t)$ is WSS.

From Eq. (*5.119*) (Prob. 5.16), we have

$$f_\Theta(\theta) = \begin{cases} \dfrac{1}{2\pi} & -\pi \le \theta \le \pi \\ 0 & \text{otherwise} \end{cases}$$

Thus

$$\mu_X(t) = E[X(t)] = \int_{-\infty}^{\infty} A\cos(\omega t + \theta)\, f_\Theta(\theta)\, d\theta$$

$$= \frac{A}{2\pi}\int_{-\pi}^{\pi} \cos(\omega t + \theta)\, d\theta = 0 \tag{6.75}$$

$$R_{XX}(t, t+\tau) = E[X(t)X(t+\tau)]$$

$$= \frac{A^2}{2\pi}\int_{-\pi}^{\pi} \cos(\omega t + \theta)\cos[\omega(t+\tau) + \theta]\, d\theta$$

$$= \frac{A^2}{2\pi}\int_{-\pi}^{\pi} \frac{1}{2}[\cos\omega\tau + \cos(2\omega t + 2\theta + \omega\tau)]\, d\theta$$

$$= \frac{A^2}{2}\cos\omega\tau \tag{6.76}$$

Since the mean of $X(t)$ is a constant and the autocorrelation of $X(t)$ is a function of time difference only, we conclude that $X(t)$ is WSS.

Note that $R_{XX}(\tau)$ is periodic with the period $T_0 = 2\pi/\omega$. A WSS random process is called *periodic* if its autocorrelation is periodic.

6.2. Consider a random process $X(t)$ given by

$$X(t) = A\cos(\omega t + \theta) \tag{6.77}$$

where ω and θ are constants and A is a random variable. Determine whether $X(t)$ is WSS.

$$\mu_X(t) = E[X(t)] = E[A\cos(\omega t + \theta)]$$

$$= \cos(\omega t + \theta)\, E[A] \tag{6.78}$$

which indicates that the mean of $X(t)$ is not constant unless $E[A] = 0$.

$$R_{XX}(t, t+\tau) = E[X(t)X(t+\tau)]$$

$$= E[A^2\cos(\omega t + \theta)\cos[\omega(t+\tau) + \theta]]$$

$$= \tfrac{1}{2}[\cos\omega\tau + \cos(2\omega t + 2\theta + \omega\tau)]E[A^2] \tag{6.79}$$

Thus, we see that the autocorrelation of $X(t)$ is not a function of the time difference τ only, and the process $X(t)$ is not WSS.

6.3. Consider a random process $Y(t)$ defined by

$$Y(t) = \int_0^t X(\tau)\, d\tau \tag{6.80}$$

where $X(t)$ is given by

$$X(t) = A\cos\omega t \tag{6.81}$$

where ω is constant and $A = N[0; \sigma^2]$.

(*a*) Determine the pdf of $Y(t)$ at $t = t_k$.

(*b*) Is $Y(t)$ WSS?

(*a*)
$$Y(t_k) = \int_0^{t_k} A\cos\omega\tau\, d\tau = \frac{\sin\omega t_k}{\omega} A \tag{6.82}$$

Then from the result of Prob. 5.24, we see that $Y(t_k)$ is a gaussian random variable with

$$E[Y(t_k)] = \frac{\sin \omega t_k}{\omega} E[A] = 0 \tag{6.83}$$

and

$$\sigma_Y^2 = \operatorname{var}[Y(t_k)] = \left(\frac{\sin \omega t_k}{\omega}\right)^2 \sigma^2 \tag{6.84}$$

Hence, by Eq. (*5.103*), the pdf of $Y(t_k)$ is

$$f_Y(y) = \frac{1}{\sqrt{2\pi}\,\sigma_Y} e^{-y^2/(2\sigma_Y^2)} \tag{6.85}$$

(*b*) From Eqs. (*6.83*) and (*6.84*), the mean and variance of $Y(t)$ depend on time $t(t_k)$, so $Y(t)$ is not WSS.

6.4. Consider a random process $X(t)$ given by

$$X(t) = A \cos \omega t + B \sin \omega t \tag{6.86}$$

where ω is constant and A and B are random variables.

(*a*) Show that the condition

$$E[A] = E[B] = 0 \tag{6.87}$$

is necessary for $X(t)$ to be stationary.

(*b*) Show that $X(t)$ is WSS if and only if the random variables A and B are uncorrelated with equal variance, that is,

$$E[AB] = 0 \tag{6.88}$$

and

$$E[A^2] = E[B^2] = \sigma^2 \tag{6.89}$$

(*a*) $\mu_X(t) = E[X(t)] = E[A] \cos \omega t + E[B] \sin \omega t$ must be independent of t for $X(t)$ to be stationary. This is possible only if $\mu_X(t) = 0$, that is,

$$E[A] = E[B] = 0$$

(*b*) If $X(t)$ is WSS, then from Eq. (*6.17*)

$$E[X^2(0)] = E\left[X^2\left(\frac{\pi}{2\omega}\right)\right] = R_{XX}(0) = \sigma_X^2$$

But

$$X(0) = A \qquad \text{and} \qquad X\left(\frac{\pi}{2\omega}\right) = B$$

Thus

$$E[A^2] = E[B^2] = \sigma_X^2 = \sigma^2$$

Using the above result, we obtain

$$\begin{aligned} R_{XX}(t, t+\tau) &= E[X(t)X(t+\tau)] \\ &= E[(A \cos \omega t + B \sin \omega t)[A \cos \omega(t+\tau) + B \sin \omega(t+\tau)]] \\ &= \sigma^2 \cos \omega\tau + E[AB] \sin(2\omega t + \omega\tau) \end{aligned} \tag{6.90}$$

which will be a function of τ only if $E[AB] = 0$.

Conversely, if $E[AB] = 0$ and $E[A^2] = E[B^2] = \sigma^2$, then from the result of part (*a*) and Eq. (*6.90*), we have

$$\mu_X(t) = 0$$

$$R_{XX}(t, t+\tau) = \sigma^2 \cos \omega\tau = R_{XX}(\tau)$$

Hence, $X(t)$ is WSS.

6.5. A random process $X(t)$ is said to be *covariance-stationary* if the covariance of $X(t)$ depends only on the time difference $\tau = t_2 - t_1$, that is,

$$C_{XX}(t, t+\tau) = C_{XX}(\tau) \tag{6.91}$$

Let $X(t)$ be given by

$$X(t) = (A+1)\cos t + B \sin t$$

where A and B are independent random variables for which

$$E[A] = E[B] = 0 \qquad \text{and} \qquad E[A^2] = E[B^2] = 1$$

Show that $X(t)$ is not WSS, but it is covariance-stationary.

$$\begin{aligned}\mu_X(t) &= E[X(t)] = E[(A+1)\cos t + B\sin t)] \\ &= E[A+1]\cos t + E[B]\sin t \\ &= \cos t\end{aligned}$$

which depends on t. Thus $X(t)$ cannot be WSS.

$$\begin{aligned}R_{XX}(t_1, t_2) &= E[X(t_1)X(t_2)] \\ &= E[[(A+1)\cos t_1 + B\sin t_1][(A+1)\cos t_2 + B\sin t_2]] \\ &= E\left[(A+1)^2\right]\cos t_1 \cos t_2 + E[B^2]\sin t_1 \sin t_2 \\ &\quad + E[(A+1)B](\cos t_1 \sin t_2 + \sin t_1 \cos t_2)\end{aligned}$$

Now

$$\begin{aligned}E\left[(A+1)^2\right] &= E[A^2 + 2A + 1] = E[A^2] + 2E[A] + 1 = 2 \\ E[(A+1)B] &= E[AB] + E[B] = E[A]E[B] + E[B] = 0 \\ E[B^2] &= 1\end{aligned}$$

Substituting these values into the expression of $R_{XX}(t_1, t_2)$, we obtain

$$\begin{aligned}R_{XX}(t_1, t_2) &= 2\cos t_1 \cos t_2 + \sin t_1 \sin t_2 \\ &= \cos(t_2 - t_1) + \cos t_1 \cos t_2\end{aligned}$$

From Eq. (6.9), we have

$$\begin{aligned}C_{XX}(t_1, t_2) &= R_{XX}(t_1, t_2) - \mu_X(t_1)\mu_X(t_2) \\ &= \cos(t_2 - t_1) + \cos t_1 \cos t_2 - \cos t_1 \cos t_2 \\ &= \cos(t_2 - t_1)\end{aligned}$$

Thus, $X(t)$ is covariance-stationary.

6.6. Show that the process $X(t)$ defined in Eq. (6.74) (Prob. 6.1) is ergodic in both the mean and the autocorrelation.

From Eq. (6.20), we have

$$\begin{aligned}\bar{x} = \langle x(t) \rangle &= \lim_{T \to \infty} \frac{1}{T} \int_{-T/2}^{T/2} A\cos(\omega t + \theta)\, dt \\ &= \frac{A}{T_0} \int_{-T_0/2}^{T_0/2} \cos(\omega t + \theta)\, dt = 0\end{aligned} \tag{6.92}$$

where $T_0 = 2\pi/\omega$.

From Eq. (*6.21*), we have

$$\begin{aligned}\bar{R}_{XX}(\tau) &= \langle x(t)x(t+\tau)\rangle \\ &= \lim_{T\to\infty}\frac{1}{T}\int_{-T/2}^{T/2} A^2\cos(\omega t+\theta)\cos[\omega(t+\tau)+\theta]\,dt \\ &= \frac{A^2}{T_0}\int_{-T_0/2}^{T_0/2}\frac{1}{2}[\cos\omega\tau+\cos(2\omega t+2\theta+\omega\tau)]\,dt \\ &= \frac{A^2}{2}\cos\omega\tau \end{aligned} \tag{6.93}$$

Thus, we have

$$\mu_X(t) = E[X(t)] = \langle x(t)\rangle = \bar{x}$$

$$R_{XX}(\tau) = E[X(t)X(t+\tau)] = \langle x(t)X(t+\tau)\rangle = \bar{R}_{XX}(\tau)$$

Hence, by definitions (*6.24*) and (*6.25*), we conclude that $X(t)$ is ergodic in both the mean and the autocorrelation.

CORRELATIONS AND POWER SPECTRA

6.7. Show that if $X(t)$ is WSS, then

$$E\left[[X(t+\tau)-X(t)]^2\right] = 2[R_{XX}(0)-R_{XX}(\tau)] \tag{6.94}$$

where $R_{XX}(\tau)$ is the autocorrelation of $X(t)$.

Using the linearity of E (the expectation operator) and Eqs. (*6.15*) and (*6.17*), we have

$$\begin{aligned}E\left[[X(t+\tau)-X(t)]^2\right] &= E\left[X^2(t+\tau)-2X(t+\tau)X(t)+X^2(t)\right] \\ &= E\left[X^2(t+\tau)\right]-2E[X(t+\tau)X(t)]+E\left[X^2(t)\right] \\ &= R_{XX}(0)-2R_{XX}(\tau)+R_{XX}(0) \\ &= 2[R_{XX}(0)-R_{XX}(\tau)]\end{aligned}$$

6.8. Let $X(t)$ be a WSS random process. Verify Eqs. (*6.26*) and (*6.27*), that is,

(*a*) $$R_{XX}(-\tau) = R_{XX}(\tau)$$

(*b*) $$|R_{XX}(\tau)| \le R_{XX}(0)$$

(*a*) From Eq. (*6.15*)

$$R_{XX}(\tau) = E[X(t)X(t+\tau)]$$

Setting $t+\tau = t'$, we have

$$\begin{aligned}R_{XX}(\tau) &= E[X(t'-\tau)X(t')] \\ &= E[X(t')X(t'-\tau)] = R_{XX}(-\tau)\end{aligned}$$

(*b*)

$$E\left[[X(t)\pm X(t+\tau)]^2\right] \ge 0$$

or $$E\left[X^2(t)\pm 2X(t)X(t+\tau)+X^2(t+\tau)\right] \ge 0$$

or $$E\left[X^2(t)\right] \pm 2E[X(t)X(t+\tau)]+E\left[X^2(t+\tau)\right] \ge 0$$

or $$2R_{XX}(0) \pm 2R_{XX}(\tau) \ge 0$$

Hence $$R_{XX}(0) \ge |R_{XX}(\tau)|$$

6.9. Show that the power spectrum of a (real) random process $X(t)$ is real, and verify Eq. (*6.39*), that is,

$$S_{XX}(-\omega) = S_{XX}(\omega)$$

From Eq. (*6.36*) and by expanding the exponential, we have

$$\begin{aligned} S_{XX}(\omega) &= \int_{-\infty}^{\infty} R_{XX}(\tau) e^{-j\omega\tau}\, d\tau \\ &= \int_{-\infty}^{\infty} R_{XX}(\tau)(\cos\omega\tau - j\sin\omega\tau)\, d\tau \\ &= \int_{-\infty}^{\infty} R_{XX}(\tau)\cos\omega\tau\, d\tau - j\int_{-\infty}^{\infty} R_{XX}(\tau)\sin\omega\tau\, d\tau \end{aligned} \tag{6.95}$$

Since $R_{XX}(-\tau) = R_{XX}(\tau)$ [Eq. (*6.26*)], the imaginary term in Eq. (*6.95*) then vanishes and we obtain

$$S_{XX}(\omega) = \int_{-\infty}^{\infty} R_{XX}(\tau)\cos\omega\tau\, d\tau \tag{6.96}$$

which indicates that $S_{XX}(\omega)$ is real.

Since the cosine is an even function of its argument, that is, $\cos(-\omega\tau) = \cos\omega\tau$, it follows that

$$S_{XX}(-\omega) = S_{XX}(\omega)$$

which indicates that the power spectrum of $X(t)$ is an even function of frequency.

6.10. A class of modulated random signal $Y(t)$ is defined by

$$Y(t) = AX(t)\cos(\omega_c t + \Theta) \tag{6.97}$$

where $X(t)$ is the random message signal and $A\cos(\omega_c t + \Theta)$ is the carrier. The random message signal $X(t)$ is a zero-mean stationary random process with autocorrelation $R_{XX}(\tau)$ and power spectrum $S_{XX}(\omega)$. The carrier amplitude A and the frequency ω_c are constants, and phase Θ is a random variable uniformly distributed over $[0, 2\pi]$. Assuming that $X(t)$ and Θ are independent, find the mean, autocorrelation, and power spectrum of $Y(t)$.

$$\begin{aligned} \mu_Y(t) &= E[Y(t)] = E[AX(t)\cos(\omega_c t + \Theta)] \\ &= AE[X(t)]E[\cos(\omega_c t + \Theta)] = 0 \end{aligned}$$

since $X(t)$ and Θ are independent and $E[X(t)] = 0$.

$$\begin{aligned} R_{YY}(t, t+\tau) &= E[Y(t)Y(t+\tau)] \\ &= E\left[A^2 X(t)X(t+\tau)\cos(\omega_c t + \Theta)\cos[\omega_c(t+\tau) + \Theta]\right] \\ &= \frac{A^2}{2} E[X(t)X(t+\tau)]E[\cos\omega_c\tau + \cos(2\omega_c t + \omega_c\tau + 2\Theta)] \\ &= \frac{A^2}{2} R_{XX}(\tau)\cos\omega_c\tau = R_{YY}(\tau) \end{aligned} \tag{6.98}$$

Since the mean of $Y(t)$ is a constant and the autocorrelation of $Y(t)$ depends only on the time difference τ, $Y(t)$ is WSS. Thus,

$$S_{YY}(\omega) = \mathscr{F}[R_{YY}(\tau)] = \frac{A^2}{2}\mathscr{F}[R_{XX}(\tau)\cos\omega_c\tau]$$

By Eqs. (*6.36*) and (*1.42*)

$$\mathscr{F}[R_{XX}(\tau)] = S_{XX}(\omega)$$

$$\mathscr{F}(\cos\omega_c\tau) = \pi\delta(\omega - \omega_c) + \pi\delta(\omega + \omega_c)$$

Then, using the convolution theorem (*1.54*) and Eq. (*1.52*), we obtain

$$S_{YY}(\omega) = \frac{A^2}{4\pi} S_{XX}(\omega) * [\pi\delta(\omega - \omega_c) + \pi\delta(\omega + \omega_c)]$$

$$= \frac{A^2}{4}[S_{XX}(\omega - \omega_c) + S_{XX}(\omega + \omega_c)] \tag{6.99}$$

6.11. Consider a random process $X(t)$ that assumes the values $\pm A$ with equal probability. A typical sample function of $X(t)$ is shown in Fig. 6-7. The average number of polarity switches (zero crossings) per unit time is α. The probability of having exactly k crossings in time τ is given by the Poisson distribution [Eq. (*5.98*)]

$$P(Z = k) = e^{-\alpha\tau} \frac{(\alpha\tau)^k}{k!} \tag{6.100}$$

where Z is the random variable representing the number of zero crossings. The process $X(t)$ is known as the *telegraph* signal. Find the autocorrelation and the power spectrum of $X(t)$.

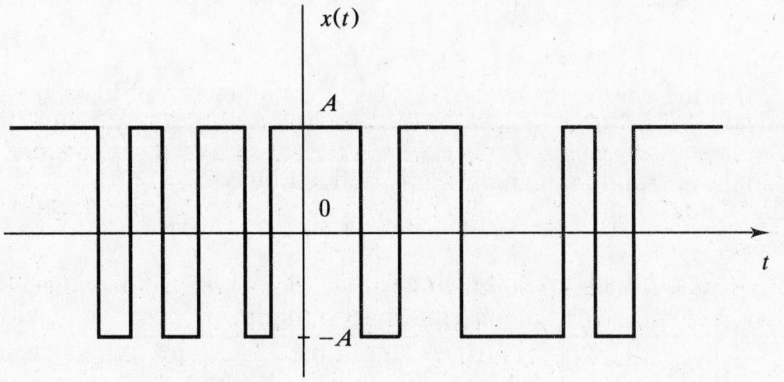

Fig. 6-7 Telegraph signal

If τ is any positive time interval, then

$$\begin{aligned} R_{XX}(t, t+\tau) &= E[X(t)X(t+\tau)] \\ &= A^2 P[X(t) \text{ and } X(t+\tau) \text{ have same signs}] \\ &\quad + (-A^2) P[X(t) \text{ and } X(t+\tau) \text{ have different signs}] \\ &= A^2 P[Z \text{ even in } (t, t+\tau)] - A^2 P[Z \text{ odd in } (t, t+\tau)] \\ &= A^2 \sum_{k \text{ even}} e^{-\alpha\tau} \frac{(\alpha\tau)^k}{k!} - A^2 \sum_{k \text{ odd}} e^{-\alpha\tau} \frac{(\alpha\tau)^k}{k!} \\ &= A^2 e^{-\alpha\tau} \sum_{k=0}^{\infty} \frac{(\alpha\tau)^k}{k!}(-1)^k \\ &= A^2 e^{-\alpha\tau} \sum_{k=0}^{\infty} \frac{(-\alpha\tau)^k}{k!} = A^2 e^{-\alpha\tau} e^{-\alpha\tau} = A^2 e^{-2\alpha\tau} \end{aligned} \tag{6.101}$$

which indicates that the autocorrelation depends only on the time difference τ. By Eq. (*6.26*), the complete solution that includes $\tau < 0$ is given by

$$R_{XX}(\tau) = A^2 e^{-2\alpha|\tau|} \tag{6.102}$$

which is sketched in Fig. 6-8(*a*).

Taking the Fourier transform of both sides of Eq. (*6.102*), we see that the power spectrum of $X(t)$ is [Eq. (*1.46*)]

$$S_{XX}(\omega) = A^2 \frac{4\alpha}{\omega^2 + (2\alpha)^2} \tag{6.103}$$

which is sketched in Fig. 6-8(b).

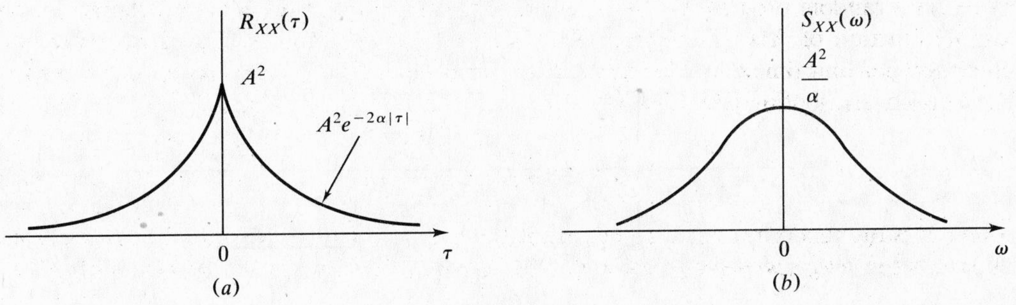

Fig. 6-8

6.12. Consider a random binary process $X(t)$ consisting of a random sequence of binary symbols 1 and 0. A typical sample function of $X(t)$ is shown in Fig. 6-9. It is assumed that

1. The symbols 1 and 0 are represented by pulses of amplitude $+A$ and $-A$ V, respectively, and duration T_b s.
2. The two symbols 1 and 0 are equally likely, and the presence of a 1 or 0 in any one interval is independent of the presence in all other intervals.
3. The pulse sequence is not synchronized, so that the starting time t_d of the first pulse after $t = 0$ is equally likely to be anywhere between 0 to T_b. That is, t_d is the sample value of a random variable T_d uniformly distributed over $[0, T_b]$.

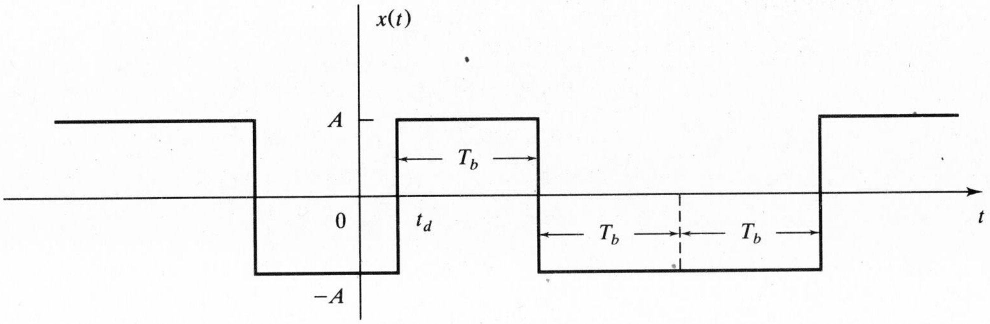

Fig. 6-9 Random binary signal

Find the autocorrelation and power spectrum of $X(t)$.

The random binary process $X(t)$ can be represented by

$$X(t) = \sum_{k=-\infty}^{\infty} A_k p(t - kT_b - T_d) \tag{6.104}$$

where $\{A_k\}$ is a sequence of independent random variables with $P[A_k = A] = P[A_k = -A] = \frac{1}{2}$, $p(t)$ is a unit amplitude pulse of duration T_b, and T_d is a random variable uniformly distributed over $[0, T_b]$.

$$\mu_X(t) = E[X(t)] = E[A_k] = \frac{1}{2}A + \frac{1}{2}(-A) = 0 \tag{6.105}$$

Let $t_2 > t_1$. When $t_2 - t_1 > T_b$, then t_1 and t_2 must fall in different pulse intervals [Fig. 6-10(a)] and the random variables $X(t_1)$ and $X(t_2)$ are therefore independent. We thus have

$$R_{XX}(t_1, t_2) = E[X(t_1)X(t_2)] = E[X(t_1)]E[X(t_2)] = 0 \tag{6.106}$$

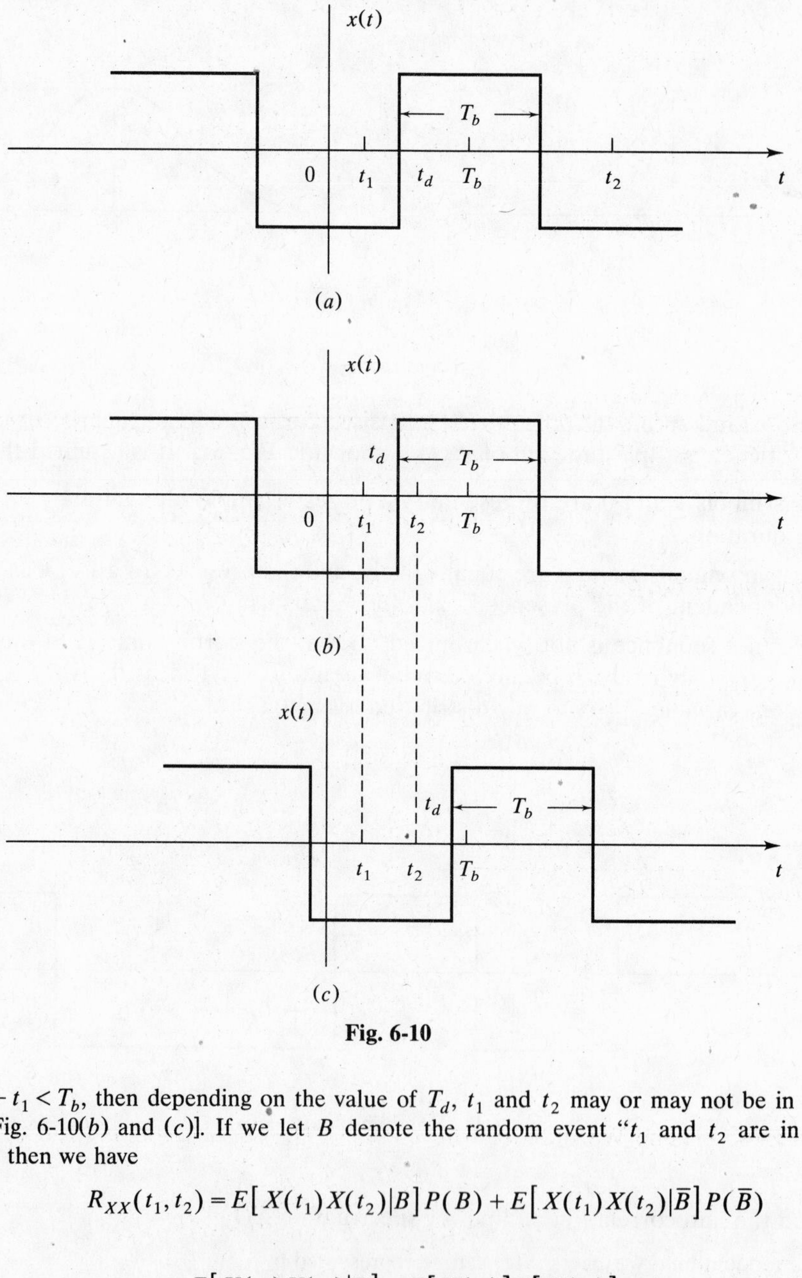

Fig. 6-10

When $t_2 - t_1 < T_b$, then depending on the value of T_d, t_1 and t_2 may or may not be in the same pulse interval [Fig. 6-10(b) and (c)]. If we let B denote the random event "t_1 and t_2 are in adjacent pulse intervals," then we have

$$R_{XX}(t_1, t_2) = E[X(t_1)X(t_2)|B]P(B) + E[X(t_1)X(t_2)|\bar{B}]P(\bar{B})$$

Now

$$E[X(t_1)X(t_2)|B] = E[X(t_1)]E[X(t_2)] = 0$$
$$E[X(t_1)X(t_2)|\bar{B}] = A^2$$

Since $P(B)$ will be the same when t_1 and t_2 fall in any time range of length T_b, it suffices to consider the case $0 < t < T_b$, as shown in Fig. 6-10(*b*). From Fig. 6-10(*b*),

$$P(B) = P(t_1 < T_d < t_2)$$
$$= \int_{t_1}^{t_2} f_{T_d}(t_d)\, dt_d = \int_{t_1}^{t_2} \frac{1}{T_b}\, dt_d = \frac{t_2 - t_1}{T_b}$$

From Eq. (*5.19*), we have

$$P(\bar{B}) = 1 - P(B) = 1 - \frac{t_2 - t_1}{T_b}$$

Thus

$$R_{XX}(t_1, t_2) = A^2\left(1 - \frac{t_2 - t_1}{T_b}\right) = R_{XX}(\tau) \tag{6.107}$$

where $\tau = t_2 - t_1$.

Since $R_{XX}(-\tau) = R_{XX}(\tau)$, we conclude that

$$R_{XX}(\tau) = \begin{cases} A^2\left(1 - \dfrac{|\tau|}{T_b}\right) & |\tau| \le T_b \\ 0 & |\tau| > T_b \end{cases} \tag{6.108}$$

which is plotted in Fig. 6-11(*a*).

From Eqs. (*6.105*) and (*6.108*), we see that $X(t)$ is WSS. Thus, from Eq. (*6.36*), the power spectrum of $X(t)$ is (see Prob. 1.55)

$$S_{XX}(\omega) = A^2 T_b \left[\frac{\sin(\omega T_b/2)}{\omega T_b/2}\right]^2 \tag{6.109}$$

which is plotted in Fig. 6-11(*b*).

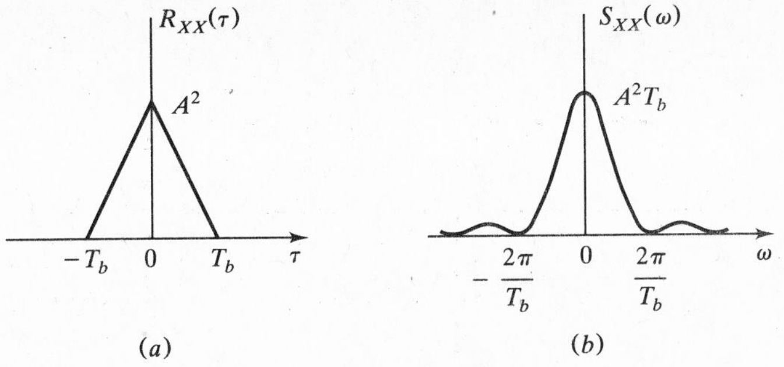

Fig. 6-11

6.13. Let $X(t)$ and $Y(t)$ be WSS random processes. Verify Eqs. (*6.29*) and (*6.30*), that is,

(*a*) $$R_{XY}(-\tau) = R_{YX}(\tau)$$

(*b*) $$|R_{XY}(\tau)| \le \sqrt{R_{XX}(0)R_{YY}(0)}$$

(*a*) By Eq. (*6.18*)

$$R_{XY}(-\tau) = E[X(t)Y(t-\tau)]$$

Setting $t - \tau = t'$, we obtain

$$R_{XY}(-\tau) = E[X(t'+\tau)Y(t')] = E[Y(t')X(t'+\tau)] = R_{YX}(\tau)$$

(*b*) From the Cauchy-Schwarz inequality Eq. (*5.165*) (Prob. 5.48) it follows that

$$\{E[X(t)Y(t+\tau)]\}^2 \le E[X^2(t)]E[Y^2(t+\tau)]$$

or

$$[R_{XY}(\tau)]^2 \le R_{XX}(0)R_{YY}(0)$$

Thus

$$|R_{XY}(\tau)| \le \sqrt{R_{XX}(0)R_{YY}(0)}$$

6.14. Let $X(t)$ and $Y(t)$ be both zero-mean and WSS random processes. Consider the random process $Z(t)$ defined by

$$Z(t) = X(t) + Y(t) \tag{6.110}$$

(*a*) Determine the autocorrelation and the power spectrum of $Z(t)$ if $X(t)$ and $Y(t)$ are jointly WSS.

(*b*) Repeat part (*a*) if $X(t)$ and $Y(t)$ are orthogonal.

(*c*) Show that if $X(t)$ and $Y(t)$ are orthogonal, then the mean square of $Z(t)$ is equal to the sum of the mean squares of $X(t)$ and $Y(t)$.

(*a*) The autocorrelation of $Z(t)$ is given by

$$\begin{aligned} R_{ZZ}(t_1,t_2) &= E[Z(t_1)Z(t_2)] \\ &= E[[X(t_1)+Y(t_1)][X(t_2)+Y(t_2)]] \\ &= E[X(t_1)X(t_2)] + E[X(t_1)Y(t_2)] \\ &\quad + E[Y(t_1)X(t_2)] + E[Y(t_1)Y(t_2)] \\ &= R_{XX}(t_1,t_2) + R_{XY}(t_1,t_2) + R_{YX}(t_1,t_2) + R_{YY}(t_1,t_2) \end{aligned} \tag{6.111}$$

If $X(t)$ and $Y(t)$ are jointly WSS, then we have

$$R_{ZZ}(\tau) = R_{XX}(\tau) + R_{XY}(\tau) + R_{YX}(\tau) + R_{YY}(\tau) \tag{6.112}$$

where $\tau = t_2 - t_1$.

Taking the Fourier transform of both sides of Eq. (*6.112*), we obtain

$$S_{ZZ}(\omega) = S_{XX}(\omega) + S_{XY}(\omega) + S_{YX}(\omega) + S_{YY}(\omega) \tag{6.113}$$

(*b*) If $X(t)$ and $Y(t)$ are orthogonal [Eq. (*6.34*)],

$$R_{XY}(\tau) = R_{YX}(\tau) = 0$$

Then Eqs. (*6.112*) and (*6.113*) become

$$R_{ZZ}(\tau) = R_{XX}(\tau) + R_{YY}(\tau) \tag{6.114}$$

and

$$S_{ZZ}(\omega) = S_{XX}(\omega) + S_{YY}(\omega) \tag{6.115}$$

(*c*) From Eqs. (*6.114*) and (*6.17*),

$$R_{ZZ}(0) = R_{XX}(0) + R_{YY}(0)$$

or

$$E[Z^2(t)] = E[X^2(t)] + E[Y^2(t)] \tag{6.116}$$

which indicates that the mean square of $Z(t)$ is equal to the sum of the mean squares of $X(t)$ and $Y(t)$.

6.15. Two random processes $X(t)$ and $Y(t)$ are given by

$$X(t) = A\cos(\omega t + \Theta) \tag{6.117a}$$

$$Y(t) = A\sin(\omega t + \Theta) \tag{6.117b}$$

where A and ω are constants and Θ is a uniform random variable over $[0, 2\pi]$. Find the cross-correlation of $X(t)$ and $Y(t)$, and verify Eq. (*6.29*).

From Eq. (*6.18*), the cross-correlation of $X(t)$ and $Y(t)$ is

$$\begin{aligned} R_{XY}(t, t+\tau) &= E[X(t)Y(t+\tau)] \\ &= E\left[A^2 \cos(\omega t + \Theta) \sin[\omega(t+\tau) + \Theta]\right] \\ &= \frac{A^2}{2} E[\sin(2\omega t + \omega\tau + 2\Theta) - \sin(-\omega\tau)] \\ &= \frac{A^2}{2} \sin \omega\tau = R_{XY}(\tau) \end{aligned} \tag{6.118a}$$

Similarly,

$$\begin{aligned} R_{YX}(t, t+\tau) &= E[Y(t)X(t+\tau)] \\ &= E\left[A^2 \sin(\omega t + \Theta) \cos[\omega(t+\tau) + \Theta]\right] \\ &= \frac{A^2}{2} E[\sin(2\omega t + \omega\tau + 2\Theta) + \sin(-\omega\tau)] \\ &= -\frac{A^2}{2} \sin \omega\tau = R_{YX}(\tau) \end{aligned} \tag{6.118b}$$

From Eqs. (*6.118a*) and (*6.118b*)

$$R_{XY}(-\tau) = \frac{A^2}{2} \sin \omega(-\tau) = -\frac{A^2}{2} \sin \omega\tau = R_{YX}(\tau)$$

which verifies Eq. (*6.29*).

6.16. Let $X(t)$ and $Y(t)$ be defined by

$$X(t) = A \cos \omega t + B \sin \omega t \tag{6.119a}$$

$$Y(t) = B \cos \omega t - A \sin \omega t \tag{6.119b}$$

where ω is constant and A and B are independent random variables both having zero mean and variance σ^2. Find the cross-correlation of $X(t)$ and $Y(t)$.

The cross-correlation of $X(t)$ and $Y(t)$ is

$$\begin{aligned} R_{XY}(t_1, t_2) &= E[X(t_1)Y(t_2)] \\ &= E[(A \cos \omega t_1 + B \sin \omega t_1)(B \cos \omega t_2 - A \sin \omega t_2)] \\ &= E[AB](\cos \omega t_1 \cos \omega t_2 - \sin \omega t_1 \sin \omega t_2) \\ &\quad - E[A^2] \cos \omega t_1 \sin \omega t_2 + E[B^2] \sin \omega t_1 \cos \omega t_2 \end{aligned}$$

Since

$$E[AB] = E[A]E[B] = 0 \qquad E[A^2] = E[B^2] = \sigma^2$$

we have

$$\begin{aligned} R_{XY}(t_1, t_2) &= \sigma^2(\sin \omega t_1 \cos \omega t_2 - \cos \omega t_1 \sin \omega t_2) \\ &= \sigma^2 \sin \omega(t_1 - t_2) \end{aligned}$$

or

$$R_{XY}(\tau) = -\sigma^2 \sin \omega\tau \tag{6.120}$$

where $\tau = t_2 - t_1$.

TRANSMISSION OF RANDOM PROCESSES THROUGH LINEAR SYSTEMS

6.17. A WSS random process $X(t)$ is applied to the input of an LTI system with impulse response $h(t) = 3e^{-2t}u(t)$. Find the mean value of the output $Y(t)$ of the system if $E[X(t)] = 2$.

By Eq. (*1.45*), the frequency response $H(\omega)$ of the system is

$$H(\omega) = \mathscr{F}[h(t)] = 3\frac{1}{j\omega + 2}$$

Then, by Eq. (*6.50*), the mean value of $Y(t)$ is

$$\mu_Y(t) = E[Y(t)] = \mu_X H(0) = 2\left(\frac{3}{2}\right) = 3$$

6.18. Let $Y(t)$ be the output of an LTI system with impulse response $h(t)$, when $X(t)$ is applied as input. Show that

(*a*)
$$R_{XY}(t_1, t_2) = \int_{-\infty}^{\infty} h(\beta) R_{XX}(t_1, t_2 - \beta)\, d\beta \tag{6.121}$$

(*b*)
$$R_{YY}(t_1, t_2) = \int_{-\infty}^{\infty} h(\alpha) R_{XY}(t_1 - \alpha, t_2)\, d\alpha \tag{6.122}$$

(*a*) Using Eq. (*6.47*), we have

$$\begin{aligned} R_{XY}(t_1, t_2) &= E[X(t_1)Y(t_2)] \\ &= E\left[X(t_1)\int_{-\infty}^{\infty} h(\beta) X(t_2 - \beta)\, d\beta\right] \\ &= \int_{-\infty}^{\infty} h(\beta) E[X(t_1)X(t_2 - \beta)]\, d\beta \\ &= \int_{-\infty}^{\infty} h(\beta) R_{XX}(t_1, t_2 - \beta)\, d\beta \end{aligned}$$

(*b*) Similarly,

$$\begin{aligned} R_{YY}(t_1, t_2) &= E[Y(t_1)Y(t_2)] \\ &= E\left[\int_{-\infty}^{\infty} h(\alpha) X(t_1 - \alpha)\, d\alpha\, Y(t_2)\right] \\ &= \int_{-\infty}^{\infty} h(\alpha) E[X(t_1 - \alpha)Y(t_2)]\, d\alpha \\ &= \int_{-\infty}^{\infty} h(\alpha) R_{XY}(t_1 - \alpha, t_2)\, d\alpha \end{aligned}$$

6.19. Let $X(t)$ be a WSS random input process to an LTI system with impulse response $h(t)$, and let $Y(t)$ be the corresponding output process. Show that

(*a*) $$R_{XY}(\tau) = h(\tau) * R_{XX}(\tau) \tag{6.123}$$

(*b*) $$R_{YY}(\tau) = h(-\tau) * R_{XY}(\tau) \tag{6.124}$$

(*c*) $$S_{XY}(\omega) = H(\omega) S_{XX}(\omega) \tag{6.125}$$

(*d*) $$S_{YY}(\omega) = H^*(\omega) S_{XY}(\omega) \tag{6.126}$$

where $*$ denotes the convolution and $H^*(\omega)$ is the complex conjugate of $H(\omega)$.

(*a*) If $X(t)$ is WSS, then Eq. (*6.121*) of Prob. 6.18 becomes

$$R_{XY}(t_1, t_2) = \int_{-\infty}^{\infty} h(\beta) R_{XX}(t_2 - t_1 - \beta)\, d\beta \tag{6.127}$$

which indicates that $R_{XY}(t_1, t_2)$ is a function of the time difference $\tau = t_2 - t_1$ only. Hence, Eq. (6.127) yields

$$R_{XY}(\tau) = \int_{-\infty}^{\infty} h(\beta) R_{XX}(\tau - \beta)\, d\beta = h(\tau) * R_{XX}(\tau)$$

(b) Similarly, if $X(t)$ is WSS, then Eq. (6.122) becomes

$$R_{YY}(t_1, t_2) = \int_{-\infty}^{\infty} h(\alpha) R_{XY}(t_2 - t_2 + \alpha)\, d\alpha$$

or

$$R_{YY}(\tau) = \int_{-\infty}^{\infty} h(\alpha) R_{XY}(\tau + \alpha)\, d\alpha = h(-\tau) * R_{XY}(\tau)$$

(c) Taking the Fourier transform of both sides of Eq. (6.123) and using Eqs. (6.41) and (1.53), we obtain

$$S_{XY}(\omega) = H(\omega) S_{XX}(\omega)$$

(d) Similarly, taking the Fourier transform of both sides of Eq. (6.124) and using Eqs. (6.36), (1.53), and (1.33), we obtain

$$S_{YY}(\omega) = H^*(\omega) S_{XY}(\omega)$$

Note that by combining Eqs. (6.125) and (6.126), we obtain Eq. (6.53), that is,

$$S_{YY}(\omega) = H^*(\omega) H(\omega) S_{XX}(\omega) = |H(\omega)|^2 S_{XX}(\omega)$$

6.20. Let $X(t)$ and $Y(t)$ be the wide-sense stationary random input process and random output process, respectively, of a quadrature phase-shifting filter ($-\pi/2$ rad phase shifter of Prob. 1.47). Show that

(a) $$R_{XX}(\tau) = R_{YY}(\tau) \tag{6.128}$$

(b) $$R_{XY}(\tau) = \hat{R}_{XX}(\tau) \tag{6.129}$$

where $\hat{R}_{XX}(\tau)$ is the Hilbert transform of $R_{XX}(\tau)$.

(a) The Hilbert transform $\hat{X}(t)$ of $X(t)$ was defined in Prob. 1.47 as the output of a quadrature phase-shifting filter with

$$h(t) = \frac{1}{\pi t} \qquad H(\omega) = -j\,\text{sgn}(\omega)$$

Since $|H(\omega)|^2 = 1$, we conclude that if $X(t)$ is a WSS random signal, then $Y(t) = \hat{X}(t)$ and by Eq. (6.53)

$$S_{YY}(\omega) = |H(\omega)|^2 S_{XX}(\omega) = S_{XX}(\omega)$$

Hence

$$R_{YY}(\tau) = \mathscr{F}^{-1}[S_{YY}(\omega)] = \mathscr{F}^{-1}[S_{XX}(\omega)] = R_{XX}(\tau)$$

(b) Using Eqs. (6.123) and (1.139) (Prob. 1.47), we have

$$R_{XY}(\tau) = h(\tau) * R_{XX}(\tau) = \frac{1}{\pi t} * R_{XX}(\tau) = \hat{R}_{XX}(\tau)$$

6.21. A WSS random process $X(t)$ with autocorrelation

$$R_{XX}(\tau) = A e^{-a|\tau|}$$

where A and a are real positive constants, is applied to the input of an LTI system with impulse response

$$h(t) = e^{-bt} u(t)$$

where b is a real positive constant. Find the autocorrelation of the output $Y(t)$ of the system.

Using Eq. (*1.45*), we see that the frequency response $H(\omega)$ of the system is

$$H(\omega) = \mathscr{F}[h(t)] = \frac{1}{j\omega + b}$$

So
$$|H(\omega)|^2 = \frac{1}{\omega^2 + b^2}$$

Using Eq. (*1.46*), we see that the power spectral density of $X(t)$ is

$$S_{XX}(\omega) = \mathscr{F}[R_{XX}(\tau)] = A\frac{2a}{\omega^2 + a^2}$$

By Eq. (*6.53*), the power spectral density of $Y(t)$ is

$$\begin{aligned} S_{YY}(\omega) &= |H(\omega)|^2 S_{XX}(\omega) \\ &= \left(\frac{1}{\omega^2 + b^2}\right)\left(\frac{2aA}{\omega^2 + a^2}\right) \\ &= \frac{aA}{(a^2 - b^2)b}\left(\frac{2b}{\omega^2 + b^2}\right) - \frac{A}{a^2 - b^2}\left(\frac{2a}{\omega^2 + a^2}\right) \end{aligned}$$

Taking the inverse Fourier transform of both sides of the above equation and using Eq. (*1.46*), we obtain

$$R_{YY}(\tau) = \frac{aA}{(a^2 - b^2)b}e^{-b|\tau|} - \frac{A}{a^2 - b^2}e^{-a|\tau|}$$

6.22. Verify Eq. (*6.38*); that is, the power spectrum of any WSS process $X(t)$ is real and

$$S_{XX}(\omega) \geq 0$$

for every ω.

The realness of the power spectrum of $X(t)$ was shown in Prob. 6.9. Consider an ideal bandpass filter with frequency response (Fig. 6-12)

$$H(\omega) = \begin{cases} 1 & \omega_1 \leq |\omega| \leq \omega_2 \\ 0 & \text{otherwise} \end{cases}$$

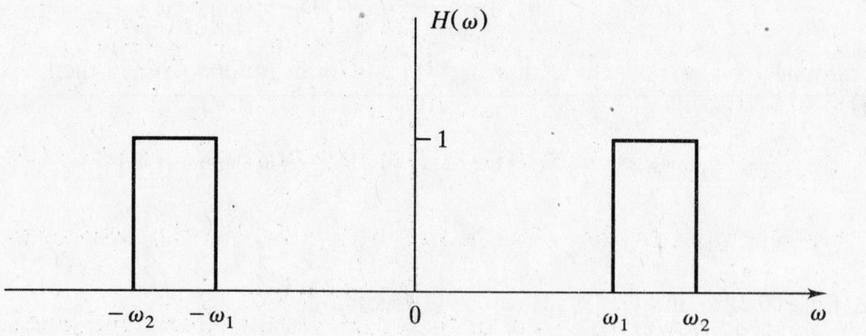

Fig. 6-12

with a random process $X(t)$ as its input. From Eq. (*6.53*) it follows that the power spectrum $S_{YY}(\omega)$ of the resulting output $Y(t)$ equals

$$S_{YY}(\omega) = \begin{cases} S_{XX}(\omega) & \omega_1 \leq |\omega| \leq \omega_2 \\ 0 & \text{otherwise} \end{cases}$$

Hence, from Eq. (*6.55*), we have

$$E\left[Y^2(t)\right] = \frac{1}{2\pi}\int_{-\infty}^{\infty} S_{YY}(\omega)\,d\omega = 2\left(\frac{1}{2\pi}\right)\int_{\omega_1}^{\omega_2} S_{XX}(\omega)\,d\omega \geq 0 \qquad (6.130)$$

which indicates that the area of $S_{XX}(\omega)$ in any interval of ω is nonnegative. This is possible only if $S_{XX}(\omega) \geq 0$ for every ω.

6.23. Consider a WSS process $X(t)$ with autocorrelation $R_{XX}(\tau)$ and power spectrum $S_{XX}(\omega)$. Let $X'(t) = dX(t)/dt$. Show that

(a) $$R_{XX'}(\tau) = \frac{dR_{XX}(\tau)}{d\tau} \tag{6.131}$$

(b) $$R_{X'X'}(\tau) = -\frac{d^2R_{XX}(\tau)}{d\tau^2} \tag{6.132}$$

(c) $$S_{X'X'}(\omega) = \omega^2 S_{XX}(\omega) \tag{6.133}$$

A system with frequency response $H(\omega) = j\omega$ is a differentiator (Fig. 6-13). Thus, if $X(t)$ is its input, then its output is $Y(t) = X'(t)$ [see Eq. (*1.35*)].

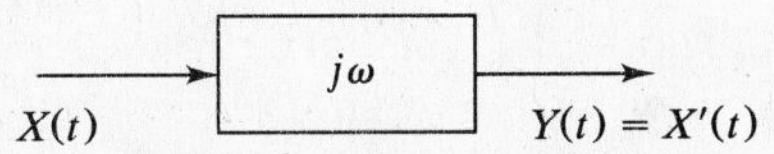

Fig. 6-13 Differentiator

(a) From Eq. (*6.125*)

$$S_{XX'}(\omega) = H(\omega)S_{XX}(\omega) = j\omega S_{XX}(\omega)$$

Taking the inverse Fourier transform of both sides, we obtain

$$R_{XX'}(\tau) = \frac{dR_{XX}(\tau)}{d\tau}$$

(b) From Eq. (*6.126*)

$$S_{X'X'}(\omega) = H^*(\omega)S_{XX'}(\omega) = -j\omega S_{XX'}(\omega)$$

Again taking the inverse Fourier transform of both sides and using the result of part (a), we have

$$R_{X'X'}(\tau) = -\frac{dR_{XX'}(\tau)}{d\tau} = -\frac{d^2R_{XX}(\tau)}{d\tau^2}$$

(c) From Eq. (*6.53*)

$$S_{X'X'}(\omega) = |H(\omega)|^2 S_{XX}(\omega) = |j\omega|^2 S_{XX}(\omega) = \omega^2 S_{XX}(\omega)$$

6.24. Suppose that the input to the differentiator of Fig. 6-13 is the zero-mean random telegraph signal of Prob. 6.11.

(a) Determine the power spectrum of the differentiator output and plot it.

(b) Determine the mean-square value of the differentiator output.

(a) From Eq. (*6.103*) of Prob. 6.11

$$S_{XX}(\omega) = A^2 \frac{4\alpha}{\omega^2 + (2\alpha)^2}$$

For the differentiator $H(\omega) = j\omega$, and from Eq. (*6.53*), we have

$$S_{YY}(\omega) = |H(\omega)|^2 S_{XX}(\omega) = A^2 \frac{4\alpha\omega^2}{\omega^2 + (2\alpha)^2} \tag{6.134}$$

which is plotted in Fig. 6-14.

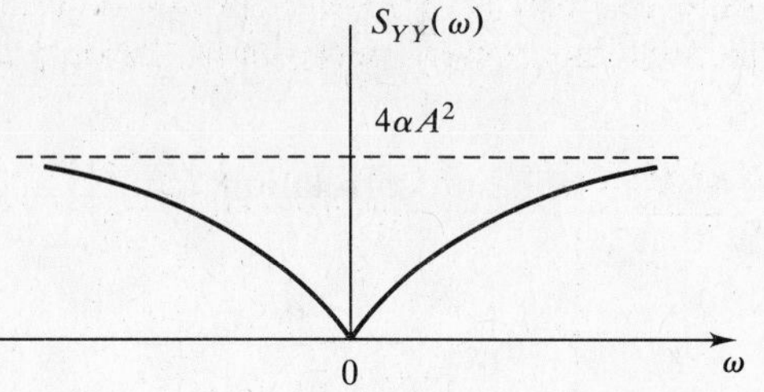

Fig. 6-14

(*b*) From Eq. (*6.55*) or Fig. 6-14

$$E[Y^2(t)] = \frac{1}{2\pi}\int_{-\infty}^{\infty} S_{YY}(\omega)\,d\omega = \infty$$

6.25. Suppose the random telegraph signal of Prob. 6.11 is the input to an ideal bandpass filter with unit gain and narrow bandwidth W_B $(=2\pi B)$ $(\ll \omega_c)$ centered at $\omega_c = 2\alpha$. Find the dc component and the average power of the output.

From Eqs. (*6.53*) and (*6.103*) and Fig. 6-8(*b*), the resulting output power spectrum

$$S_{YY}(\omega) = |H(\omega)|^2 S_{XX}(\omega)$$

is shown in Fig. 6-15. Since $H(0) = 0$, from Eq. (*6.50*) we see that

$$\mu_Y = H(0)\mu_X = 0$$

Hence the dc component of the output is zero.

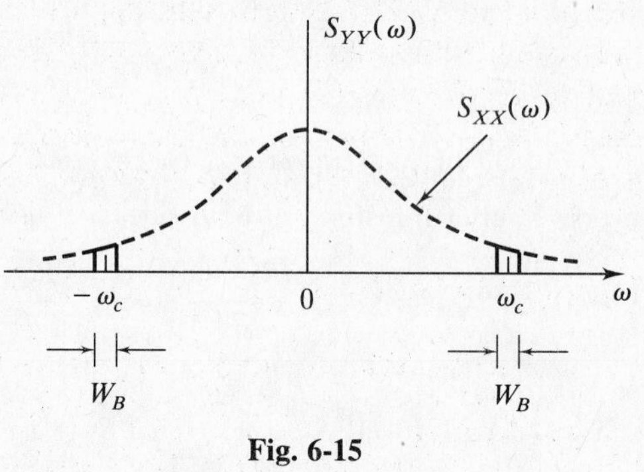

Fig. 6-15

From Eq. (*6.103*) (Prob. 6.11)

$$S_{XX}(\pm\omega_c) = A^2 \frac{4\alpha}{(2\alpha)^2 + (2\alpha)^2} = \frac{A^2}{2\alpha}$$

Since $W_B \ll \omega_c$,

$$S_{YY}(\omega) \approx \begin{cases} \dfrac{A^2}{2\alpha} & |\omega - \omega_c| < \dfrac{W_B}{2} \\ 0 & \text{otherwise} \end{cases}$$

The average output power is

$$E[Y^2(t)] = \frac{1}{2\pi}\int_{-\infty}^{\infty} S_{YY}(\omega)\,d\omega$$

$$\approx \frac{1}{2\pi}(2W_B)\left(\frac{A^2}{2\alpha}\right) = \frac{A^2 W_B}{2\pi\alpha} = \frac{A^2 B}{\alpha} \qquad (6.135)$$

6.26. Suppose that a WSS random process $X(t)$ with power spectrum $S_{XX}(\omega)$ is the input to the filter shown in Fig. 6-16. Find the power spectrum of the output process $Y(t)$.

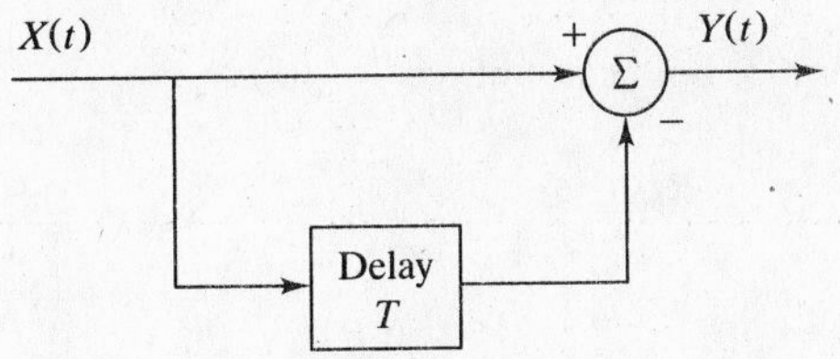

Fig. 6-16

From Fig. 6-16, $Y(t)$ can be expressed as

$$Y(t) = X(t) - X(t-T) \tag{6.136}$$

From Eq. (*1.77*) the impulse response of the filter is

$$h(t) = \delta(t) - \delta(t-T)$$

and by Eqs. (*1.38*) and (*1.39*) the frequency response of the filter is

$$H(\omega) = 1 - e^{-j\omega T}$$

Thus, by Eq. (*6.53*) the output power spectrum is

$$\begin{aligned} S_{YY}(\omega) = |H(\omega)|^2 S_{XX}(\omega) &= |1 - e^{-j\omega T}|^2 S_{XX}(\omega) \\ &= \left[(1-\cos\omega T)^2 + \sin^2\omega T\right] S_{XX}(\omega) \\ &= 2(1-\cos\omega T) S_{XX}(\omega) \end{aligned} \tag{6.137}$$

6.27. Suppose that $X(t)$ is the input to an LTI system with impulse response $h_1(t)$ and that $Y(t)$ is the input to another LTI system with impulse response $h_2(t)$. It is assumed that $X(t)$ and $Y(t)$ are jointly wide-sense stationary. Let $V(t)$ and $Z(t)$ denote the random processes at the respective system outputs (Fig. 6-17). Find the cross-correlation and cross spectral density of $V(t)$ and $Z(t)$ in terms of the cross-correlation and cross spectral density of $X(t)$ and $Y(t)$.

Using Eq. (*6.47*), we have

$$\begin{aligned} R_{VZ}(t_1, t_2) &= E[V(t_1)Z(t_2)] \\ &= E\left[\int_{-\infty}^{\infty} X(t_1-\alpha)h_1(\alpha)\,d\alpha \int_{-\infty}^{\infty} Y(t_2-\beta)h_2(\beta)\,d\beta\right] \\ &= \int_{-\infty}^{\infty}\int_{-\infty}^{\infty} h_1(\alpha)h_2(\beta)E[X(t_1-\alpha)Y(t_2-\beta)]\,d\alpha\,d\beta \\ &= \int_{-\infty}^{\infty}\int_{-\infty}^{\infty} h_1(\alpha)h_2(\beta)R_{XY}(t_1-\alpha, t_2-\beta)\,d\alpha\,d\beta \\ &= \int_{-\infty}^{\infty}\int_{-\infty}^{\infty} h_1(\alpha)h_2(\beta)R_{XY}(t_2-t_1+\alpha-\beta)\,d\alpha\,d\beta \end{aligned} \tag{6.138}$$

since $X(t)$ and $Y(t)$ are jointly WSS.

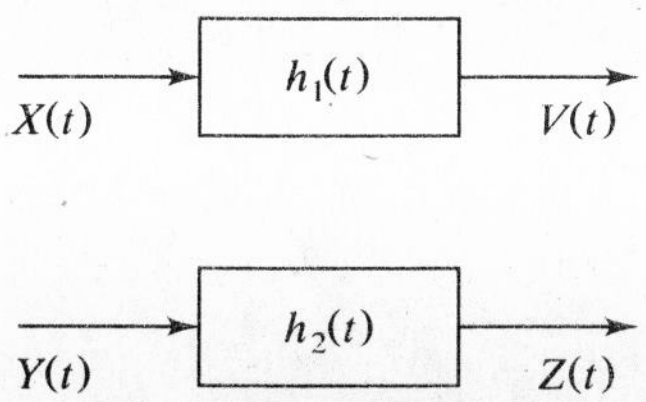

Fig. 6-17

Equation (*6.138*) indicates that $R_{VZ}(t_1, t_2)$ depends only on the time difference $\tau = t_2 - t_1$. Thus

$$R_{VZ}(\tau) = \int_{-\infty}^{\infty}\int_{-\infty}^{\infty} h_1(\alpha)h_2(\beta)R_{XY}(\tau + \alpha - \beta)\,d\alpha\,d\beta \tag{6.139}$$

Taking the Fourier transform of both sides of Eq. (*6.139*), we obtain

$$S_{VZ}(\omega) = \int_{-\infty}^{\infty} R_{VZ}(\tau)e^{-j\omega\tau}\,d\tau$$

$$= \int_{-\infty}^{\infty}\int_{-\infty}^{\infty}\int_{-\infty}^{\infty} h_1(\alpha)h_2(\beta)R_{XY}(\tau + \alpha - \beta)e^{-j\omega\tau}\,d\alpha\,d\beta\,d\tau$$

Let $\tau + \alpha - \beta = \lambda$, or equivalently $\tau = \lambda - \alpha + \beta$. Then

$$S_{VZ}(\omega) = \int_{-\infty}^{\infty}\int_{-\infty}^{\infty}\int_{-\infty}^{\infty} h_1(\alpha)h_2(\beta)R_{XY}(\lambda)e^{-j\omega(\lambda-\alpha+\beta)}\,d\alpha\,d\beta\,d\lambda$$

$$= \int_{-\infty}^{\infty} h_1(\alpha)e^{j\omega\alpha}\,d\alpha \int_{-\infty}^{\infty} h_2(\beta)e^{-j\omega\beta}\,d\beta \int_{-\infty}^{\infty} R_{XY}(\lambda)e^{-j\omega\lambda}\,d\lambda$$

$$= H_1(-\omega)H_2(\omega)S_{XY}(\omega)$$

$$= H_1^*(\omega)H_2(\omega)S_{XY}(\omega) \tag{6.140}$$

where $H_1(\omega)$ and $H_2(\omega)$ are the frequency responses of the respective systems in Fig. 6-17.

SPECIAL CLASSES OF RANDOM PROCESSES

6.28. Show that if a gaussian random process is WSS, then it is SSS.

If the gaussian process $X(t)$ is WSS, then

$$\mu_i = E[X(t_i)] = \mu \ (= \text{constant}) \qquad \text{for all } t_i$$

and

$$R_{XX}(t_i, t_j) = R_{XX}(t_j - t_i)$$

Therefore, in the expression for the joint probability density of Eq. (*6.58*) and Eqs. (*6.59*), (*6.60*), and (*6.61*),

$$\mu_1 = \mu_2 = \cdots = \mu_n = \mu \to E[X(t_i)] = E[X(t_i + c)]$$

$$C_{ij} = C_{XX}(t_i, t_j) = R_{XX}(t_i, t_j) - \mu_i\mu_j$$

$$= R_{XX}(t_j - t_i) - \mu^2 = C_{XX}(t_i + c, t_j + c)$$

for any c. It then follows that

$$f_{\mathbf{X}(t_i)}(\mathbf{x}) = f_{\mathbf{X}(t_i + c)}(\mathbf{x})$$

for any c. Therefore, $X(t)$ is SSS by Eq. (*6.11*).

6.29. Let $\mathbf{X}$ be an n-dimensional gaussian random vector [Eq. (*6.56*)] with independent components. Show that the multivariate gaussian joint density function is given by

$$f_{\mathbf{X}}(\mathbf{x}) = \frac{1}{(2\pi)^{n/2}\prod_{i=1}^{n}\sigma_i} \exp\left[-\frac{1}{2}\sum_{i=1}^{n}\left(\frac{x_i - \mu_i}{\sigma_i}\right)^2\right] \tag{6.141}$$

where $\mu_i = E[X_i]$ and $\sigma_i^2 = \text{var}(X_i)$.

The multivariate gaussian density function is given by Eq. (*6.58*). Since $X_i = X(t_i)$ are independent, we have

$$C_{ij} = \begin{cases} \sigma_i^2 & i = j \\ 0 & i \neq j \end{cases} \tag{6.142}$$

Thus, from Eq. (*6.60*) the covariance matrix **C** becomes

$$\mathbf{C} = \begin{bmatrix} \sigma_1^2 & 0 & \cdots & 0 \\ 0 & \sigma_2^2 & \cdots & 0 \\ \cdots & \cdots & \cdots & \cdots \\ 0 & 0 & \cdots & \sigma_n^2 \end{bmatrix} \tag{6.143}$$

It therefore follows that

$$|\det \mathbf{C}|^{1/2} = \sigma_1 \sigma_2 \cdots \sigma_n = \prod_{i=1}^{n} \sigma_i \tag{6.144}$$

and

$$\mathbf{C}^{-1} = \begin{bmatrix} \dfrac{1}{\sigma_1^2} & 0 & \cdots & 0 \\ 0 & \dfrac{1}{\sigma_2^2} & \cdots & 0 \\ \cdots & \cdots & \cdots & \cdots \\ 0 & 0 & \cdots & \dfrac{1}{\sigma_n^2} \end{bmatrix} \tag{6.145}$$

Then we can write

$$(\mathbf{x} - \boldsymbol{\mu})^T \mathbf{C}^{-1} (\mathbf{x} - \boldsymbol{\mu}) = \sum_{i=1}^{n} \left(\frac{x_i - \mu_i}{\sigma_i} \right)^2 \tag{6.146}$$

Substituting Eqs. (*6.144*) and (*6.146*) into Eq. (*6.58*), we obtain Eq. (*6.141*).

6.30. Let $X_1 = X(t_1)$ and $X_2 = X(t_2)$ be jointly gaussian random variables, each with a zero mean and a variance σ^2. Show that the joint bivariate gaussian density function is given by

$$f_{X_1X_2}(x_1, x_2) = \frac{1}{2\pi\sigma^2\sqrt{1-\rho^2}} \exp\left[-\frac{1}{2} \frac{x_1^2 - 2\rho x_1 x_2 + x_2^2}{\sigma^2(1-\rho^2)} \right] \tag{6.147}$$

where ρ is the correlation coefficient of X_1 and X_2 given by $\rho = C_{12}/\sigma^2$ [Eq. (*5.80*)].

Substituting $C_{11} = C_{22} = \sigma^2$ and $C_{12} = C_{21} = \rho\sigma^2$ into Eq. (*6.60*), we have

$$\mathbf{C} = \begin{bmatrix} \sigma^2 & \rho\sigma^2 \\ \rho\sigma^2 & \sigma^2 \end{bmatrix} = \sigma^2 \begin{bmatrix} 1 & \rho \\ \rho & 1 \end{bmatrix}$$

$$|\det \mathbf{C}|^{1/2} = \sigma^2 \sqrt{(1-\rho^2)}$$

$$\mathbf{C}^{-1} = \frac{1}{\sigma^2(1-\rho^2)} \begin{bmatrix} 1 & -\rho \\ -\rho & 1 \end{bmatrix}$$

Since $\boldsymbol{\mu} = \mathbf{0}$, $\mathbf{x} - \boldsymbol{\mu} = \mathbf{x}$, and

$$\mathbf{x}^T\mathbf{C}^{-1}\mathbf{x} = \frac{1}{\sigma^2(1-\rho^2)}[x_1\ x_2]\begin{bmatrix} 1 & -\rho \\ -\rho & 1 \end{bmatrix}\begin{bmatrix} x_1 \\ x_2 \end{bmatrix}$$

$$= \frac{1}{\sigma^2(1-\rho^2)}(x_1^2 - 2\rho x_1 x_2 + x_2^2)$$

Substituting these results in Eq. (*6.58*), we obtain Eq. (*6.147*).

Note that Eq. (*6.147*) can be obtained from Eq. (*5.122*) (Prob. 5.22) by setting $X = X_1$, $Y = X_2$, $\mu_X = \mu_Y = 0$, and $\sigma_X = \sigma_Y = \sigma$.

6.31. The relation between the input $X(t)$ and output $Y(t)$ of a diode is expressed as

$$Y(t) = X^2(t) \tag{6.148}$$

Let $X(t)$ be a zero-mean stationary gaussian random process with autocorrelation

$$R_{XX}(\tau) = e^{-\alpha|\tau|} \qquad \alpha > 0$$

Find the output mean $\mu_Y(t)$, the output autocorrelation $R_{YY}(\tau)$, and the output power spectral density $S_{YY}(\omega)$.

$$\mu_Y(t) = E[Y(t)] = E[X^2(t)] = R_{XX}(0) \tag{6.149}$$

$$R_{YY}(t_1, t_2) = E[Y(t_1)Y(t_2)] = E[X^2(t_1)X^2(t_2)]$$

Since $X(t_1)$ and $X(t_2)$ are zero-mean jointly gaussian random variables, by Eq. (*5.175*) (Prob. 5.54)

$$E[X^2(t_1)X^2(t_2)] = E[X^2(t_1)]E[X^2(t_2)] + 2\{E[X(t_1)X(t_2)]\}^2 \tag{6.150}$$

Since $X(t)$ is stationary

$$E[X^2(t_1)] = E[X^2(t_2)] = R_{XX}(0)$$

and

$$E[X(t_1)X(t_2)] = R_{XX}(t_2 - t_1) = R_{XX}(\tau)$$

Hence

$$R_{YY}(t_1, t_2) = R_{YY}(\tau) = [R_{XX}(0)]^2 + 2[R_{XX}(\tau)]^2 \tag{6.151}$$

and using Eqs. (*1.40*) and (*1.54*), we have

$$S_{YY}(\omega) = \mathscr{F}[R_{YY}(\tau)] = 2\pi[R_{XX}(0)]^2\delta(\omega) + \frac{1}{\pi}S_{XX}(\omega) * S_{XX}(\omega) \tag{6.152}$$

Now, for the given input autocorrelation, by Eqs. (*6.149*) and (*6.151*),

$$\mu_Y(t) = R_{XX}(0) = 1$$

and

$$R_{YY}(\tau) = 1 + 2e^{-2\alpha|\tau|}$$

By using Eqs. (*1.40*) and (*1.46*), the output power spectral density is

$$S_{YY}(\omega) = \mathscr{F}[R_{YY}(\tau)] = 2\pi\delta(\omega) + \frac{8\alpha}{\omega^2 + 4\alpha^2}$$

6.32. The input $X(t)$ to the *RC* filter shown in Fig. 6-18 is a white noise process.

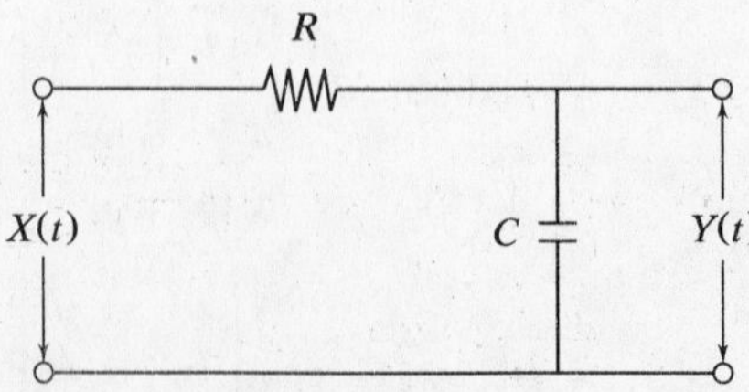

Fig. 6-18 *RC* filter

(*a*) Determine the power spectrum of the output process $Y(t)$.
(*b*) Determine the autocorrelation and the mean-square value of $Y(t)$.

From Prob. 1.46 the frequency response of the RC filter is

$$H(\omega)=\frac{1}{1+j\omega RC}$$

(*a*) From Eqs. (*6.62*) and (*6.53*)

$$S_{XX}(\omega)=\frac{\eta}{2}$$

$$S_{YY}(\omega)=|H(\omega)|^2 S_{XX}(\omega)=\frac{1}{1+(\omega RC)^2}\frac{\eta}{2} \tag{6.153}$$

(*b*) Rewriting Eq. (*6.153*) as

$$S_{YY}(\omega)=\frac{\eta}{2}\frac{1}{2RC}\frac{2[1/(RC)]}{\omega^2+[1/(RC)]^2}$$

and using the Fourier transform pair Eq. (*1.46*), we obtain

$$R_{YY}(\tau)=\frac{\eta}{2}\frac{1}{2RC}e^{-|\tau|/(RC)} \tag{6.154}$$

Finally, from Eq. (*6.154*)

$$E\left[Y^2(t)\right]=R_{YY}(0)=\frac{\eta}{4RC} \tag{6.155}$$

6.33. The input $X(t)$ to an ideal bandpass filter having the frequency response characteristic shown in Fig. 6-19 is a white noise process. Determine the total noise power at the output of the filter.

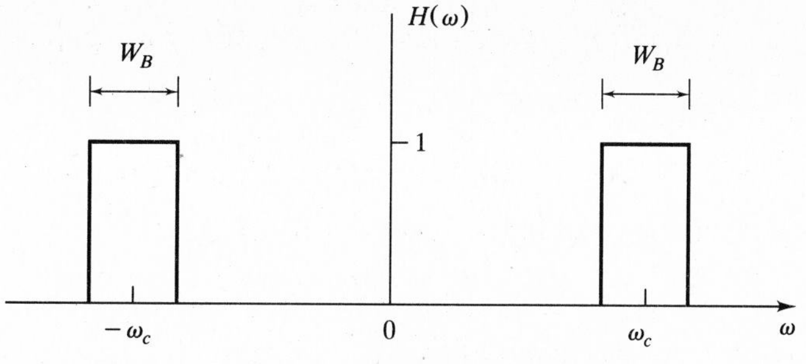

Fig. 6-19

$$S_{XX}(\omega)=\frac{\eta}{2}$$

$$S_{YY}(\omega)=|H(\omega)|^2 S_{XX}(\omega)=\frac{\eta}{2}|H(\omega)|^2$$

The total noise power at the output of the filter is

$$\begin{aligned}E\left[Y^2(t)\right]&=\frac{1}{2\pi}\int_{-\infty}^{\infty}S_{YY}(\omega)\,d\omega=\frac{1}{2\pi}\frac{\eta}{2}\int_{-\infty}^{\infty}|H(\omega)|^2\,d\omega\\&=\frac{\eta}{2}\frac{1}{2\pi}(2W_B)=\eta B\end{aligned} \tag{6.156}$$

where $B=W_B/(2\pi)$ (in hertz).

6.34. The equivalent noise bandwidth of a system is defined as

$$B_{eq} = \frac{1}{2\pi} \frac{\int_0^\infty |H(\omega)|^2 \, d\omega}{|H(\omega)|^2_{max}} \quad \text{Hz} \tag{6.157}$$

where $|H(\omega)|^2_{max} = \max |H(\omega)|^2$.

(*a*) Determine the equivalent noise bandwidth of the ideal bandpass filter shown in Fig. 6-19.

(*b*) Determine the equivalent noise bandwidth of the low-pass *RC* filter shown in Fig. 6-18, and compare the result with the 3-dB bandwidth obtained in Prob. 1.49.

(*a*) For the ideal bandpass filter shown in Fig. 6-19, we have $\max |H(\omega)|^2 = 1$ and

$$B_{eq} = \frac{1}{2\pi} \int_0^\infty |H(\omega)|^2 \, d\omega = \frac{W_B}{2\pi} = B \quad \text{Hz} \tag{6.158}$$

(*b*) For the low-pass *RC* filter shown in Fig. 6-18 we have

$$|H(\omega)|^2 = \frac{1}{1 + (\omega RC)^2}$$

and

$$\max |H(\omega)|^2 = |H(0)|^2 = 1$$

Thus

$$B_{eq} = \frac{1}{2\pi} \int_0^\infty \frac{d\omega}{1 + (\omega RC)^2} = \frac{1}{2\pi} \frac{1}{2} \int_{-\infty}^\infty \frac{d\omega}{1 + (\omega RC)^2}$$

$$= \frac{1}{4RC} \quad \text{Hz} \tag{6.159}$$

From Prob. 1.49 the 3-dB bandwidth of the low-pass *RC* filter is

$$B_{3\,dB} = \frac{1}{2\pi} W_{3\,dB} = \frac{1}{2\pi RC} \quad \text{Hz}$$

Thus

$$B_{eq} = \frac{\pi}{2} B_{3\,dB} \approx 1.57 B_{3\,dB}$$

6.35. A narrowband white noise $X(t)$ has the power spectrum shown in Fig. 6-20(*a*). Represent $X(t)$ in terms of quadrature components. Derive the power spectra of $X_c(t)$ and $X_s(t)$, and show that

$$E[X_c^2(t)] = E[X_s^2(t)] = E[X^2(t)] \tag{6.160}$$

From Eq. (*6.67*)

$$X(t) = X_c(t) \cos \omega_c t - X_s(t) \sin \omega_c t$$

From Eq. (*6.70*) and Fig. 6-20(*a*), we have

$$S_{X_cX_c}(\omega) = S_{X_sX_s}(\omega) = \begin{cases} \eta & |\omega| \le W \ (= 2\pi B) \\ 0 & \text{otherwise} \end{cases}$$

which is plotted in Fig. 6-20(*b*). From Fig. 6-20(*a*)

$$E[X^2(t)] = \frac{1}{2\pi} \int_{-\infty}^\infty S_{XX}(\omega) \, d\omega$$

$$= \frac{1}{2\pi} (2) \int_{\omega_c - W}^{\omega_c + W} \frac{\eta}{2} \, d\omega = \frac{1}{2\pi} \eta 2W = 2\eta B$$

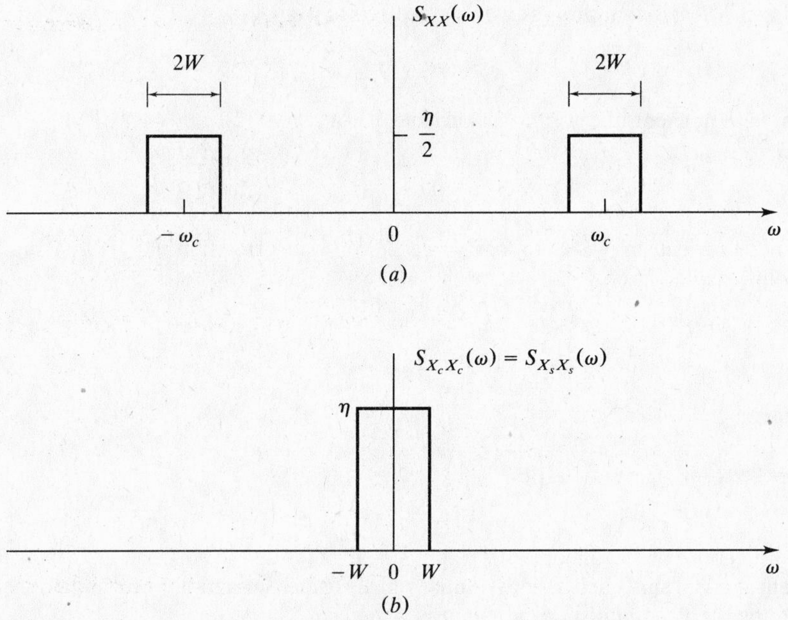

Fig. 6-20

From Fig. 6-20(*b*)

$$E[X_c^2(t)] = E[X_s^2(t)] = \frac{1}{2\pi}(2)\int_0^W \eta\,d\omega = \frac{1}{2\pi}\eta(2W) = 2\eta B$$

Hence

$$E[X^2(t)] = E[X_c^2(t)] = E[X_s^2(t)] = \frac{1}{2\pi}\eta(2W) = 2\eta B \tag{6.161}$$

Supplementary Problems

6.36. Consider a random process $X(t)$ defined by

$$X(t) = \cos \Omega t$$

where Ω is a random variable uniformly distributed over $[0, \omega_0]$. Determine whether $X(t)$ is stationary.

Ans. Nonstationary

Hint: Examine specific sample functions of $X(t)$ for different frequencies, say, $\Omega = \pi/2$, π, and 2π.

6.37. Consider the random process $X(t)$ defined by

$$X(t) = A \cos \omega t$$

where ω is a constant and A is a random variable uniformly distributed over $[0, 1]$. Find the autocorrelation and autocovariance of $X(t)$.

Ans. $R_{XX}(t_1, t_2) = \frac{1}{3}\cos t_1 \cos t_2$

$C_{XX}(t_1, t_2) = \frac{1}{12}\cos t_1 \cos t_2$

6.38. Let $X(t)$ be a WSS random process with autocorrelation

$$R_{XX}(\tau) = Ae^{-\alpha|\tau|}$$

Find the second moment of the random variable $Y = X(5) - X(2)$.

Ans. $2A(1 - e^{-3\alpha})$

6.39. Let $X(t)$ be a zero-mean WSS random process with autocorrelation $R_{XX}(\tau)$. A random variable Y is formed by integrating $X(t)$:

$$Y = \frac{1}{2T}\int_{-T}^{T} X(t)\,dt$$

Find the mean and variance of Y.

Ans. $\mu_Y = 0; \sigma_Y^2 = \dfrac{1}{T}\displaystyle\int_0^{2T} R_{XX}(\tau)\left(1 - \frac{\tau}{2T}\right)d\tau$

6.40. A sample function of a random telegraph signal $X(t)$ is shown in Fig. 6-21. This signal makes independent random shifts between two equally likely values, A and 0. The number of shifts per unit time is governed by the Poisson distribution with parameter α.

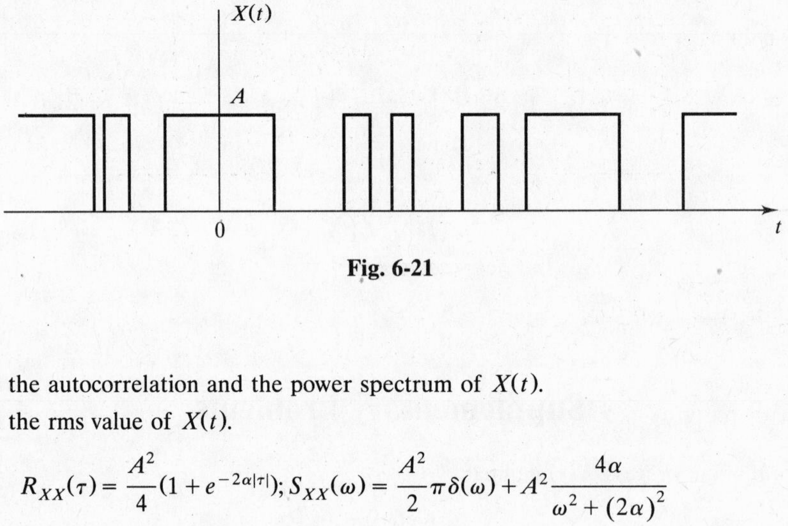

Fig. 6-21

(*a*) Find the autocorrelation and the power spectrum of $X(t)$.

(*b*) Find the rms value of $X(t)$.

Ans. (*a*) $R_{XX}(\tau) = \dfrac{A^2}{4}(1 + e^{-2\alpha|\tau|}); S_{XX}(\omega) = \dfrac{A^2}{2}\pi\delta(\omega) + A^2\dfrac{4\alpha}{\omega^2 + (2\alpha)^2}$

(*b*) $\dfrac{A}{2}$

6.41. Suppose that $X(t)$ is a gaussian process with

$$\mu_X = 2 \qquad R_{XX}(\tau) = 5e^{-0.2|\tau|}$$

Find the probability that $X(4) \le 1$.

Ans. 0.159

6.42. The output of a filter is given by

$$Y(t) = X(t+T) - X(t-T)$$

where $X(t)$ is a WSS process with power spectrum $S_{XX}(\omega)$ and T is a constant. Find the power spectrum of $Y(t)$.

Ans. $S_{YY}(\omega) = 4\sin^2 \omega T S_{XX}(\omega)$

6.43. Let $\hat{X}(t)$ be the Hilbert transform of a WSS process $X(t)$. Show that

$$R_{X\hat{X}}(0) = E\left[X(t)\hat{X}(t)\right] = 0$$

Hint: Use relation (*b*) of Prob. 6.20 and definition (*1.139*) of Prob. 1.47.

6.44. When a metallic resistor R is at temperature T, random electron motion produces a noise voltage $V(t)$ at the open-circuited terminals. This voltage $V(t)$ is known as the *thermal noise*. Its power spectrum $S_{VV}(\omega)$ is practically constant for $f \le 10^{12}$ Hz and is given by

$$S_{VV}(\omega) = 2kTR$$

where k = Boltzmann constant = $1.37(10^{-23})$, joules per kelvin (J/K)
T = absolute temperature, kelvins (K)
R = resistance, ohms (Ω)

Calculate the thermal noise voltage (rms value) across the simple RC circuit shown in Fig. 6-22 with $R = 1$ kilohm (kΩ), $C = 1$ microfarad (μF), at $T = 27°$ C.

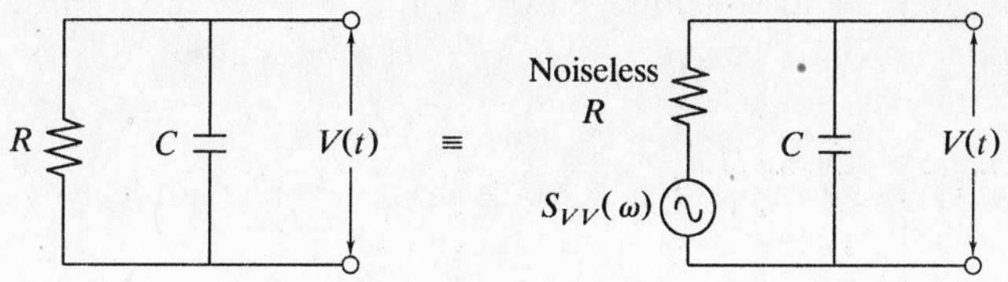

Fig. 6-22

Ans. $V_{\text{rms}} = \sqrt{\dfrac{kT}{C}} \approx 0.2\ \mu\text{V}$

Chapter 7

Performance of Communication Systems in the Presence of Noise

7.1 INTRODUCTION

The presence of noise degrades the performance of analog and digital communication systems. The extent to which noise affects the performance of communication systems is measured by the output signal-to-noise power ratio or the probability of error. The signal-to-noise ratio is used to measure the performance of analog communication systems, whereas the probability of error is used as a performance measure of digital communication systems.

In the following analysis, we are concerned mainly with the additive noise that accompanies the signal at the input to the receiver.

7.2 ADDITIVE NOISE AND SIGNAL-TO-NOISE RATIO

A schematic of a communication system is shown in Fig. 7-1. It is assumed that the input of the transmitter is modeled by the random process $X(t)$, the channel introduces no distortion other than additive random noise, and the receiver is a linear system. At the receiver input, we have a signal mixed with noise. The signal and the noise power at the receiver input are S_i and N_i, respectively. Since the receiver is linear, the receiver output $Y_o(t)$ can be written as

$$Y_o(t) = X_o(t) + n_o(t) \tag{7.1}$$

where $X_o(t)$ and $n_o(t)$ are the signal and noise components at the receiver output, respectively.

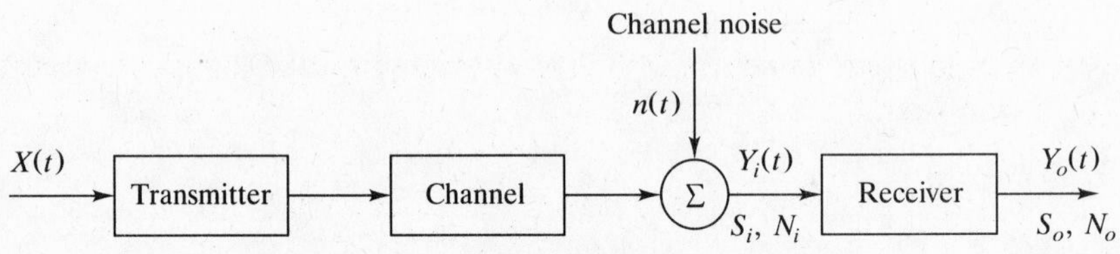

Fig. 7-1 A communication system model

We make two further assumptions about additive noise:

1. The noise is zero-mean white gaussian with power spectral density $S_{nn}(\omega) = \eta/2$.
2. The noise is uncorrelated with $X(t)$.

Under the above specified assumptions, we have

$$E\left[Y_o^2(t)\right] = E\left[X_o^2(t)\right] + E\left[n_o^2(t)\right] = S_o + N_o \tag{7.2}$$

where $S_o = E[X_o^2(t)]$ and $N_o = E[n_o^2(t)]$ and they are the average signal and noise power at the receiver output, respectively. The *output signal-to-noise ratio* $(S/N)_o$ is defined as

$$\left(\frac{S}{N}\right)_o = \frac{S_o}{N_o} = \frac{E\left[X_o^2(t)\right]}{E\left[n_o^2(t)\right]} \tag{7.3}$$

In analog communication systems, the quality of the received signal is determined by this parameter. Note that this ratio is meaningful only when Eq. (7.1) holds.

7.3 NOISE IN BASEBAND COMMUNICATION SYSTEMS

In baseband communication systems, the signal is transmitted directly without any modulation. The results obtained for baseband systems serve as a basis for comparing with other systems. Figure 7-2 shows a simple analog baseband system. For a baseband system, the receiver is a low-pass filter which passes the message while reducing the noise at the output. Obviously, the filter should reject all noise frequency components that fall outside the message band. We assume that the low-pass filter is ideal with bandwidth $W(=2\pi B)$.

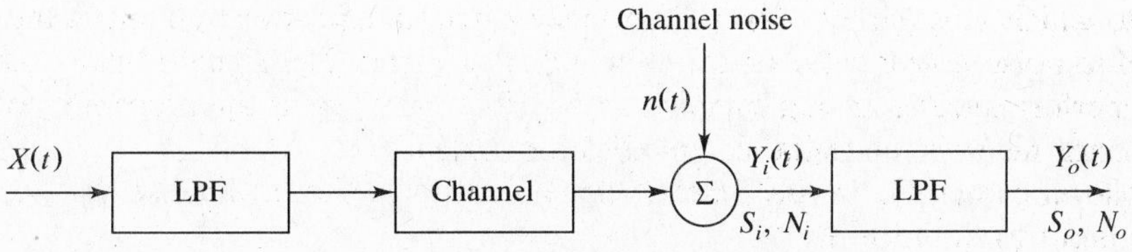

Fig. 7-2 A baseband system

It is assumed that the message signal $X(t)$ is a zero-mean ergodic random process band-limited to W with power spectral density $S_{XX}(\omega)$. The channel is assumed to be distortionless over the message band so that

$$X_o(t) = X(t - t_d) \tag{7.4}$$

where t_d is the time delay of the system. The average output signal power S_o is

$$\begin{aligned} S_o &= E[X_o^2(t)] = E[X^2(t-t_d)] \\ &= \frac{1}{2\pi}\int_{-W}^{W} S_{XX}(\omega)\,d\omega = S_X = S_i \end{aligned} \tag{7.5}$$

where S_X is the average signal power and S_i is the signal power at the input of the receiver. The average output noise power N_o is

$$N_o = E[n_o^2(t)] = \frac{1}{2\pi}\int_{-W}^{W} S_{nn}(\omega)\,d\omega$$

For the case of additive white noise, $S_{nn}(\omega) = \eta/2$, and

$$N_o = \frac{1}{2\pi}\int_{-W}^{W} \frac{\eta}{2}\,d\omega = \eta\frac{W}{2\pi} = \eta B \tag{7.6}$$

The white noise assumption simplifies the calculations and provides essential aspects of analysis. The output signal-to-noise ratio is

$$\left(\frac{S}{N}\right)_o = \frac{S_o}{N_o} = \frac{S_i}{\eta B} \tag{7.7}$$

Let

$$\frac{S_i}{\eta B} = \gamma \tag{7.8}$$

Then

$$\left(\frac{S}{N}\right)_o = \gamma \tag{7.9}$$

The parameter γ is directly proportional to S_i. Hence, comparing various systems for the output SNR for a given S_i is the same as comparing these systems for the output SNR for a given γ.

7.4 NOISE IN AMPLITUDE MODULATION SYSTEMS

A block diagram of a continuous-wave communication system is shown in Fig. 7-3. The receiver front end (RF/IF stages) is modeled as an ideal bandpass filter with a bandwidth $2W$ centered at ω_c. This bandpass filter also known as a *predetection filter*, limits the amount of noise outside the band that reaches the detector ("out-of-band noise"). Predetection bandpass filtering produces

$$Y_i(t) = X_c(t) + n_i(t) \tag{7.10}$$

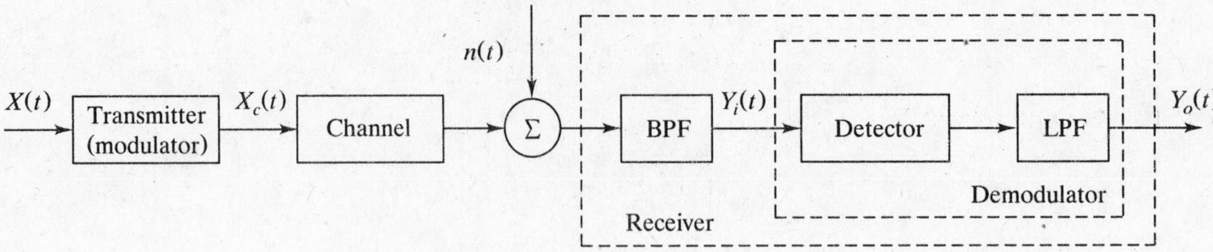

Fig. 7-3 A CW communication system

where $n_i(t)$ is the narrowband noise, which can be expressed as [Eq. (*6.67*)]

$$n_i(t) = n_c(t)\cos\omega_c t - n_s(t)\sin\omega_c t \tag{7.11}$$

If the power spectral density of $n(t)$ is $\eta/2$, then (Prob. 6.35)

$$E\left[n_c^2(t)\right] = E\left[n_s^2(t)\right] = E\left[n_i^2(t)\right] = 2\eta B \tag{7.12}$$

A. Synchronous Detection:

1. *DSB Systems:*

In a DSB system, the transmitted signal $X_c(t)$ has the form[1]

$$X_c(t) = A_c X(t)\cos\omega_c t \tag{7.13}$$

The demodulator portion of the DSB system is shown in Fig. 7-4. The input of the receiver is

$$\begin{aligned} Y_i(t) &= A_c X(t)\cos\omega_c t + n_i(t) \\ &= \left[A_c X(t) + n_c(t)\right]\cos\omega_c t - n_s(t)\sin\omega_c t \end{aligned} \tag{7.14}$$

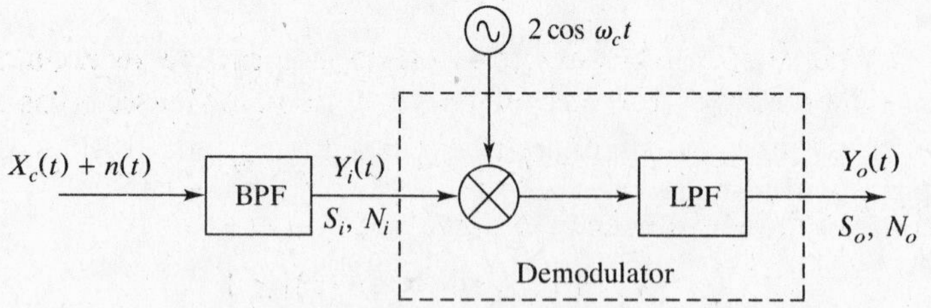

Fig. 7-4 DSB modulation

Multiplying $Y_i(t)$ by $2\cos\omega_c t$ and using a low-pass filter, we obtain

$$Y_o(t) = A_c X(t) + n_c(t) = X_o(t) + n_o(t) \tag{7.15}$$

---

[1]We need a random phase Θ in the carrier to make $X_c(t)$ stationary (see Prob. 6.10). The random phase [which is independent of $X(t)$] does not affect the final results, hence is neglected here.

where $$X_o(t) = A_c X(t) \quad \text{and} \quad n_o(t) = n_c(t)$$

We see that the output signal and noise are additive and the quadrature noise component $n_s(t)$ has been rejected by the demodulator. Now

$$S_o = E[X_o^2(t)] = E[A_c^2 X^2(t)] = A_c^2 E[X^2(t)] = A_c^2 S_X \tag{7.16a}$$

$$N_o = E[n_o^2(t)] = E[n_c^2(t)] = E[n_i^2(t)] = 2\eta B \tag{7.16b}$$

and the output SNR is

$$\left(\frac{S}{N}\right)_o = \frac{S_o}{N_o} = \frac{A_c^2 S_X}{2\eta B} \tag{7.17}$$

The input signal power S_i is given by

$$S_i = E[X_c^2(t)] = \frac{1}{2} A_c^2 S_X \tag{7.18}$$

Thus, from Eqs. (*7.17*) and (*7.8*) we obtain

$$\left(\frac{S}{N}\right)_o = \frac{S_i}{\eta B} = \gamma \tag{7.19}$$

which indicates that insofar as noise is concerned, DSB with ideal synchronous detection has the same performance as baseband system.

The SNR at the input of the detector is

$$\left(\frac{S}{N}\right)_i = \frac{S_i}{N_i} = \frac{S_i}{2\eta B} \tag{7.20}$$

and $$\frac{(S/N)_o}{(S/N)_i} = \alpha_d = 2 \tag{7.21}$$

The ratio α_d is known as the *detector gain* and is often used as a *figure of merit* for the demodulation.

2. ***SSB Systems:***

Similar calculations for an SSB system using synchronous detection yield the same noise performance as for a baseband or DSB system (Prob. 7.4).

3. ***AM Systems:***

In ordinary AM (or simply AM) systems, AM signals can be demodulated by synchronous detector or by envelope detector. The modulated signal in an AM system has the form

$$X_c(t) = A_c[1 + \mu X(t)] \cos \omega_c t \tag{7.22}$$

where μ is the modulation index of the AM signal and

$$\mu \le 1 \quad \text{and} \quad |X(t)| \le 1$$

If the synchronous detector includes an ideal dc suppressor, the receiver output $Y_o(t)$ will be (Fig. 7-4)

$$Y_o(t) = A_c \mu X(t) + n_c(t) = X_o(t) + n_o(t) \tag{7.23}$$

where

$$X_o(t) = A_c \mu X(t) \quad \text{and} \quad n_o(t) = n_c(t)$$

so $$\left(\frac{S}{N}\right)_o = \frac{S_o}{N_o} = \frac{A_c^2 \mu^2 S_X}{2\eta B} \tag{7.24}$$

The input signal power S_i is

$$S_i = \frac{1}{2} E\left[A_c^2[1+\mu X(t)]^2\right]$$

Since $X(t)$ is assumed to have a zero mean,

$$S_i = \frac{1}{2} A_c^2 (1+\mu^2 S_X) \tag{7.25}$$

Thus

$$S_o = A_c^2 \mu^2 S_X = \frac{2\mu^2 S_X}{1+\mu^2 S_X} S_i$$

and

$$\left(\frac{S}{N}\right)_o = \frac{S_o}{N_o} = \frac{\mu^2 S_X}{1+\mu^2 S_X}\left(\frac{S_i}{\eta B}\right) = \frac{\mu^2 S_X}{1+\mu^2 S_X}\gamma \tag{7.26}$$

Because $\mu^2 S_x \le 1$, we have

$$\left(\frac{S}{N}\right)_o \le \frac{\gamma}{2} \tag{7.27}$$

which indicates that the output SNR in AM is at least 3 dB worse that that in DSB and SSB systems.

B. Envelope Detection and Threshold Effect:

An ordinary AM signal is usually demodulated by envelope detection. With reference to Fig. 7-3, the input to the detector is

$$\begin{aligned} Y_i(t) &= X_c(t) + n_i(t) \\ &= \{A_c[1+\mu X(t)] + n_c(t)\}\cos\omega_c t - n_s(t)\sin\omega_c t \end{aligned} \tag{7.28}$$

We can analyze the effect of the noise by considering a phasor representation of $Y_i(t)$

$$Y_i(t) = \text{Re}\left[Y(t)e^{j\omega_c t}\right] \tag{7.29}$$

where

$$Y(t) = A_c[1+\mu X(t)] + n_c(t) + jn_s(t) \tag{7.30}$$

The phase diagram is shown in Fig. 7-5. From Fig. 7-5, we see that $Y_i(t)$ can be written as

$$Y_i(t) = V(t)\cos\left[\omega_c(t) + \phi(t)\right] \tag{7.31}$$

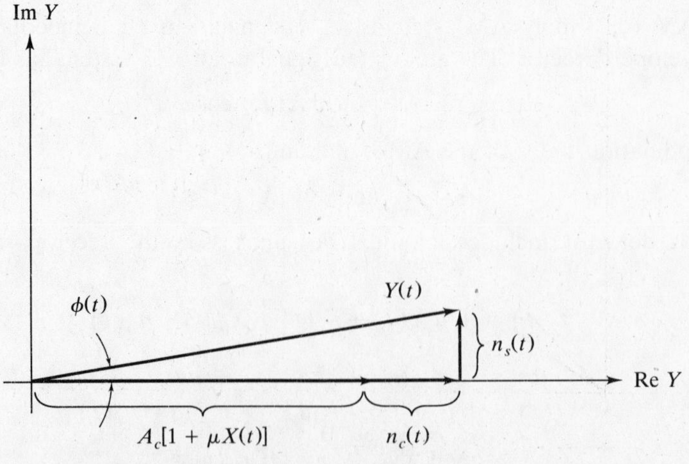

Fig. 7-5 Phasor diagram for AM when $(S/N)_i \gg 1$

where

$$V(t) = \sqrt{\{A_c[1+\mu X(t)] + n_c(t)\}^2 + n_s^2(t)} \tag{7.32a}$$

$$\phi(t) = \tan^{-1}\frac{n_s(t)}{A_c[1+\mu X(t)] + n_c(t)} \tag{7.32b}$$

1. Large-SNR (Signal Dominance) Case:

When $(S/N)_i \gg 1$, $A_c[1+\mu X(t)] \gg n_i(t)$, and hence $A_c[1+\mu X(t)] \gg n_c(t)$ and $n_s(t)$ for almost all t. Under this condition, the envelope $V(t)$ can be approximated by

$$V(t) \approx A_c[1+\mu X(t)] + n_c(t) \tag{7.33}$$

An ideal envelope detector reproduces the envelope $V(t)$ minus its dc component, so

$$Y_o(t) = A_c \mu X(t) + n_c(t) \tag{7.34}$$

which is identical to that of a synchronous detector [Eq. (*7.23*)]. The output SNR is then as given in Eq. (*7.26*), that is,

$$\left(\frac{S}{N}\right)_o = \frac{\mu^2 S_X}{1+\mu^2 S_X}\gamma \tag{7.35}$$

Therefore for AM, when $(S/N)_i \gg 1$, the performance of the envelope detector is identical to that of the synchronous detector.

2. Small-SNR (Noise Dominance) Case:

When $(S/N)_i \ll 1$, the envelope of the resultant signal is primarily dominated by the envelope of the noise signal (Fig. 7-6). From the diagram of Fig. 7-6, the envelope of the resultant signal is approximated by

$$V(t) \approx V_n(t) + A_c[1+\mu X(t)]\cos\phi_n(t) \tag{7.36}$$

where $V_n(t)$ and $\phi_n(t)$ are the envelope and the phase of the noise $n_i(t)$. Equation (*7.36*) indicates that the output contains no term proportional to $X(t)$ and that noise is multiplicative. The signal $X(t)$ is multiplied by noise in the form of $\cos\phi_n(t)$ which is random. Thus the message signal $X(t)$

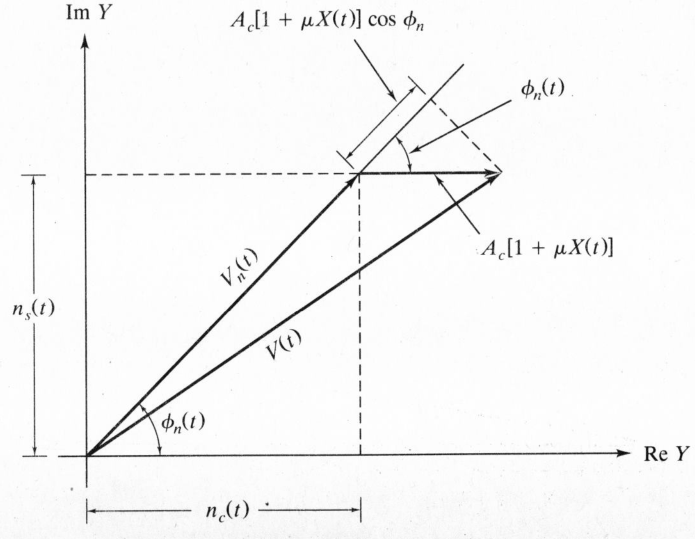

Fig. 7-6 Phasor diagram for AM when $(S/N)_i \ll 1$

is badly mutilated, and its information has been lost. Under these circumstances, it is meaningless to talk about output SNR.

The loss or mutilation of the message at low predetection SNR is called the *threshold effect*. The name comes about because there is some value of $(S/N)_i$ above which signal distortion due to noise is negligible and below which system performance deteriorates rapidly. The threshold occurs when $(S/N)_i$ is about 10 dB or less (Prob. 7.6).

7.5 NOISE IN ANGLE MODULATION SYSTEMS

With reference to Fig. 7-3, in angle modulation systems the transmitted signal $X_c(t)$ has the form

$$X_c(t) = A_c \cos[\omega_c t + \phi(t)] \tag{7.37}$$

where

$$\phi(t) = \begin{cases} k_p X(t) & \text{for PM} \\ k_f \int_{-\infty}^{t} X(\tau)\, d\tau & \text{for FM} \end{cases} \tag{7.38}$$

Figure 7-7 shows a model for the angle demodulation system. The predetection filter bandwidth B_T is approximately $2(D+1)B$, where D is the deviation ratio and B is the bandwidth of the message signal [Eq. (3.27)]. The detector input is

$$\begin{aligned} Y_i(t) &= X_c(t) + n_i(t) \\ &= A_c \cos[\omega_c t + \phi(t)] + n_i(t) \end{aligned} \tag{7.39}$$

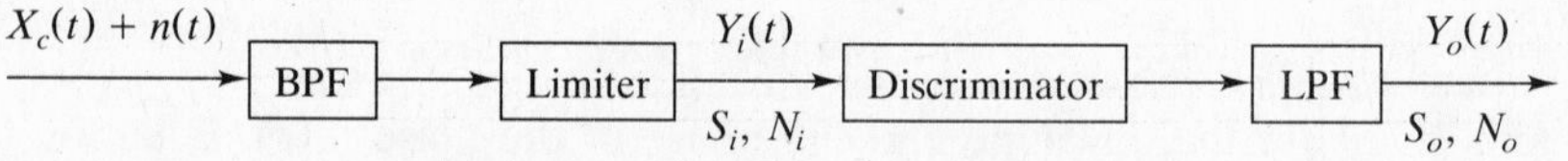

Fig. 7-7 Angle demodulation system

The carrier amplitude remains constant, therefore

$$S_i = E[X_c^2(t)] = \frac{1}{2} A_c^2 \tag{7.40a}$$

and

$$N_i = \eta B_T \tag{7.40b}$$

Hence,

$$\left(\frac{S}{N}\right)_i = \frac{A_c^2}{2\eta B_T} \tag{7.41}$$

which is independent of $X(t)$.

The $(S/N)_i$ of Eq. (7.41) is often called the *carrier-to-noise ratio* (CNR).

Because $n_i(t)$ is narrowband, we write

$$n_i(t) = v_n(t) \cos[\omega_c t + \phi_n(t)] \tag{7.42}$$

where $v_n(t)$ is *Rayleigh*-distributed and $\phi_n(t)$ is uniformly distributed in $(0, 2\pi)$ (Prob. 5.33). Then $Y_i(t)$ can be written as

$$Y_i(t) = V(t) \cos[\omega_c t + \theta(t)] \tag{7.43}$$

where

$$V(t) = \left\{ \left[A_c \cos \phi + v_n(t) \cos \phi_n(t) \right]^2 + \left[A_c \sin \phi + v_n(t) \sin \phi_n(t) \right]^2 \right\}^{1/2}$$

and

$$\theta(t) = \tan^{-1} \frac{A_c \sin \phi + v_n(t) \sin \phi_n(t)}{A_c \cos \phi + v_n(t) \cos \phi_n(t)}$$

The limiter suppresses any amplitude variation $V(t)$. Hence, in angle modulation, SNRs are derived from consideration of $\theta(t)$ only. The expression for $\theta(t)$ is too complicated for analysis without some simplification. The detector is assumed to be ideal. The output of the detector is

$$Y_o(t) = \begin{cases} \theta(t) & \text{for PM} \\ \dfrac{d\theta(t)}{dt} & \text{for FM} \end{cases} \tag{7.44}$$

A. Signal Dominance Case:

A phasor diagram (Fig. 7-8) for this case is obtained from

$$Y_i(t) = \text{Re}\left[Y(t) e^{j\omega_c t} \right] \tag{7.45}$$

where

$$Y(t) = A_c e^{j\phi(t)} + v_n(t) e^{j\phi_n(t)} \tag{7.46}$$

and $v_n(t) \ll A_c$ for almost all t. From Fig. 7-8 the length L of arc AB is

$$L = Y(t)\left[\theta(t) - \phi(t)\right] \tag{7.47}$$

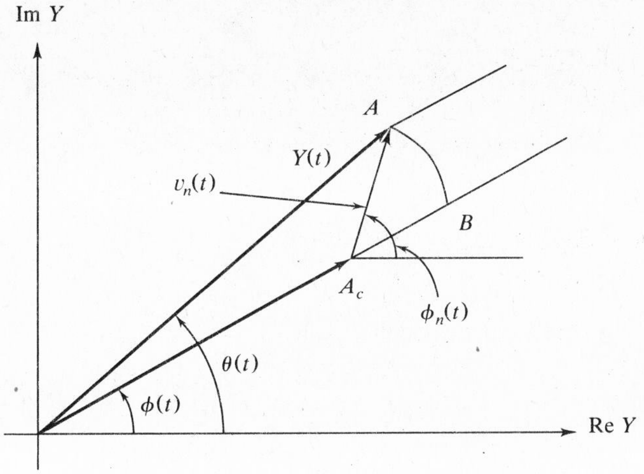

Fig. 7-8 Phasor diagram for angle modulation

and

$$Y(t) \approx A_c + v_n(t) \cos\left[\phi_n(t) - \phi(t)\right] \approx A_c$$

$$L \approx v_n(t) \sin\left[\phi_n(t) - \phi(t)\right]$$

Hence, from Eq. (7.47), we obtain

$$\theta(t) \approx \phi(t) + \frac{v_n(t)}{A_c} \sin\left[\phi_n(t) - \phi(t)\right] \tag{7.48}$$

For the purpose of computing an output SNR, replacing $\phi_n(t) - \phi(t)$ with $\phi_n(t)$ will not affect the result.[2] Thus, we write

$$\theta(t) \approx \phi(t) + \frac{v_n(t)}{A_c} \sin \phi_n(t)$$

$$= \phi(t) + \frac{n_s(t)}{A_c} \tag{7.49}$$

From Eqs. (7.44) and (7.38) the detector output is

$$Y_o(t) = \theta(t) = k_p X(t) + \frac{n_s(t)}{A_c} \qquad \text{for PM} \tag{7.50}$$

$$Y_o(t) = \frac{d\theta(t)}{dt} = k_f X(t) + \frac{n_s'(t)}{A_c} \qquad \text{for FM} \tag{7.51}$$

B. $(S/N)_o$ in PM Systems:

From Eq. (7.50)

$$S_o = E\left[k_p^2 X^2(t)\right] = k_p^2 E\left[X^2(t)\right] = k_p^2 S_X \tag{7.52}$$

$$N_o = E\left[\frac{1}{A_c^2} n_s^2(t)\right] = \frac{1}{A_c^2} E\left[n_s^2(t)\right] = \frac{1}{A_c^2}(2\eta B) \tag{7.53}$$

Hence,

$$\left(\frac{S}{N}\right)_o = \frac{k_p^2 A_c^2 S_X}{2\eta B} \tag{7.54}$$

From Eqs. (7.8) and (7.40a)

$$\gamma = \frac{S_i}{\eta B} = \frac{A_c^2}{2\eta B} \tag{7.55}$$

and then Eq. (7.54) can be expressed as

$$\left(\frac{S}{N}\right)_o = k_p^2 S_X \gamma \tag{7.56}$$

c. $(S/N)_o$ in FM Systems:

From Eq. (7.51)

$$S_o = E\left[k_f^2 X^2(t)\right] = k_f^2 E\left[X^2(t)\right] = k_f^2 S_X \tag{7.57}$$

$$N_o = E\left[\frac{1}{A_c^2}\left[n_s'(t)\right]^2\right] = \frac{1}{A_c^2} E\left[\left[n_s'(t)\right]^2\right] \tag{7.58}$$

By using Eq. (6.133) (Prob. 6.23), the power spectral density of $n_s'(t)$ is given by

$$S_{n_s' n_s'}(\omega) = \omega^2 S_{n_s n_s}(\omega) = \begin{cases} \omega^2 \eta & \text{for } |\omega| < W(=2\pi B) \\ 0 & \text{otherwise} \end{cases} \tag{7.59}$$

---

[2]This is based on the observation that in the sense of ensemble averages $\phi_n - \phi$ differs from ϕ_n only by a shift of the mean value.

Then

$$N_o = \frac{1}{A_c^2}\frac{1}{2\pi}\int_{-W}^{W}\omega^2\eta\, d\omega = \frac{2}{3}\frac{\eta}{A_c^2}\frac{W^3}{2\pi} \tag{7.60}$$

Hence,

$$\left(\frac{S}{N}\right)_o = \frac{3A_c^2(2\pi)k_f^2 S_X}{2\eta W^3} \tag{7.61}$$

Using Eq. (*7.55*), we can express Eq. (*7.61*) as

$$\left(\frac{S}{N}\right)_o = 3\left(\frac{k_f^2 S_X}{W^2}\right)\left(\frac{A_c^2}{2\eta B}\right) = 3\left(\frac{k_f^2 S_X}{W^2}\right)\gamma \tag{7.62}$$

Since $\Delta\omega = |k_f X(t)|_{\max} = k_f$ $[|X(t)| \le 1]$, Eq. (*7.62*) can be rewritten as

$$\left(\frac{S}{N}\right)_o = 3\left(\frac{\Delta\omega}{W}\right)^2 S_X\gamma = 3D^2 S_X\gamma. \tag{7.63}$$

where D is the deviation ratio, defined in Eq. (*3.26*).

Equation (*7.60*) indicates that the output noise power is inversely proportional to the mean carrier power $A_c^2/2$ in FM. This effect of a decrease in output noise power as the carrier power increases is called *noise quieting*.

D. Threshold Effects in Angle Modulation Systems:

When $A_c^2 \ll E[n_i^2(t)]$, the resulting phasor [Eq. (*7.46*)] is dominated by the term $v_n(t)e^{j\phi_n(t)}$. For this case, the phase of the detector input is

$$\theta(t) \approx \phi_n(t) + \frac{A_c}{v_n(t)}\sin[\phi(t) - \phi_n(t)] \tag{7.64}$$

The noise now dominates, and the message signal has been corrupted by the noise beyond the possibility of recovery—an effect similar to the one we observed in an AM system using envelope detection. This threshold effect is illustrated in Fig. 7-9.

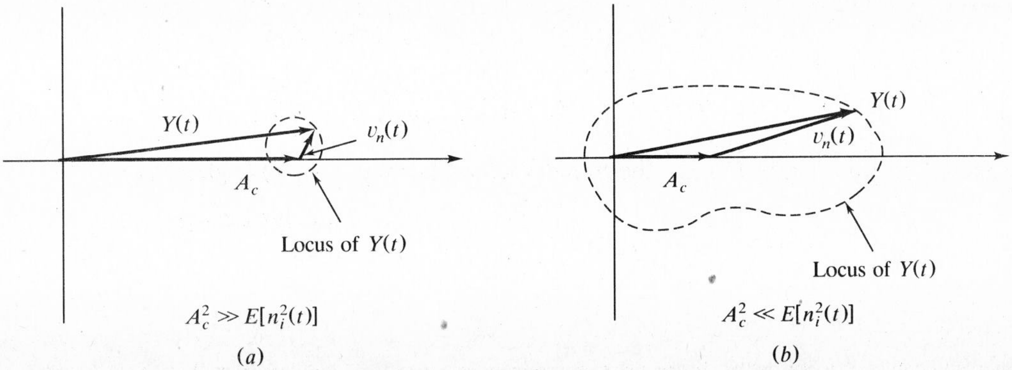

Fig. 7-9 Threshold effect in FM systems

If $A_c^2 \gg E[n_i^2(t)]$, we may expect $v_n(t) \ll A_c$ most of the time, and the tip of the resultant phasor $Y(t)$ traces randomly around the end of the carrier phasor, as illustrated in Fig. 7-9(a). The variation in phase is relatively small; however, if the noise becomes large enough, the tip of the resultant phasor $Y(t)$ may move away from the endpoint of the carrier phase and may even occasionally encircle the origin [Fig. 7-9(b)], where the phase of $Y(t)$ rapidly changes by 2π [Fig. 7-10(a)]. The output of the frequency discriminator then will contain a spike every time an encirclement occurs. This is illustrated in Fig. 7-10(b).

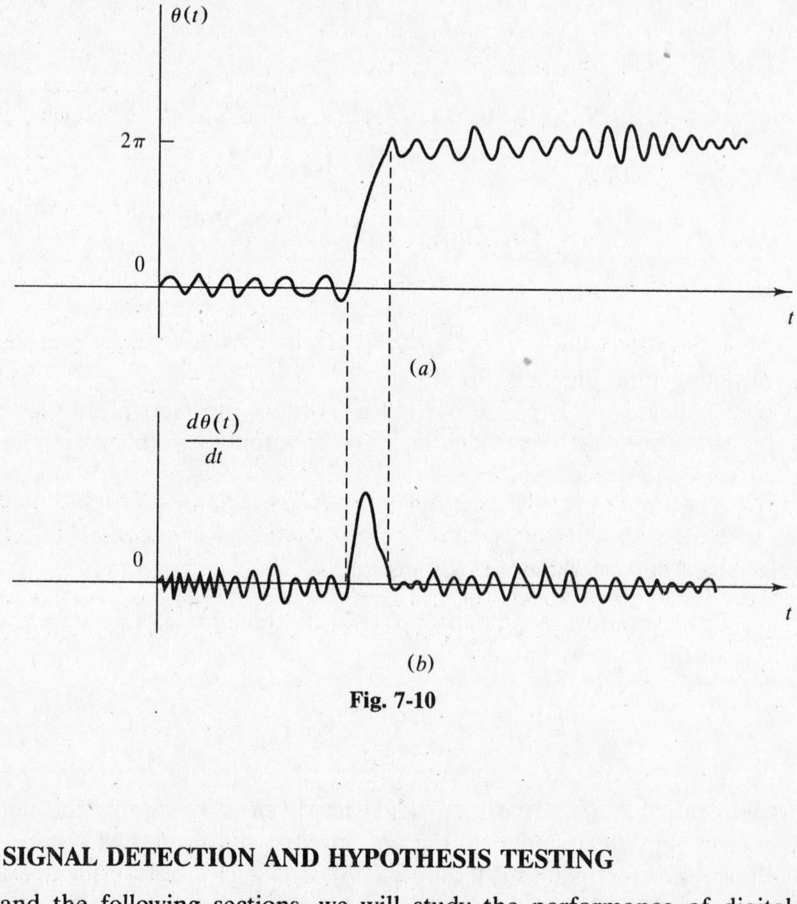

(a)

(b)

Fig. 7-10

7.6 BINARY SIGNAL DETECTION AND HYPOTHESIS TESTING

In this and the following sections, we will study the performance of digital communication systems in the presence of the additive noise as measured by the probability of error. We assume throughout a distortionless channel, so the received signal is free of intersymbol interference (ISI). We also assume *additive white gaussian noise* (AWGN) with zero mean value, independent of the signal.

Figure 7-11 portrays the operations of a binary receiver. The transmitted signal over a symbol interval $(0, T)$ is represented by

$$s_i(t) = \begin{cases} s_1(t) & 0 \le t \le T \quad \text{for 1} \\ s_2(t) & 0 \le t \le T \quad \text{for 0} \end{cases} \tag{7.65}$$

The received signal $r(t)$ by the receiver is represented by

$$r(t) = s_i(t) + n(t) \qquad i = 1, 2 \qquad 0 \le t \le T \tag{7.66}$$

where $n(t)$ is a zero-mean AWGN.

There are two separate steps involved in signal detection. The first step consists of reducing the received signal $r(t)$ to a single number $z(T)$. This operation can be performed by a linear filter followed by a sampler, as shown in block 1 of Fig. 7-11. The output of receiver (block 1), sampled at $t = T$, yields

$$z(T) = a_i(T) + n_o(T) \qquad i = 1, 2 \tag{7.67a}$$

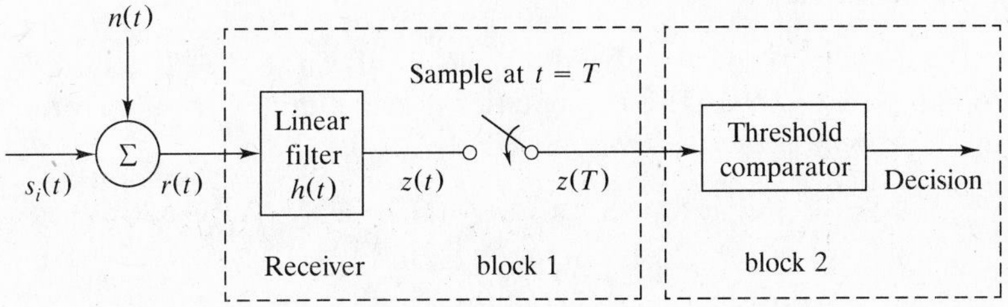

Fig. 7-11 Digital signal detection

where $a_i(T)$ is the signal component of $z(T)$ and $n_o(T)$ is the noise component. We often write Eq. (*7.67a*) as

$$z = a_i + n_o \qquad i = 1, 2 \tag{7.67b}$$

Note that the noise component n_o is a zero-mean gaussian random variable, and thus z is a gaussian random variable with a mean of either a_1 or a_2 depending on whether $s_1(t)$ or $s_2(t)$ was sent. The sample z is sometimes called the *test statistic*.

The second step of the signal detection process consists of comparing the test statistic z to a threshold level λ in block 2 (threshold comparator) of Fig. 7-11. The final step in block 2 is to make the decision.

$$z \underset{H_2}{\overset{H_1}{\gtrless}} \lambda \tag{7.68}$$

where H_1 and H_2 are the two possible hypotheses. Choosing H_1 is equivalent to deciding that signal $s_1(t)$ was sent, and choosing H_2 is equivalent to deciding that signal $s_2(t)$ was sent. Equation (*7.68*) indicates that hypothesis H_1 is chosen if $z > \lambda$, and hypothesis H_2 is chosen if $z < \lambda$. If $z = \lambda$, the decision can be an arbitrary one.

7.7 PROBABILITY OF ERROR AND MAXIMUM LIKELIHOOD DETECTOR

A. Probability of Error:

For the binary signal detection system, there are two ways in which errors can occur. That is, given that signal $s_1(t)$ was transmitted, an error results if hypothesis H_2 is chosen; or given that signal $s_2(t)$ was transmitted, an error results if hypothesis H_1 is chosen. Thus the probability of error P_e is expressed as [Eq. (*5.30*)]

$$P_e = P(H_2|s_1)P(s_1) + P(H_1|s_2)P(s_2) \tag{7.69}$$

where $P(s_1)$ and $P(s_2)$ are the a priori probabilities that $s_1(t)$ and $s_2(t)$, respectively, are transmitted.

When symbols 1 and 0 occur with equal probability, that is, $P(s_1) = P(s_2) = \frac{1}{2}$,

$$P_e = \tfrac{1}{2}[P(H_2|s_1) + P(H_1|s_2)] \tag{7.70}$$

B. Maximum Likelihood Detector:

A popular criterion for choosing the threshold λ of Eq. (*7.68*) is based on minimizing the probability of error of Eq. (*7.69*). The computation for this minimum error value of $\lambda = \lambda_0$ starts with forming the following *likelihood ratio test* (Prob. 7.17)

$$\Lambda(z) = \frac{f(z|s_1)}{f(z|s_2)} \underset{H_2}{\overset{H_1}{\gtrless}} \frac{P(s_2)}{P(s_1)} \tag{7.71}$$

where $f(z|s_i)$ is the conditional pdf known as the *likelihood* of s_i. The ratio $\Lambda(z)$ is known as the *likelihood ratio*. Equation (*7.71*) states that we should choose hypothesis H_1 if the likelihood ratio $\Lambda(z)$ is greater than the ratio of a priori probabilities. If $P(s_1) = P(s_2)$, Eq. (*7.71*) reduces to

$$\Lambda(z) = \frac{f(z|s_1)}{f(z|s_2)} \underset{H_2}{\overset{H_1}{\gtrless}} 1 \tag{7.72a}$$

or

$$f(z|s_1) \underset{H_2}{\overset{H_1}{\gtrless}} f(z|s_2) \tag{7.72b}$$

If $P(s_1) = P(s_2)$ and the likelihoods $f(z|s_i)$ $(i = 1, 2)$ are symmetric, then Eq. (*7.71*) yields the criterion (Prob. 7.18)

$$z \underset{H_2}{\overset{H_1}{\gtrless}} \lambda_0 \tag{7.73}$$

where

$$\lambda_0 = \frac{a_1 + a_2}{2} \tag{7.74}$$

It can be shown that the threshold λ_0 represented by Eq. (*7.74*) is the *optimum threshold* for minimizing the error of probability (Prob. 7.19). The criterion of Eq. (*7.73*) is known as the *minimum error criterion*. A detector that minimizes the error probability (for the case where the signal classes are equally likely) is also known as a *maximum likelihood detector*.

C. Probability of Error with Gaussian Noise:

The pdf of the gaussian random noise n_o in Eq. (*7.67b*) is [Eq. (*5.103*)]

$$f_{n_o}(\xi) = \frac{1}{\sqrt{2\pi}\,\sigma_{n_o}} e^{-\xi^2/(2\sigma_{n_o}^2)} \tag{7.75}$$

where $\sigma_{n_o}^2$ is the noise variance. It follows from Eqs. (*7.67b*) and (*7.75*) that

$$f(z|s_1)=\frac{1}{\sqrt{2\pi}\,\sigma_{n_o}}e^{-(z-a_1)^2/(2\sigma_{n_o}^2)} \tag{7.76a}$$

$$f(z|s_2)=\frac{1}{\sqrt{2\pi}\,\sigma_{n_o}}e^{-(z-a_2)^2/(2\sigma_{n_o}^2)} \tag{7.76b}$$

which are illustrated in Fig. 7-12.

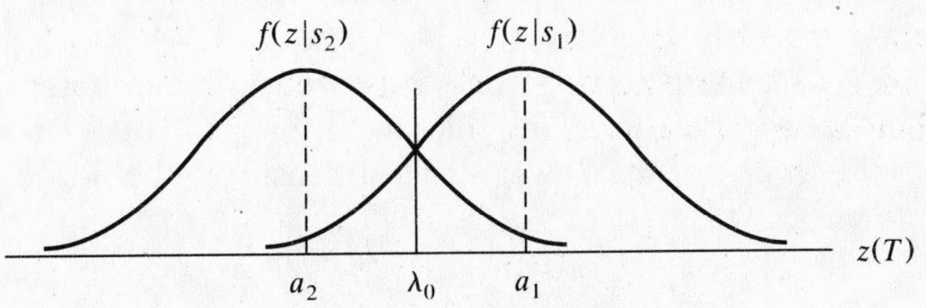

Fig. 7-12 Condition pdf

Now

$$P(H_2|s_1)=\int_{-\infty}^{\lambda_0} f(z|s_1)\,dz \tag{7.77a}$$

$$P(H_1|s_2)=\int_{\lambda_0}^{\infty} f(z|s_2)\,dz \tag{7.77b}$$

Because of the symmetry of $f(z|s_i)$, Eq. (*7.70*) reduces to

$$P_e=P(H_2|s_1)=P(H_1|s_2) \tag{7.78}$$

Thus, the probability of error P_e is numerically equal to the area under the "tail" of either likelihood function $f(z|s_1)$ or $f(z|s_2)$ falling on the "incorrect" side of the threshold;

$$P_e=\int_{\lambda_0}^{\infty} f(z|s_2)\,dz \tag{7.79}$$

where $\lambda_0=(a_1+a_2)/2$ is the optimum threshold [Eq. (*7.74*)]. Using Eq. (*7.76b*), we have

$$P_e=\int_{\lambda_0}^{\infty}\frac{1}{\sqrt{2\pi}\,\sigma_{n_o}}e^{-(z-a_2)^2/(2\sigma_{n_o}^2)}\,dz$$

Let $y=(z-a_2)/\sigma_{n_o}$. Then $\sigma_{n_o}\,dy=dz$ and

$$P_e=\int_{(a_1-a_2)/(2\sigma_{n_o})}^{\infty}\frac{1}{\sqrt{2\pi}}e^{-y^2/2}\,dy=Q\left(\frac{a_1-a_2}{2\sigma_{n_o}}\right) \tag{7.80}$$

where $Q(\cdot)$ is the complementary error function, or the Q function defined in Eq. (*5.105*). The values of the Q function are tabulated in App. C.

7.8 OPTIMUM DETECTION

In this section we consider optimizing the linear filter in the receiver (block 1) of Fig. 7-11 by minimizing the probability of error P_e.

A. The Matched Filter:

A matched filter is a linear filter designed to provide the maximum output SNR for a given transmitted signal. Consider that a known signal $s(t)$ plus AWGN $n(t)$ is the input to an LTI filter followed by a sampler, as shown in Fig. 7-11. Let $a(t)$ be the output of the filter. Then from Eq. (7.67a), at $t = T$, we have

$$\left(\frac{S}{N}\right)_o = \frac{a^2(T)}{E[n_o^2(T)]} = \frac{a^2(T)}{\sigma_{n_o}^2} \tag{7.81}$$

We wish to find the filter frequency response $H_0(\omega)$ that maximizes Eq. (7.81). It can be shown that (Prob. 7.23)

$$\left(\frac{S}{N}\right)_o \leq \frac{2}{\eta}\frac{1}{2\pi}\int_{-\infty}^{\infty}|S(\omega)|^2\,d\omega = \frac{2E}{\eta} \tag{7.82}$$

where $S(\omega) = \mathscr{F}[s(t)]$, $\eta/2$ is the power spectral density of the input noise, and E is the energy of the input signal $s(t)$. Note that the right-hand side of this inequality does not depend on $H(\omega)$ but only on the input signal energy and the power spectral density of the noise. Thus,

$$\left(\frac{S}{N}\right)_{o_{\max}} = \frac{2E}{\eta} \tag{7.83}$$

The equality in Eq. (7.82) holds only if the optimum filter frequency response $H_0(\omega)$ is employed such that (Prob. 7.23)

$$H(\omega) = H_0(\omega) = S^*(\omega)e^{-j\omega T} \tag{7.84}$$

where * denotes the complex conjugate.

The impulse response $h(t)$ of this optimum filter is [see Eqs. (1.30) and (1.33)]

$$h(t) = \mathscr{F}^{-1}[H(\omega)] = \begin{cases} s(T-t) & 0 \leq t \leq T \\ 0 & \text{otherwise} \end{cases} \tag{7.85}$$

Equation (7.85) and Fig. 7-13 illustrate the matched filter's basic property: The impulse response of the matched filter is a delayed version of the mirror image of the signal form.

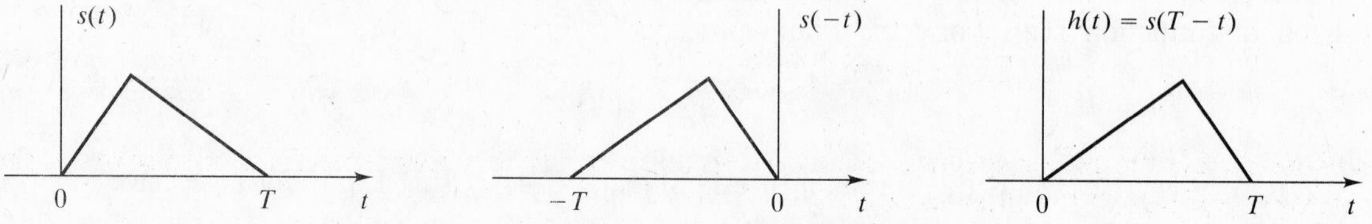

Fig. 7-13 Matched filter characteristics

B. Correlator:

The output $z(t)$ of a causal filter can be expressed as [Eq. (1.82)]

$$z(t) = r(t) * h(t) = \int_0^t r(\tau)h(t-\tau)\,d\tau \tag{7.86}$$

Substituting $h(t)$ of Eq. (7.85) into Eq. (7.86), we obtain

$$z(t) = \int_0^t r(\tau)s[T-(t-\tau)]\,d\tau \tag{7.87}$$

When $t = T$, we have

$$z(T) = \int_0^T r(\tau)s(\tau)\,d\tau \tag{7.88}$$

The operation of Eq. (7.88) is known as the *correlation* of $r(t)$ and $s(t)$.

Since the matched filter output and the correlator output are identical at the sampling time $t = T$, the matched filter and correlator depicted in Fig. 7-14 are used interchangeably.

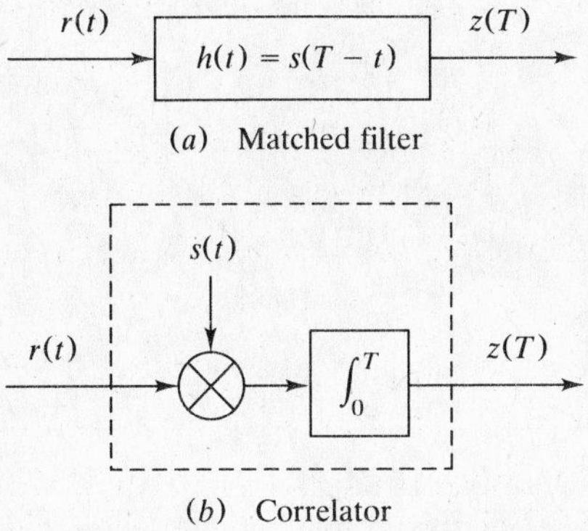

Fig. 7-14 Equivalence of matched filter and correlator

C. Optimum Detection:

To minimize P_e of Eq. (7.80), we need to determine the linear filter that maximizes $(a_1 - a_2)/(2\sigma_{n_o})$ or, equivalently, that maximizes

$$\frac{(a_1 - a_2)^2}{\sigma_{n_o}^2} \tag{7.89}$$

where $a_1 - a_2$ is the difference of the signal components at the filter output, at time $t = T$; hence $(a_1 - a_2)^2$ is the instantaneous power of the difference signal, and $\sigma_{n_o}^2$ is the average output noise power.

Consider a filter that is matched to the input signal $s_1(t) - s_2(t)$. From Eqs. (7.81) and (7.83), we have

$$\left(\frac{S}{N}\right)_o = \frac{(a_1 - a_2)^2}{\sigma_{n_o}^2} = \frac{E_d}{\eta/2} = \frac{2E_d}{\eta} \tag{7.90}$$

where $\eta/2$ is the power spectral density of the noise at the filter input and E_d is the energy of the difference signal at the filter input:

$$E_d = \int_0^T [s_1(t) - s_2(t)]^2 \, dt \tag{7.91}$$

Hence, using Eqs. (7.80) and (7.90), we obtain

$$P_e = Q\left(\frac{a_1 - a_2}{2\sigma_{n_o}}\right) = Q\left(\sqrt{\frac{E_d}{2\eta}}\right) \tag{7.92}$$

7.9 ERROR PROBABILITY PERFORMANCE OF BINARY TRANSMISSION SYSTEMS

By using Eq. (7.92), the probabilities of error for various binary transmission systems are given in the following.

A. Unipolar Baseband Signaling:

$$s_i(t) = \begin{cases} s_1(t) = A & 0 \le t \le T \\ s_2(t) = 0 & 0 \le t \le T \end{cases} \tag{7.93}$$

The probability of error P_e is

$$P_e = Q\left(\sqrt{\frac{A^2T}{2\eta}}\right) = Q\left(\sqrt{\frac{E_b}{\eta}}\right) \tag{7.94}$$

where $E_b = A^2T/2$ is the average signal energy per bit.

B. Bipolar Baseband Signaling:

$$s_i(t) = \begin{cases} s_1(t) = +A & 0 \le t \le T \\ s_2(t) = -A & 0 \le t \le T \end{cases} \tag{7.95}$$

The probability of error P_e is (Prob. 7.28)

$$P_e = Q\left(\sqrt{\frac{2A^2T}{\eta}}\right) = Q\left(\sqrt{\frac{2E_b}{\eta}}\right) \tag{7.96}$$

where $E_b = A^2T$ is the average signal energy per bit.

C. Amplitude-Shift Keying (or On-Off Keying):

$$s_i(t) = \begin{cases} s_1(t) = A\cos\omega_c t & 0 \le t \le T \\ s_2(t) = 0 & 0 \le t \le T \end{cases} \tag{7.97}$$

with T an integer times $1/f_c$. The probability of error P_e is

$$P_e = Q\left(\sqrt{\frac{A^2T}{4\eta}}\right) = Q\left(\sqrt{\frac{E_b}{\eta}}\right) \tag{7.98}$$

where $E_b = A^2T/4$ is the average signal energy per bit.

D. Phase-Shift Keying:

$$s_i(t) = \begin{cases} s_1(t) = A\cos\omega_c t & 0 \le t \le T \\ s_2(t) = A\cos(\omega_c t + \pi) & \\ \quad = -A\cos\omega_c t & 0 \le t \le T \end{cases} \tag{7.99}$$

with T an integer times $1/f_c$. The probability of error P_e is (Prob. 7.30)

$$P_e = Q\left(\sqrt{\frac{A^2T}{\eta}}\right) = Q\left(\sqrt{\frac{2E_b}{\eta}}\right) \tag{7.100}$$

where $E_b = A^2T/2$ is the average signal energy per bit.

E. Frequency-Shift Keying:

$$s_i(t) = \begin{cases} s_1(t) = A\cos\omega_1 t & 0 \le t \le T \\ s_2(t) = A\cos\omega_2 t & 0 \le t \le T \end{cases} \tag{7.101}$$

If we assume $\omega_1 T \gg 1$, $\omega_2 T \gg 1$, and $(\omega_1 - \omega_2)T \gg 1$, then the probability of error P_e is (Prob. 7.33).

$$P_e \approx Q\left(\sqrt{\frac{A^2T}{2\eta}}\right) = Q\left(\sqrt{\frac{E_b}{\eta}}\right) \tag{7.102}$$

where $E_b = A^2T/2$ is the average signal energy per bit.

Solved Problems

NOISE IN BASEBAND SYSTEMS

7.1. Consider an analog baseband communication system with additive white noise. The transmission channel is assumed to be distortionless and the power spectral density of white noise $\eta/2$ is 10^{-9} watt per hertz (W/Hz). The signal to be transmitted is an audio signal with 4-kHz bandwidth. At the receiver end, an *RC* low-pass filter with a 3-dB bandwidth of 8 kHz is used to limit the noise power at the output. Calculate the output noise power.

From Prob. 1.49 the frequency response $H(\omega)$ of an *RC* low-pass filter with a 3-dB bandwidth of 8 kHz is given by

$$H(\omega) = \frac{1}{1 + j\omega/\omega_0}$$

where $\omega_0 = 2\pi(8000)$. Using Eqs. (*6.40*) and (*6.53*), we see that the output noise power N_o is

$$\begin{aligned} N_o = E\left[n_o^2(t)\right] &= \frac{1}{2\pi}\int_{-\infty}^{\infty} \frac{\eta}{2}|H(\omega)|^2\,d\omega \\ &= \frac{\eta}{2}\frac{1}{2\pi}\int_{-\infty}^{\infty} \frac{1}{1 + (\omega/\omega_0)^2}\,d\omega \\ &= \frac{1}{4}\eta\omega_0 = \frac{1}{4}2(10^{-9})(2\pi)(8)(10^3)\text{W} = 25.2\ \mu\text{W} \end{aligned}$$

7.2. Consider an analog baseband communication system with additive white noise having power spectral density $\eta/2$ and a distorting channel having the frequency response

$$H_c(\omega) = \frac{1}{1 + j\omega/W}$$

The distortion is equalized by a receiver filter having the frequency response

$$H_{eq}(\omega) = \begin{cases} \dfrac{1}{H_c(\omega)} & 0 \le |\omega| \le W \\ 0 & \text{otherwise} \end{cases}$$

Obtain an expression for the output SNR.

$$\begin{aligned} S_o &= \frac{1}{2\pi}\int_{-\infty}^{\infty} |H_c(\omega)|^2 |H_{eq}(\omega)|^2 S_{XX}(\omega)\,d\omega \\ &= \frac{1}{2\pi}\int_{-W}^{W} S_{XX}(\omega)\,d\omega = S_X \\ N_o &= \frac{1}{2\pi}\int_{-\infty}^{\infty} \frac{\eta}{2}|H_{eq}(\omega)|^2\,d\omega \\ &= \frac{\eta}{2\pi}\int_0^W \left[1 + \left(\frac{\omega}{W}\right)^2\right] d\omega = \frac{\eta}{2\pi}\left(\frac{4}{3}W\right) = \frac{4}{3}\eta B \end{aligned}$$

Thus

$$\left(\frac{S}{N}\right)_o = \frac{S_o}{N_o} = \frac{S_X}{\frac{4}{3}\eta B} = \frac{3}{4}\frac{S_X}{\eta B} \tag{7.103}$$

NOISE IN AMPLITUDE MODULATION SYSTEMS

7.3. A DSB signal with additive white noise is demodulated by a synchronous detector (Fig. 7-4) with a phase error ϕ. Taking the local oscillator signal to be $2\cos(\omega_c t + \phi)$, show that

$$\left(\frac{S}{N}\right)_o = \gamma \cos^2 \phi \tag{7.104}$$

where γ is defined by Eq. (*7.8*).

From Eq. (*7.14*)

$$Y_i(t) = [A_c X(t) + n_c(t)] \cos \omega_c t - n_s(t) \sin \omega_c t$$

Multiplying $Y_i(t)$ by $2\cos(\omega_c t + \phi)$ and using a low-pass filter, we obtain

$$\begin{aligned} Y_o(t) &= A_c X(t) \cos \phi + n_c(t) \cos \phi + n_s(t) \sin \phi \\ &= X_o(t) + n_o(t) \end{aligned}$$

where

$$X_o(t) = A_c X(t) \cos \phi$$

$$n_o(t) = n_c(t) \cos \phi + n_s(t) \sin \phi$$

So

$$S_o = E[X_o^2(t)] = A_c^2 \cos^2 \phi \, E[X^2(t)] = A_c^2 (\cos^2 \phi) S_X$$

$$\begin{aligned} N_o = E[n_o^2(t)] &= E[n_c^2(t) \cos^2 \phi + n_s^2(t) \sin^2 \phi + n_c(t) n_s(t) \sin 2\phi] \\ &= E[n_c^2(t)] \cos^2 \phi + E[n_s^2(t)] \sin^2 \phi \\ &= E[n_i^2(t)](\cos^2 \phi + \sin^2 \phi) = E[n_i^2(t)] = 2\eta B \end{aligned}$$

since $E[n_c(t) n_s(t)] = 0$ [Eq. (*6.73*)].

Thus, by Eqs. (*7.17*), (*7.18*), and (*7.19*), we obtain

$$\left(\frac{S}{N}\right)_o = \frac{A_c^2 S_X \cos^2 \phi}{2\eta B} = \frac{\frac{1}{2} A_c^2 S_X}{\eta B} \cos^2 \phi = \gamma \cos^2 \phi$$

where γ is [Eq. (*7.8*)]

$$\gamma = \frac{\frac{1}{2} A_c^2 S_X}{\eta B} = \frac{S_i}{\eta B}$$

7.4. Show that the performance of an SSB system using synchronous detection is equivalent to the performance of both DSB and baseband systems.

The SSB signal $X_c(t)$ can be expressed as [Eq. (*2.11*)]

$$X_c(t) = A_c[X(t) \cos \omega_c t + \hat{X}(t) \sin \omega_c t] \tag{7.105}$$

where $\hat{X}(t)$ denotes the Hilbert transform of $X(t)$. Note that Eq. (*7.105*) represents the lower-sideband SSB signal and that the minimum bandwidth of the predetection filter is W for a single-band signal.

Refer to Fig. 7-4; the input of the receiver is

$$\begin{aligned} Y_i(t) &= X_c(t) + n_i(t) \\ &= [A_c X(t) + n_c(t)] \cos \omega_c t + [A_c \hat{X}(t) - n_s(t)] \sin \omega_c t \end{aligned} \tag{7.106}$$

Synchronous detection rejects the quadrature components of both the signal and noise, yielding

$$Y_o(t) = A_c X(t) + n_c(t) = X_o(t) + n_o(t) \tag{7.107}$$

where

$$X_o(t) = A_c X(t) \quad \text{and} \quad n_o(t) = n_c(t)$$

The output signal power is

$$S_o = E[X_o^2(t)] = A_c^2 E[X^2(t)] = A_c^2 S_X$$

The power spectral density of $n_c(t)$ is illustrated in Fig. 7-15 for the case of lower-sideband SSB.

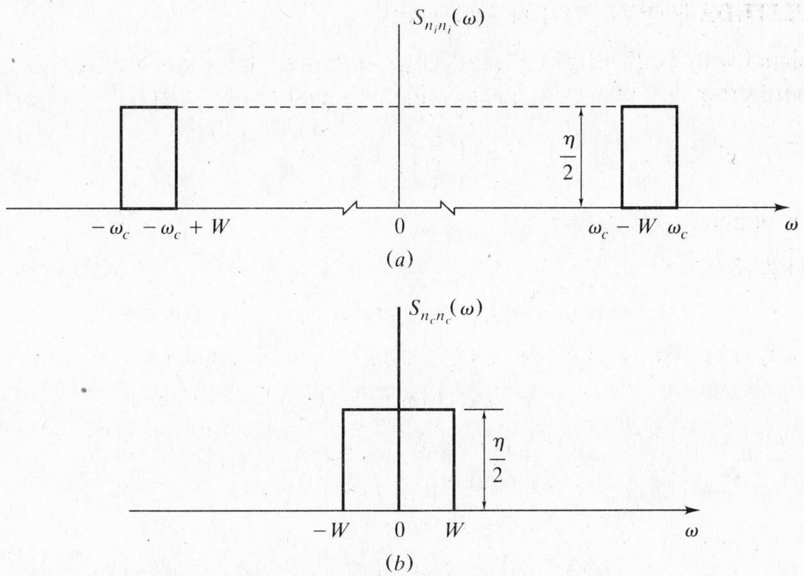

Fig. 7-15

The output noise power is

$$N_o = E\left[n_c^2(t)\right] = \frac{\eta W}{2\pi} = \eta B$$

Thus, the output SNR is

$$\left(\frac{S}{N}\right)_o = \frac{A_c^2 S_X}{\eta B} \tag{7.108}$$

The input signal power is

$$S_i = E\left[X_c^2(t)\right] = A_c^2\left\{\frac{1}{2}E\left[X^2(t)\right] + \frac{1}{2}E\left[\hat{X}^2(t)\right]\right\}$$
$$= A_c^2 E\left[X^2(t)\right] = A_c^2 S_X$$

since [Prob. 6.43 and Eq. (*6.128*) of Prob. 6.20]

$$E\left[X(t)\hat{X}(t)\right] = 0 \qquad \text{and} \qquad E\left[X^2(t)\right] = E\left[\hat{X}^2(t)\right]$$

We obtain

$$\left(\frac{S}{N}\right)_o = \frac{S_i}{\eta B} = \gamma \tag{7.109}$$

which indicates that insofar as noise is concerned, SSB with ideal synchronous detection has the same performance as both DSB and baseband systems.

7.5. Assuming sinusoidal modulation, show that in an AM system with envelope detection the output SNR is given by

$$\left(\frac{S}{N}\right)_o = \frac{\mu^2}{2 + \mu^2}\gamma \tag{7.110}$$

where μ is the modulation index for AM.

For sinusoidal modulation

$$X(t) = \cos \omega_m t$$

So
$$S_X = E[X^2(t)] = \frac{1}{2}$$

Thus, from Eq. (7.35), we obtain

$$\left(\frac{S}{N}\right)_o = \frac{\mu^2 S_X}{1+\mu^2 S_X}\gamma = \frac{\mu^2(\frac{1}{2})}{1+\mu^2(\frac{1}{2})}\gamma = \frac{\mu^2}{2+\mu^2}\gamma$$

Note that with 100 percent modulation ($\mu = 1$)

$$\left(\frac{S}{N}\right)_o = \frac{1}{3}\gamma \tag{7.111}$$

7.6. The *threshold level* (or *value*) in an AM system with envelope detection is usually defined as that value of $(S/N)_i$ for which $V_n \ll A_c$ with probability 0.99. Here A_c and V_n are the amplitudes of the carrier and narrowband noise, respectively. Show that if $\mu = 1$ and $S_X = E[X^2(t)] = 1$, then $P(V_n \ll A_c) = 0.99$ requires

$$\left(\frac{S}{N}\right)_i = 4\ln 10 = 9.2 \approx 10 \tag{7.112}$$

From Eq. (6.69a) and Eq. (5.148) (Prob. 5.33) the pdf of V_n is

$$f_{V_n}(v_n) = \frac{v_n}{N_i}e^{-v_n^2/(2N_i)} \qquad \text{for } v_n > 0 \tag{7.113}$$

where
$$N_i = E[n_i^2(t)] = 2\eta B$$

and
$$P(V_n \ge A_c) = \int_{A_c}^{\infty} f_{V_n}(v_n)\,dv_n$$
$$= \int_{A_c}^{\infty}\frac{v_n}{N_i}e^{-v_n^2/(2N_i)}\,dv_n = \int_{A_c^2/(2N_i)}^{\infty} e^{-\lambda}\,d\lambda = e^{-A_c^2/(2N_i)}$$

Now
$$P(V_n < A_c) = 0.99 \longrightarrow P(V_n \ge A_c) = 0.01$$

Hence,
$$e^{-A_c^2/(2N_i)} = e^{-A_c^2/(4\eta B)} = 0.01 \tag{7.114}$$

From Eq. (7.25) with $\mu = 1$ and $S_X = 1$, we have

$$S_i = \frac{1}{2}A_c^2(1+\mu^2 S_X) = A_c^2$$

and
$$\left(\frac{S}{N}\right)_i = \frac{S_i}{N_i} = \frac{A_c^2}{2\eta B}$$

Thus, Eq. (7.114) can be rewritten as

$$e^{-(1/2)(S/N)_i} = 0.01$$

from which $(S/N)_i$ at threshold is obtained at

$$\left(\frac{S}{N}\right)_{i_{\text{th}}} = 2\ln\frac{1}{0.01} = 4\ln 10 = 9.2 \approx 10\ (10\text{ dB}) \tag{7.115}$$

7.7. Consider an AM system with additive thermal noise having a power spectral density $\eta/2 = 10^{-12}$ W/Hz. Assume that the baseband message signal $X(t)$ has a bandwidth of 4 kHz and the amplitude distribution shown by Fig. 7-16. The signal is demodulated by envelope detection and appropriate postdetection filtering. Assume $\mu = 1$.

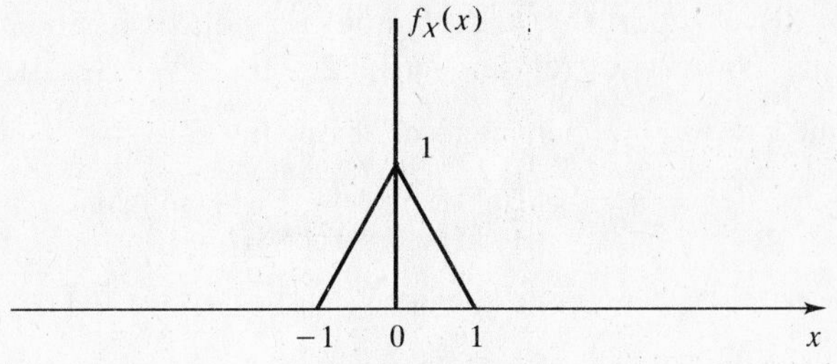

Fig. 7-16

(*a*) Find the minimum value of the carrier amplitude A_c that will yield $(S/N)_o \geq 40$ dB.

(*b*) Find the threshold value of A_c.

(*a*)
$$X_c(t) = A_c[1 + X(t)] \cos \omega_c t$$

$$S_X = E[X^2(t)] = \int_{-\infty}^{\infty} x^2 f_X(x)\, dx = 2\int_0^1 x^2(-x+1)\, dx = \frac{1}{6}$$

From Eq. (*7.26*) ($\mu = 1$) we have

$$\left(\frac{S}{N}\right)_o = \frac{S_X}{1+S_X}\gamma = \frac{\frac{1}{6}}{1+\frac{1}{6}}\gamma = \frac{1}{7}\gamma \geq 10^4$$

Hence
$$\gamma \geq 7(10^4)$$

From Eqs. (*7.8*) and (*7.25*)

$$\frac{A_c^2(1+\frac{1}{6})}{4(10^{-12})(4)(10^3)} \geq 7(10^4)$$

from which we obtain

$$A_c \geq 31(10^{-3})\ \text{V} = 31\ \text{mV}$$

Thus, the minimum value of A_c required is 31 mV.

(*b*) From Eq. (*7.115*)

$$\left(\frac{S}{N}\right)_{i_{\text{th}}} = 10$$

Therefore

$$\gamma|_{\text{at threshold}} = \gamma_{\text{th}} = 2\left(\frac{S}{N}\right)_{i_{\text{th}}} \approx 20 \qquad (7.116)$$

and from part (*a*)

$$\frac{A_c^2(1+\frac{1}{6})}{4(10^{-12})(4)(10^3)} = 20$$

from which we obtain

$$A_c = 0.52(10^{-3})\ \text{V} = 0.52\ \text{mV}$$

which is the threshold value of A_c.

7.8. Calculate the transmission bandwidth B_T and the required transmitter power S_T of DSB, SSB, and AM systems for transmitting an audio signal which has a bandwidth of 10 kHz with an output SNR of 40 dB. Assume that the channel introduces a 40-dB power loss and channel noise is AWGN with power spectral density $\eta/2 = 10^{-9}$ W/Hz. Assume $\mu^2 S_X = 0.5$ for AM.

From Chap. 2 the bandwidth requirements are easily found as

$$B_T = \begin{cases} 20 \text{ kHz} & \text{for DSB and AM} \\ 10 \text{ kHz} & \text{for SSB} \end{cases}$$

Transmitter power for DSB and SSB systems: From Eqs. (*7.19*) and (*7.109*)

$$\left(\frac{S}{N}\right)_o = \frac{S_i}{\eta B} = 10^4 \; (= 40 \text{ dB})$$

and

$$S_i = \eta B(10^4) = 2(10^{-9})(10^4)(10^4) = 0.2 \text{ W}$$

Since the channel power loss is 40 dB, the required transmitted power S_T is

$$S_T = 0.2(10^4) = 2000 \text{ W} = 2 \text{ kW}$$

For an AM system with envelope detection: From Eq. (*7.26*) with $\mu^2 S_X = 0.5$

$$\left(\frac{S}{N}\right)_o = \frac{1}{3}\left(\frac{S_i}{\eta B}\right)$$

Thus, the required transmitted power S_T is 3 times that for the DSB or SSB system, that is,

$$S_T = 6 \text{ kW}$$

7.9. Consider an AM system using a square-law detector as shown in Fig. 7-17 (Prob. 2.7). Assume that the channel noise is AWGN with power spectral density $\eta/2$. Find the output SNR.

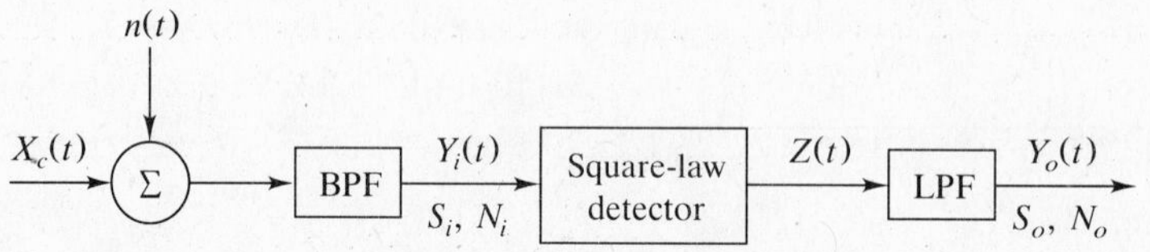

Fig. 7-17 Square-law detector

$$X_c(t) = A_c[1 + \mu X(t)] \cos \omega_c t$$

The detector input $Y_i(t)$ is

$$Y_i(t) = X_c(t) + n_i(t)$$

The detector output $Z(t)$ is

$$\begin{aligned} Z(t) &= Y_i^2(t) \\ &= \frac{A_c^2}{2}[1 + 2\mu X(t) + \mu^2 X^2(t)](1 + \cos 2\omega_c t) + n_i^2(t) + 2A_c n_i(t)[1 + \mu X(t)] \cos \omega_c t \end{aligned}$$

Since it is required that $|\mu X(t)| \ll 1$ for distortionless square-law demodulation (Prob. 2.7), we can approximate $Z(t)$ as

$$Z(t) \approx \frac{A_c^2}{2}[1 + 2\mu X(t)](1 + \cos 2\,\omega_c t) + n_i^2(t) + 2A_c n_i(t) \cos \omega_c t \tag{7.117}$$

After low-pass filtering and blocking the dc component, the output $Y_o(t)$ of the receiver is given by

$$Y_o(t) = X_o(t) + n_o(t)$$

where

$$X_o(t) = A_c^2 \mu X(t)$$

and $n_o(t)$ is equal to the result of low-pass filtering $n_i^2(t) + 2A_c n_i(t)\cos\omega_c t$ and removing dc components. So

$$S_o = E[X_o^2(t)] = A_c^4\mu^2 E[X^2(t)] = A_c^4\mu^2 S_X \tag{7.118}$$

Next, by Eq. (*6.152*) (Prob. 6.31) we have

$$S_{n_i^2 n_i^2}(\omega) = 2\pi E[n_i^2(t)]\delta(\omega) + \frac{1}{\pi} S_{n_i n_i}(\omega) * S_{n_i n_i}(\omega) \tag{7.119}$$

and from Eq. (*6.99*) of Prob. 6.10 the power spectral density of $2A_c n_i(t)\cos\omega_c t$, denoted by $S_{nic}(\omega)$, is given by

$$S_{nic}(\omega) = A_c^2[S_{n_i n_i}(\omega - \omega_c) + S_{n_i n_i}(\omega + \omega_c)] \tag{7.120}$$

These power spectral densities are shown in Fig. 7-18. The postdetection filter removes the dc component [the first term of Eq. (*7.119*)] and passes only those portions of the power spectral components that lie within $-W$ to W ($W = 2\pi B$). From Fig. 7-18 the output noise power is

$$N_o = E[n_o^2(t)] = 3\eta^2 B^2 + 2A_c^2\eta B \tag{7.121}$$

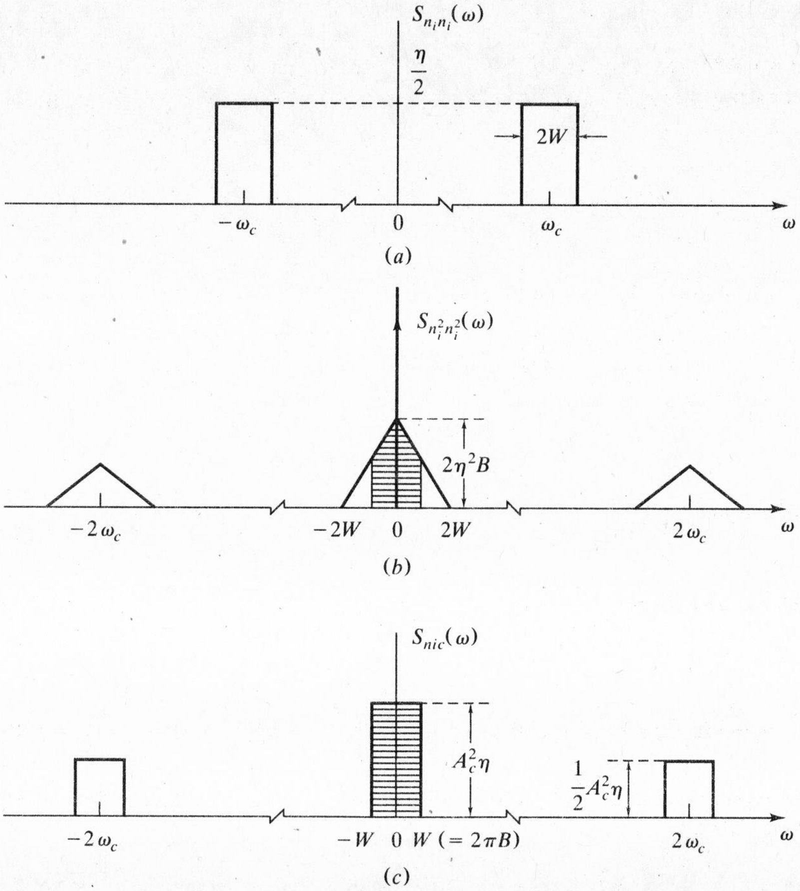

Fig. 7-18

Hence, the output SNR is

$$\left(\frac{S}{N}\right)_o = \frac{S_o}{N_o} = \frac{A_c^4 \mu^2 S_X}{3\eta^2 B^2 + 2A_c^2 \eta B} \tag{7.122}$$

NOISE IN ANGLE MODULATION SYSTEM

7.10. Consider an FM broadcast system with parameter $\Delta f = 75$ kHz and $B = 15$ kHz. Assuming $S_X = \frac{1}{2}$, find the output SNR and calculate the improvement (in dB) over the baseband system.

Substituting the given parameters into Eq. (*7.63*), we obtain

$$\left(\frac{S}{N}\right)_o = 3\left(\frac{75(10^3)}{15(10^3)}\right)^2 \left(\frac{1}{2}\right)\gamma = 37.5\gamma$$

Now
$$10 \log 37.5 = 15.7 \text{ dB}$$
which indicates that $(S/N)_o$ is about 16 dB better than the baseband system.

7.11. Show that in an FM system the output SNR, assuming sinusoidal modulation, is given by

$$\left(\frac{S}{N}\right)_o = \frac{3}{2}\beta^2\gamma \tag{7.123}$$

where β is the modulation index for FM.

For sinusoidal modulation

$$X(t) = \cos \omega_m t$$

so
$$S_X = E[X^2(t)] = \frac{1}{2}$$

From Eq. (*7.62*), we obtain

$$\left(\frac{S}{N}\right)_o = 3\left(\frac{k_f^2 S_X}{W^2}\right)\gamma = \frac{3}{2}\left(\frac{\Delta\omega}{W}\right)^2 \gamma = \frac{3}{2}\beta^2\gamma$$

since
$$\Delta\omega = |k_f X(t)|_{\max} = k_f \quad \text{and} \quad \frac{\Delta\omega}{W} = \beta$$

It is important to note that the modulation index β is determined by the bandwidth W of the postdetection low-pass filter and is not related to the sinusoidal message frequency ω_m, except insofar as this filter is chosen so as to pass the spectrum of the desired filter. For a specified bandwidth W, the sinusoidal message frequency may lie anywhere between 0 and W and would yield the same output SNR.

7.12. Show that narrowband FM offers no improvement in SNR over AM.

For AM with envelope detection and assuming 100 percent sinusoidal modulation, the output SNR is given by [Prob. 7.5, Eq. (*7.111*)]

$$\left(\frac{S}{N}\right)_o = \frac{1}{3}\gamma$$

For FM with sinusoidal modulation, the output SNR is given by [Prob. 7.11, Eq. (*7.123*)]

$$\left(\frac{S}{N}\right)_o = \frac{3}{2}\beta^2\gamma$$

Hence, we see that the use of FM offers the possibility of improved SNR over AM when

$$\frac{3}{2}\beta^2 > \frac{1}{3} \quad \text{or} \quad \beta > 0.47$$

In Sec. 3.7 we mentioned that a value of $\beta < 0.2$ is considered to define an FM signal to be narrowband. We conclude that the narrowband FM offers no improvement in SNR over AM.

Note that based on the above noise consideration, $\beta = 0.5$ is often considered as defining roughly the transition from narrowband FM to wideband FM.

7.13. An audio signal $X(t)$ is to be transmitted over a radio frequency (RF) channel with additive white noise. It is required that the output SNR be greater than 40 dB. Assume the following characteristics for $X(t)$ and the channel:

$$E[X(t)] = 0 \qquad |X(t)| \le 1 \qquad S_X = E[X^2(t)] = \tfrac{1}{2} \qquad B = 15 \text{ kHz}$$

Power spectral density of white noise $\eta/2 = 10^{-10}$ W/Hz

Power loss in channel = 50 dB

Calculate the transmission bandwidth B_T and the required average transmitted power S_T for

(*a*) DSB modulation

(*b*) AM with 100 percent modulation and envelope detection

(*c*) PM with $k_p = 3$

(*d*) FM with a deviation ratio $D = 5$

(*a*) For DSB modulation

$$B_T = 2B = 30 \text{ kHz}$$

From Eqs. (*7.8*) and (*7.19*), we have

$$\left(\frac{S}{N}\right)_o = \gamma = \frac{S_i}{\eta B} = \frac{S_i}{2(10^{-10})(15)(10^3)} \ge 10^4 \ (= 40 \text{ dB})$$

or

$$S_i \ge 3(10^{-2}) \text{ W}$$

Hence,

$$S_T = S_i(10^5) \ge 3000 \text{ W} = 3 \text{ kW}$$

(*b*) For AM with $\mu = 1$ and envelope detection

$$B_T = 2B = 30 \text{ kHz}$$

From Eq. (*7.35*), we have

$$\left(\frac{S}{N}\right)_o = \frac{1}{3}\gamma$$

Hence,

$$S_T \ge 9 \text{ kW}$$

(*c*) For PM with $k_p = 3$, from Eqs. (*3.21*) and (*3.25*),

$$B_T + 2(k_p + 1)B = 120 \text{ kHz}$$

From Eq. (*7.56*), we have

$$\left(\frac{S}{N}\right)_o = k_p^2 S_X \gamma = (3^2)\left(\frac{1}{2}\right)\frac{S_i}{2(10^{-10})(15)(10^3)} \ge 10^4$$

or

$$S_i \ge \frac{2}{3}(10^{-2}) \text{ W}$$

Hence,

$$S_T = S_i(10^5) \ge 667 \text{ W}$$

(*d*) For FM with $D = 5$, from Eq. (*3.27*)

$$B_T = 2(D + 1)B = 180 \text{ kHz}$$

From Eq. (*7.63*), we have

$$\left(\frac{S}{N}\right)_o = 3D^2 S_X \gamma = 3(5^2)\left(\frac{1}{2}\right)\frac{S_i}{2(10^{-10})(15)(10^3)} \ge 10^4$$

or

$$S_i \geq \frac{2}{25}(10^{-2}) \text{ W}$$

Hence,

$$S_T = S_i(10^5) \geq 80 \text{ W}$$

7.14. The result of Prob. 7.6 indicates that the threshold level for AM is equivalent to the input SNR $(S/N)_i = 10$. Assume this conclusion is also valid for FM.

(*a*) Find the output SNR at the threshold level for FM (assuming sinusoidal modulation).

(*b*) Find the modulation index β that produces $(S/N)_o = 30$ dB at the threshold.

(*a*) From Eq. (*7.62*), we have

$$\left(\frac{S}{N}\right)_o = 3\left(\frac{k_f^2 S_X}{W^2}\right)\left(\frac{A_c^2}{2\eta B}\right) \tag{7.124}$$

For sinusoidal modulation

$$S_X = \frac{1}{2} \qquad \Delta\omega = k_f \qquad \beta = \frac{\Delta\omega}{W}$$

Thus, Eq. (*7.124*) can be rewritten as

$$\left(\frac{S}{N}\right)_o = \frac{3}{2}\beta^2\left(\frac{\frac{1}{2}A_c^2}{\eta B}\right)$$

Using Eq. (*7.41*), we obtain

$$\left(\frac{S}{N}\right)_o = \frac{3}{2}\beta^2\left(\frac{B_T}{B}\right)\left(\frac{S}{N}\right)_i$$

According to Carson's rule (*3.25*),

$$B_T = 2(\beta + 1)B$$

Thus, we obtain

$$\left(\frac{S}{N}\right)_o = 3\beta^2(\beta + 1)\left(\frac{S}{N}\right)_i \tag{7.125}$$

Setting $(S/N)_i = 10$, we see that the output SNR at the threshold level is

$$\left(\frac{S}{N}\right)_{o_{\text{th}}} = 30\beta^2(\beta + 1) \tag{7.126}$$

(*b*)

$$\left(\frac{S}{N}\right)_{o_{\text{th}}} = 30\beta^2(\beta + 1) = 10^3 \ (= 30 \text{ dB})$$

Solving for β, we obtain $\beta = 2.92$.

7.15. In commercial FM broadcasting, a scheme known as *preemphasis/deemphasis filtering* is used to improve the output SNR. In this scheme, the high-frequency component in the input signal is emphasized at the transmitter before the noise is introduced, and at the output of the FM demodulator the inverse operation is performed to deemphasize the high-frequency components. The circuits shown in Fig. 7-19 are used as preemphasis (PE) and deemphasis (DE) filters. The standard value for the time constant is $R_1C = 75$ μs, and $R_2C \gg R_1C$.

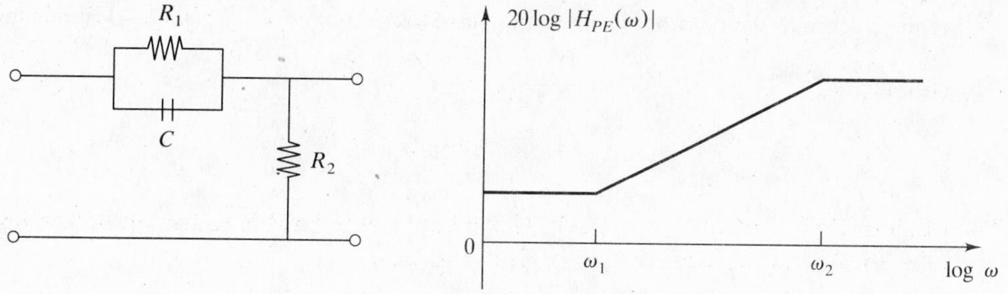

(*a*) Preemphasis filter and its frequency response

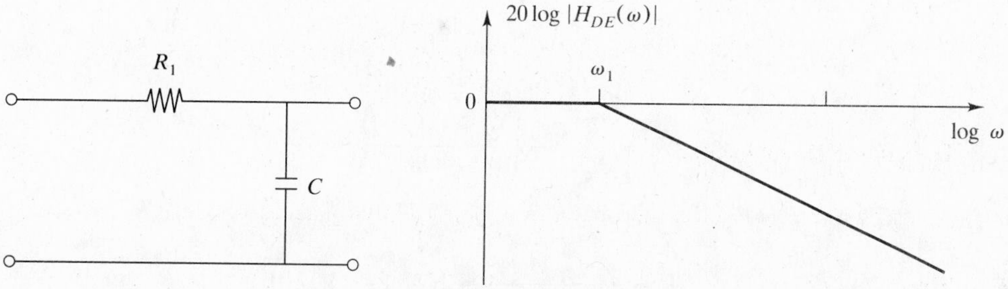

(*b*) Deemphasis filter and its frequency response

Fig. 7-19

Calculate the noise improvement factor Γ defined by

$$\Gamma = \frac{N_o}{N_o'}$$

where N_o is the output noise power without preemphasis/deemphasis filters and N_o' is the output noise power when these filters are used.

The frequency response for the preemphasis filter is

$$H_{PE}(\omega) = K\frac{1 + j\omega\tau_1}{1 + j\omega\tau_2}$$

where $$K = \frac{R_2}{R_1 + R_2} \qquad \tau_1 = R_1 C \qquad \tau_2 = \frac{R_1 R_2 C}{R_1 + R_2} \approx R_2 C$$

For $\omega \ll \omega_2$ where $\omega_2 = 1/\tau_2$, $H_{PE}(\omega)$ can be approximated by

$$H_{PE}(\omega) = K(1 + j\omega\tau_1) = K\left(1 + j\frac{\omega}{\omega_1}\right) \tag{7.127}$$

with $$\omega_1 = \frac{1}{\tau_1} \qquad \text{and} \qquad f_1 = \frac{1}{2\pi\tau_1} = \frac{1}{2\pi(75)(10^{-6})} = 2.1(10^3)\text{ Hz} = 2.1\text{ kHz}$$

The frequency response for the deemphasis filter is

$$H_{DE}(\omega) = \frac{1}{1 + j\omega\tau_1} = \frac{1}{1 + j(\omega/\omega_1)} \tag{7.128}$$

We note that

$$H_{PE}(\omega)H_{DE}(\omega) \approx K \qquad \text{for } \omega \ll \omega_2$$

which is the requirement for no distortion.

In the presence of the deemphasis filter, the output noise power N_o of Eq. (7.60) is modified to

$$N_o' = \frac{1}{A_c^2}\frac{1}{2\pi}\int_{-W}^{W}\omega^2\eta|H_{DE}(\omega)|\,d\omega$$

$$= \frac{\eta}{A_c^2\pi}\int_0^W \frac{\omega^2}{1+(\omega/\omega_1)^2}\,d\omega$$

$$= \frac{\eta\omega_1^3}{A_c^2\pi}\left(\frac{W}{\omega_1} - \tan^{-1}\frac{W}{\omega_1}\right)$$

Without the deemphasis filter the output noise power N_o is [Eq. (7.60)]

$$N_o = \frac{\eta W^3}{3A_c^2\pi}$$

Thus,

$$\Gamma = \frac{N_o}{N_o'} = \frac{1}{3}\frac{(W/\omega_1)^3}{W/\omega_1 - \tan^{-1}(W/\omega_1)} \tag{7.129}$$

If $f_1 = 2.1$ kHz and $B = 15$ kHz, then $W/\omega_1 = B/f_1 = 7.14$ and

$$\Gamma = \frac{1}{3}\frac{(7.14)^3}{7.14 - 1.43} = 21.25\ (= 13.27\text{ dB}) \tag{7.130}$$

Note that Eq. (7.130) neglects any increase in the modulation signal power resulting from the preemphasis of higher spectral components. The actual SNR improvement achieved by taking preemphasis into consideration is treated in the next problem.

7.16. Suppose that the power spectral density of the signal $X(t)$ is

$$S_{XX}(\omega) = \begin{cases} \dfrac{1}{1+(\omega/\omega_1)^2} & |\omega| < 2\pi(15)(10^3) \\ 0 & \text{otherwise} \end{cases}$$

where $\omega_1 = 2\pi(2.1)(10^3)$ rad/s.

(a) Determine the required change in modulation level with and without preemphasis for a fixed modulation power.

(b) What is the net SNR improvement under these conditions?

(a) Without preemphasis, the modulation signal power S_X is

$$S_X = \frac{1}{2\pi}\int_{-W}^{W}\frac{1}{1+(\omega/\omega_1)^2}\,d\omega = \frac{\omega_1}{\pi}\tan^{-1}\frac{W}{\omega_1}$$

With preemphasis, using Eq. (7.127) of Prob. 7.15 gives

$$S_X = \frac{1}{2\pi}\int_{-W}^{W} S_{XX}(\omega)|H_{PE}(\omega)|^2\,d\omega = \frac{1}{2\pi}\int_{-W}^{W}K\,d\omega = K\frac{W}{\pi}$$

Thus, if the modulation signal power is to remain fixed, we must have

$$K = \frac{\tan^{-1}(W/\omega_1)}{W/\omega_1}$$

With $B = 15$ kHz and $f_1 = 2.1$ kHz, we obtain

$$K = \frac{\tan^{-1}7.14}{7.14} = 0.200\ (= -6.98\text{ dB})$$

(*b*) From Eq. (*7.130*) of Prob. 7.15 the noise improvement factor with preemphasis/deemphasis is 13.27 dB. However, now the modulation level must be decreased by 6.98 dB, so the net SNR improvement is 13.27 − 6.98 = 6.29 dB.

PROBABILITY OF ERROR AND MAXIMUM LIKELIHOOD DETECTOR

7.17. Derive the likelihood ratio test given by Eq. (*7.71*), that is,

$$\Lambda(z)=\frac{f(z|s_1)}{(z|s_2)}\ \underset{H_2}{\overset{H_1}{\gtrless}}\ \frac{P(s_2)}{P(s_1)}$$

A reasonable receiver decision rule is to choose hypothesis H_1 if the a posteriori probability $P(s_1|z)$ is greater than the a posteriori probability $P(s_2|z)$. Otherwise, we should choose hypothesis H_2. (See Prob. 5.8.) Hence

$$P(s_1|z)\ \underset{H_2}{\overset{H_1}{\gtrless}}\ P(s_2|z) \tag{7.131}$$

or

$$\frac{P(s_1|z)}{P(s_2|z)}\ \underset{H_2}{\overset{H_1}{\gtrless}}\ 1 \tag{7.132}$$

The decision criterion of Eq. (*7.132*) is called the *maximum a posteriori* (MAP) criterion.

Expressing Bayes' rule [Eq. (*5.24*)] for a continuous conditional pdf [Eq. (*5.46*)], we have

$$P(s_i|z)=\frac{f(z|s_i)P(s_i)}{f(z)} \qquad i=1,2 \tag{7.133}$$

where $f(z|s_i)$ is the conditional pdf of the received sample z conditioned on the signal class s_i. Thus, by using Eq. (*7.133*), Eq. (*7.131*) yields

$$f(z|s_1)P(s_1)\ \underset{H_2}{\overset{H_1}{\gtrless}}\ f(z|s_2)P(s_2) \tag{7.134}$$

or

$$\Lambda(z)=\frac{f(z|s_1)}{f(z|s_2)}\ \underset{H_2}{\overset{H_1}{\gtrless}}\ \frac{P(s_2)}{P(s_1)}$$

7.18. Derive Eq. (*7.73*), that is,

$$z(T)\ \underset{H_2}{\overset{H_1}{\gtrless}}\ \lambda_0=\frac{a_1+a_2}{2}$$

Using Eqs. (*7.76a*) and (*7.76b*), we see that the likelihood ratio $\Lambda(z)$ defined in Eq. (*7.71*) can be written as

$$\Lambda(z)=\frac{f(z|s_1)}{f(z|s_2)}=\frac{e^{-(z-a_1)^2/(2\sigma_{n_o}^2)}}{e^{-(z-a_2)^2/(2\sigma_{n_o}^2)}}=e^{z(a_1-a_2)/\sigma_{n_o}^2-(a_1^2-a_2^2)/(2\sigma_{n_o}^2)}$$

Hence, the likelihood ratio test (*7.71*) can be expressed as

$$e^{z(a_1-a_2)/\sigma_{n_o}^2-(a_1^2-a_2^2)/(2\sigma_{n_o}^2)}\ \underset{H_2}{\overset{H_1}{\gtrless}}\ \frac{P(s_2)}{P(s_1)} \tag{7.135}$$

The inequality relationship of Eq. (*7.135*) is preserved for any monotonically increasing (or decreasing) transformation. Taking the natural logarithm of both sides, we obtain

$$\frac{z(a_1-a_2)}{\sigma_{n_o}^2}-\frac{a_1^2-a_2^2}{2\sigma_{n_o}^2}\ \underset{H_2}{\overset{H_1}{\gtrless}}\ \ln\frac{P(s_2)}{P(s_1)} \tag{7.136}$$

When $P(s_1)=P(s_2)$,

$$\ln\frac{P(s_2)}{P(s_1)}=\ln 1=0$$

Eq. (*7.136*) yields

$$z\ \underset{H_2}{\overset{H_1}{\gtrless}}\ \frac{a_1^2-a_2^2}{2(a_1-a_2)}=\frac{a_1+a_2}{2}=\lambda_0 \tag{7.137}$$

7.19. Verify that the threshold λ_0 given by Eq. (*7.137*) yields the minimum probability of error when $P(s_1)=P(s_2)$.

Assume that the threshold is set at λ (Fig. 7-20). From Eq. (*7.69*)

$$\begin{aligned}P_e&=P(H_2|s_1)P(s_1)+P(H_1|s_2)P(s_2)\\&=P(s_1)\int_{-\infty}^{\lambda}f(z|s_1)\,dz+P(s_2)\int_{\lambda}^{\infty}f(z|s_2)\,dz\\&=P(s_1)\int_{-\infty}^{\lambda}f(z|s_1)\,dz+P(s_2)\left[1-\int_{-\infty}^{\lambda}f(z|s_2)\,dz\right]\\&=P(s_2)+\int_{-\infty}^{\lambda}\left[P(s_1)f(z|s_1)-P(s_2)f(z|s_2)\right]dz\end{aligned}$$

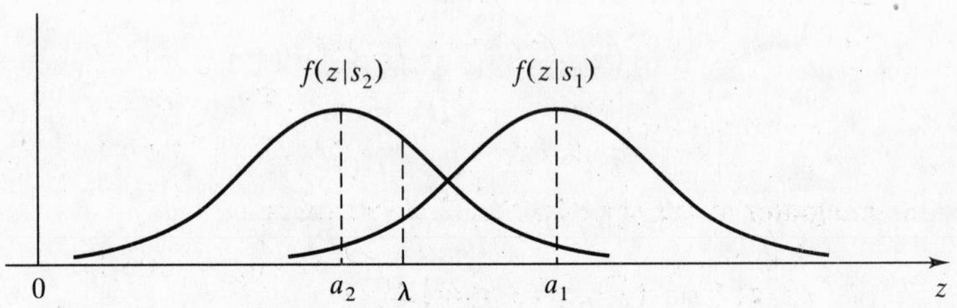

Fig. 7-20

To find the threshold λ_0 which minimizes P_e, we set

$$\frac{dP_e}{d\lambda}=0$$

which yields

$$P(s_1)f(\lambda_0|s_1)=P(s_2)f(\lambda_0|s_2) \tag{7.138}$$

or

$$\frac{f(\lambda_0|s_1)}{f(\lambda_0|s_2)}=\frac{P(s_2)}{P(s_1)} \tag{7.139}$$

Using Eqs. (*7.76a*) and (*7.76b*), we obtain

$$\frac{f(\lambda_0|s_1)}{f(\lambda_0|s_2)}=\frac{e^{-(\lambda_0-a_1)^2/(2\sigma_{n_o}^2)}}{e^{-(\lambda_0-a_2)^2/(2\sigma_{n_o}^2)}}=\frac{P(s_2)}{P(s_1)}$$

or

$$e^{\lambda_0(a_1-a_2)/\sigma_{n_o}^2-(a_1^2-a_2^2)/(2\sigma_{n_o}^2)}=\frac{P(s_2)}{P(s_1)}$$

or

$$\frac{\lambda_0(a_1-a_2)}{\sigma_{n_o}^2}+\frac{a_1^2-a_2^2}{2\sigma_{n_o}^2}=\ln\frac{P(s_2)}{P(s_1)}$$

From which we obtain

$$\lambda_0=\frac{1}{2}(a_1+a_2)+\frac{\sigma_{n_o}^2}{a_1-a_2}\ln\frac{P(s_2)}{P(s_1)} \tag{7.140}$$

When $P(s_1)=P(s_2)$, Eq. (7.140) reduces to

$$\lambda_0=\frac{1}{2}(a_1+a_2)$$

7.20. A bipolar binary signal $s_i(t)$ is a $+A$-V or $-A$-V pulse during the interval $(0,T)$. The linear filter shown in Fig. 7-11 is an integrator, as shown in Fig. 7-21. (This detector is known as an *integrate-and-dump detector*.) Assuming $\sigma_{n_o}^2=0.1$, determine the optimum detection threshold λ_0 if the a priori probabilities are

(*a*) $P(s_1)=0.5$

(*b*) $P(s_1)=0.7$

(*c*) $P(s_1)=0.2$

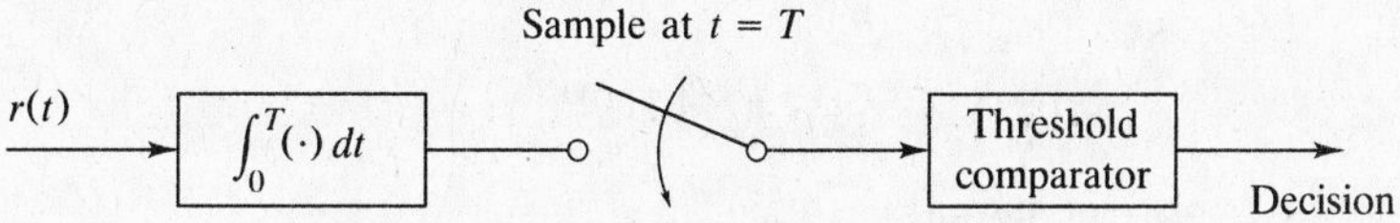

Fig. 7-21 Integrate-and-dump detector

The received signal $r(t)$ is

$$r(t)=s_i(t)+n(t)$$

where

$$s_i(t)=\begin{cases} s_1(t)=+A & 0\le t\le T & \text{for } 1\\ s_2(t)=-A & 0\le t\le T & \text{for } 0\end{cases}$$

The output of the integrator at the end of a signaling interval is

$$z(T)=\int_0^T[s_i(t)+n(t)]\,dt=\begin{cases} a_1+n_o & \text{for } 1\\ a_2+n_o & \text{for } 0\end{cases} \tag{7.141}$$

where

$$a_1=\int_0^T A\,dt=AT \qquad a_2=\int_0^T(-A)\,dt=-AT \tag{7.142}$$

and n_o is a random variable defined by

$$n_o=\int_0^T n(t)\,dt \tag{7.143}$$

(*a*) $P(s_1)=P(s_2)=0.5$
From Eqs. (7.74) and (7.142), the optimum threshold is

$$\lambda_0=\frac{a_1+a_2}{2}=\frac{AT+(-AT)}{2}=0$$

(*b*) $P(s_1)=0.7$, $P(s_2)=0.3$
From Eqs. (7.140) and (7.142) the optimum threshold is

$$\lambda_0=\frac{0.1}{2AT}\ln\frac{0.3}{0.7}=-\frac{0.04}{AT}$$

(c) $P(s_1) = 0.2$, $P(s_2) = 0.8$
From Eqs. (7.140) and (7.142) the optimum threshold is

$$\lambda_0 = \frac{0.1}{2AT} \ln \frac{0.8}{0.2} = \frac{0.07}{AT}$$

7.21. Show that the probability of error P_e in the binary system of Prob. 7.20 can be expressed as

$$P_e = Q\left(\sqrt{\frac{2A^2T}{\eta}}\right) \tag{7.144}$$

From Eq. (7.143),

$$n_o = \int_0^T n(t)\,dt$$

Since n_o is obtained by a linear operation on a sample function from a gaussian process, it is a gaussian random variable.

$$E[n_o] = E\left[\int_0^T n(t)\,dt\right] = \int_0^T E[n(t)]\,dt = 0 \tag{7.145}$$

since $n(t)$ has zero mean.

$$\sigma_{n_o}^2 = E[n_o^2] = E\left[\left[\int_0^T n(t)\,dt\right]^2\right]$$

$$= \int_0^T \int_0^T E[n(t)n(\tau)]\,dt\,d\tau$$

Since $n(t)$ is a white noise, we have [Eq. (6.63)]

$$E[n(t)n(\tau)] = \frac{\eta}{2}\delta(t-\tau)$$

and
$$\sigma_{n_o}^2 = \int_0^T \int_0^T \frac{\eta}{2}\delta(t-\tau)\,dt\,d\tau = \frac{\eta}{2}\int_0^T d\tau = \frac{\eta T}{2} \tag{7.146}$$

Thus, using Eqs. (7.80) and (7.142), we obtain

$$P_e = Q\left(\frac{a_1 - a_2}{2\sigma_{n_o}}\right) = Q\left(\frac{AT}{\sqrt{\eta T/2}}\right) = Q\left(\sqrt{\frac{2A^2T}{\eta}}\right)$$

7.22. In the binary system of Prob. 7.20, $P(s_1) = P(s_2) = \frac{1}{2}$, $\eta/2 = 10^{-9}$ W/Hz, $A = 10$ mV, and the transmission rate of data (bit rate) is 10^4 b/s.

(a) Find the probability of error P_e.

(b) If the bit rate is increased to 10^5 b/s, what value of A is needed to attain the same P_e as in part (a)?

(a)
$$\frac{2A^2T}{\eta} = \frac{A^2T}{\eta/2} = \frac{(0.01)^2(10^{-4})}{10^{-9}} = 10$$

Hence, from Eq. (7.144) and Table C-1, we obtain

$$P_e = Q\left(\sqrt{\frac{2A^2T}{\eta}}\right) = Q(\sqrt{10}) = 7.8(10^{-4})$$

(b)
$$\frac{A^2T}{\eta/2} = \frac{A^2(10^{-5})}{10^{-9}} = 10$$

Solving for A, we obtain

$$A = 31.62(10^{-3})\ \text{V} = 31.62\ \text{mV}$$

THE MATCHED FILTER AND CORRELATOR

7.23. Derive Eq. (*7.82*), that is,

$$\left(\frac{S}{N}\right)_o \le \frac{2E}{\eta}$$

where E is the energy of the input signal $s(t)$ and $\eta/2$ is the power spectral density of the input noise $n(t)$ (Fig. 7-11).

Let $H(\omega)$ be the frequency response of the linear filter. Let $a(t)$ be the output signal of the filter. Then by Eq. (*1.85*)

$$a(T) = \frac{1}{2\pi}\int_{-\infty}^{\infty} H(\omega)S(\omega)e^{j\omega T}\,d\omega \tag{7.147}$$

where $S(\omega) = \mathscr{F}[s(t)]$. The output noise power N_o is [by Eq. (*6.53*)]

$$N_o = E\left[n_o^2(t)\right] = \frac{\eta}{2}\frac{1}{2\pi}\int_{-\infty}^{\infty} |H(\omega)|^2\,d\omega \tag{7.148}$$

Thus, the output SNR is

$$\left(\frac{S}{N}\right)_o = \frac{\left[1/(2\pi)^2\right]\left|\int_{-\infty}^{\infty} H(\omega)S(\omega)e^{j\omega T}\,d\omega\right|^2}{(\eta/2)[1/(2\pi)]\int_{-\infty}^{\infty} |H(\omega)|^2\,d\omega} \tag{7.149}$$

We can examine the above SNR by applying *Schwarz inequality*, which states that

$$\left|\int_{-\infty}^{\infty} f_1(\omega)f_2(\omega)\,d\omega\right|^2 \le \int_{-\infty}^{\infty} |f_1(\omega)|^2\,d\omega \int_{-\infty}^{\infty} |f_2(\omega)|^2\,d\omega \tag{7.150}$$

The equality holds if

$$f_1(\omega) = kf_2^*(\omega) \tag{7.151}$$

where k is an arbitrary real number and * denotes the complex conjugate.

If we set

$$f_1(\omega) = H(\omega) \qquad f_2(\omega) = S(\omega)e^{j\omega T}$$

we can write

$$\left|\int_{-\infty}^{\infty} H(\omega)S(\omega)e^{j\omega T}\,d\omega\right|^2 \le \int_{-\infty}^{\infty} |H(\omega)|^2\,d\omega \int_{-\infty}^{\infty} |S(\omega)|^2\,d\omega \tag{7.152}$$

Substituting Eq. (*7.152*) into Eq. (*7.149*), we obtain

$$\left(\frac{S}{N}\right)_o \le \frac{2}{\eta}\frac{1}{2\pi}\int_{-\infty}^{\infty} |S(\omega)|^2\,d\omega = \frac{2E}{\eta}$$

where

$$E = \frac{1}{2\pi}\int_{-\infty}^{\infty} |S(\omega)|^2\,d\omega$$

which is the energy of the input signal $s(t)$.

7.24. Find the output of the matched filter and determine the maximum value of $(S/N)_o$ if the input $s(t)$ is a rectangular pulse of amplitude A and duration T.

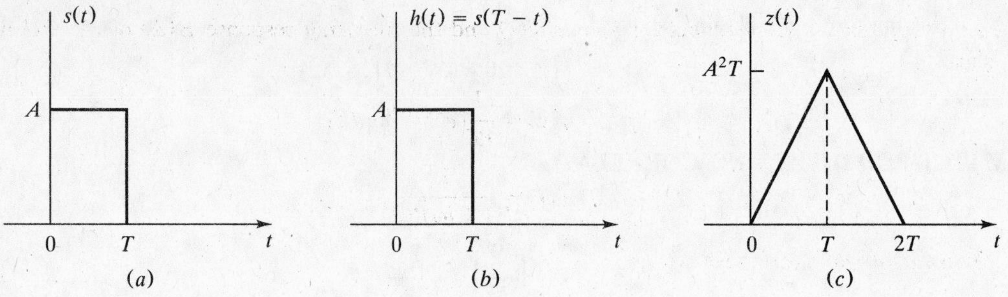

Fig. 7-22

For the given $s(t)$ [Fig. 7-22(a)] the impulse response $h(t)$ of the matched filter is [Eq. (7.85)]

$$h(t) = s(T - t) = s(t)$$

which is exactly the same as the input $s(t)$ [Fig. 7-22(b)]. Thus, the output $z(t)$ is

$$z(t) = s(t) * h(t) = \begin{cases} A^2 t & 0 \le t \le T \\ -A^2 t + 2A^2 T & T \le t \le 2T \\ 0 & \text{otherwise} \end{cases}$$

which is plotted in Fig. 7-22(c).

Note that $z(T) = A^2 T$ is the maximum value of $z(t)$. From Eq. (7.83) the maximum value of $(S/N)_o$ is

$$\left(\frac{S}{N}\right)_{o_{\max}} = \frac{2E}{\eta}$$

where

$$E = \int_{-\infty}^{\infty} s^2(t)\,dt = \int_0^T A^2\,dt = A^2 T$$

Hence,

$$\left(\frac{S}{N}\right)_{o_{\max}} = \frac{2A^2 T}{\eta} \tag{7.153}$$

7.25. Repeat Prob. 7.24 if a simple RC filter [Fig. 7-23(a)] is used instead of the matched filter.

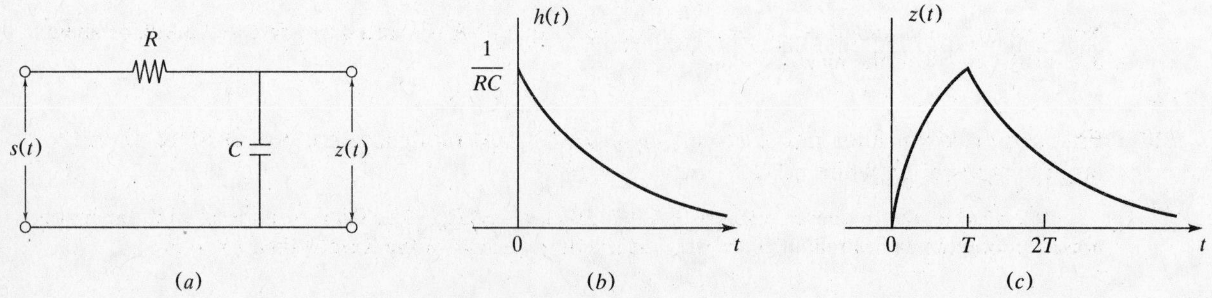

Fig. 7-23

From Prob. 1.46 the impulse response $h(t)$ and the frequency response $H(\omega)$ of the RC filter are given by

$$h(t) = \frac{1}{RC} e^{-t/(RC)} u(t)$$

$$H(\omega) = \frac{1}{1 + j\omega RC}$$

Then the output $z(t)$ is given by

$$z(t) = s(t) * h(t) = \begin{cases} 0 & t < 0 \\ A(1 - e^{-t/(RC)}) & 0 \le t \le T \\ A(1 - e^{-T/(RC)}) e^{-(t-T)/(RC)} & t > T \end{cases}$$

which is plotted in Fig. 7-23(*c*). Note that the maximum value of $z(t)$ is reached at $t = T$ and

$$z(T) = A(1 - e^{-T/(RC)})$$

The average output noise power is

$$N_o = E[n_o^2(t)] = \frac{1}{2\pi} \int_{-\infty}^{\infty} \frac{\eta}{2} \frac{d\omega}{1 + (\omega RC)^2} = \frac{\eta}{4RC} \tag{7.154}$$

Thus,

$$\left(\frac{S}{N}\right)_o = \frac{z^2(T)}{N_o} = \frac{4A^2T}{\eta} \frac{(1 - e^{-T/(RC)})^2}{T/(RC)} \tag{7.155}$$

We now maximize $(S/N)_o$ with respect to RC. Letting $x = T/(RC)$ and

$$g(x) = \frac{(1 - e^{-x})^2}{x}$$

and setting

$$g'(x) = \frac{2xe^{-x}(1 - e^{-x}) - (1 - e^{-x})^2}{x^2} = 0$$

we obtain

$$2xe^{-x} = 1 - e^{-x} \qquad \text{or} \qquad 1 + 2x = e^x$$

Solving for x, we obtain

$$x = \frac{T}{RC} \approx 1.257$$

Substituting this value into Eq. (*7.155*), we obtain

$$\left(\frac{S}{N}\right)_{o\max} = (0.815) \frac{2A^2T}{\eta} \tag{7.156}$$

Note that by using a simple RC filter, the maximum output SNR is reduced by a factor of 0.815, or about 0.89 dB from that of the matched filter.

7.26. Find the optimum filter frequency response $H_0(\omega)$ that maximizes the output SNR when the input noise is not a white noise.

Let $H(\omega)$ be the frequency response of a linear filter. Let $S_{nn}(\omega)$ be the power spectrum of the input noise. Proceeding as in Prob. 7.23, we see that the output signal of the filter at time $t = T$ is

$$a(T) = \frac{1}{2\pi} \int_{-\infty}^{\infty} H(\omega) S(\omega) e^{j\omega T} d\omega$$

The output noise power is

$$N_o = E[n_o^2(t)] = \frac{1}{2\pi}\int_{-\infty}^{\infty} S_{nn}(\omega)|H(\omega)|^2\,d\omega$$

Then the output SNR is

$$\left(\frac{S}{N}\right)_o = \frac{[1/(2\pi)]\left|\int_{-\infty}^{\infty} H(\omega)S(\omega)e^{j\omega T}\,d\omega\right|^2}{\int_{-\infty}^{\infty} S_{nn}(\omega)|H(\omega)|^2\,d\omega} \tag{7.157}$$

To find the frequency response $H(\omega)$ that maximizes Eq. (*7.157*), we apply Schwarz inequality Eq. (*7.150*). Setting

$$f_1(\omega) = \sqrt{S_{nn}(\omega)}\,H(\omega) \qquad f_2(\omega) = \frac{S(\omega)e^{j\omega T}}{\sqrt{S_{nn}(\omega)}}$$

we can write

$$\left|\int_{-\infty}^{\infty} H(\omega)S(\omega)e^{j\omega T}\,d\omega\right|^2 \le \int_{-\infty}^{\infty} S_{nn}(\omega)|H(\omega)|^2\,d\omega \int_{-\infty}^{\infty}\frac{|S(\omega)|^2}{S_{nn}(\omega)}\,d\omega \tag{7.158}$$

The equality holds if

$$\sqrt{S_{nn}(\omega)}\,H(\omega) = k\frac{S^*(\omega)e^{-j\omega T}}{\sqrt{S_{nn}(\omega)}} \tag{7.159}$$

Substituting Eq. (*7.158*) into Eq. (*7.157*), we obtain

$$\left(\frac{S}{N}\right)_o \le \frac{1}{2\pi}\int_{-\infty}^{\infty}\frac{|S(\omega)|^2}{S_{nn}(\omega)}\,d\omega \tag{7.160}$$

The maximum value of $(S/N)_o$ occurs when the equality holds in Eq. (*7.160*) or Eq. (*7.158*). Thus, from Eq. (*7.159*) the optimum filter frequency response $H_0(\omega)$ is

$$H_0(\omega) = k\frac{S^*(\omega)e^{-j\omega T}}{S_{nn}(\omega)} \tag{7.161}$$

An optimum filter given by Eq. (*7.161*) is called a *matched filter for colored noise*.

7.27. Referring to Eqs. (*7.90*) and (*7.91*), we have the output SNR of a matched filter receiver as

$$\left(\frac{S}{N}\right)_o = \frac{2E_d}{\eta} = \frac{2}{\eta}\int_0^T [s_1(t) - s_2(t)]^2\,dt \tag{7.162}$$

Now suppose that we want $s_1(t)$ and $s_2(t)$ to have the same signal energy. Show that the optimum choice of $s_2(t)$ is

$$s_2(t) = -s_1(t) \tag{7.163}$$

and the resultant output SNR is

$$\left(\frac{S}{N}\right)_o = \frac{8}{\eta}\int_0^T s_1^2(t)\,dt = \frac{8E}{\eta} \tag{7.164}$$

where E is the signal energy.

From Eq. (*7.162*)

$$\left(\frac{S}{N}\right)_o = \frac{2}{\eta}\int_0^T [s_1^2(t) + s_2^2(t) - 2s_1(t)s_2(t)]\,dt$$

$$= \frac{4E}{\eta} - \frac{4}{\eta}\int_0^T s_1(t)s_2(t)\,dt \tag{7.165}$$

Using Schwarz inequality (*7.150*), we obtain

$$\left|\int_0^T s_1(t)s_2(t)\,dt\right| \le \sqrt{\int_0^T s_1^2(t)\,dt \int_0^T s_2^2(t)\,dt} = E$$

The equality holds when

$$s_2(t) = ks_1(t)$$

Equal energy requirement implies $k = \pm 1$, and maximizing $(S/N)_o$ requires $k = -1$. Hence

$$s_2(t) = -s_1(t)$$

Substituting this relation in Eq. (*7.165*), we obtain

$$\left(\frac{S}{N}\right)_o = \frac{4E}{\eta} + \frac{4}{\eta}\int_0^T s_1^2(t)\,dt = \frac{8E}{\eta}$$

ERROR PROBABILITY PERFORMANCE OF BINARY TRANSMISSION SYSTEMS

7.28. Derive Eq. (*7.96*), that is,

$$P_e = Q\left(\sqrt{\frac{2A^2T}{\eta}}\right) = Q\left(\sqrt{\frac{2E_b}{\eta}}\right)$$

From Eq. (*7.95*)

$$s_i(t) = \begin{cases} s_1(t) = +A & 0 \le t \le T \\ s_2(t) = -A & 0 \le t \le T \end{cases}$$

Then by Eqs. (*7.91*) and (*7.92*) we obtain

$$E_d = \int_0^T [s_1(t) - s_2(t)]^2\,dt$$

$$= \int_0^T (2A)^2\,dt = 4A^2T$$

$$P_e = Q\left(\sqrt{\frac{E_d}{2\eta}}\right) = Q\left(\sqrt{\frac{2A^2T}{\eta}}\right) = Q\left(\sqrt{\frac{2E_b}{\eta}}\right)$$

where the average energy per bit is $E_b = A^2T$. Note that Eq. (*7.96*) is the same as Eq. (*7.144*) obtained in Prob. 7.21.

7.29. A bipolar binary signal $s_i(t)$ is a $+1$-V or -1-V pulse during the interval $(0, T)$. Additive white noise with power spectral density $\eta/2 = 10^{-5}$ W/Hz is added to the signal. Determine the maximum bit rate that can be sent with a bit error probability of $P_e \le 10^{-4}$.

By Eq. (*7.96*)

$$P_e = Q\left(\sqrt{\frac{2A^2T}{\eta}}\right) \le 10^{-4}$$

From Table C-1

$$Q(x) = 10^{-4} \longrightarrow x = 3.71$$

Hence,

$$\sqrt{\frac{2A^2T}{\eta}} = \sqrt{\frac{2(1)^2T}{2(10^{-5})}} = 3.71$$

from which

$$T = (3.71)^2(10^{-5}) = 13.76(10^{-5}) \text{ s}$$

Thus, the maximum bit rate R is

$$R = \frac{1}{T} = 7.26(10^3) \text{ b/s} = 7.26 \text{ kb/s}$$

7.30. Derive Eq. (*7.100*), that is,

$$P_e = Q\left(\sqrt{\frac{A^2T}{\eta}}\right) = Q\left(\sqrt{\frac{2E_b}{\eta}}\right)$$

From Eq. (*7.99*)

$$s_i(t) = \begin{cases} s_1(t) = A\cos\omega_c t & 0 \le t \le T \\ s_2(t) = A\cos(\omega_c t + \pi) & \\ \quad\;\; = -A\cos\omega_c t & 0 \le t \le T \end{cases}$$

with T an integer times $1/f_c$. Again, by Eqs. (*7.91*) and (*7.92*), we obtain

$$E_d = \int_0^T [s_1(t) - s_1(t)]^2\, dt$$

$$= \int_0^T (2A\cos\omega_c t)^2\, dt = 2A^2T$$

and

$$P_e = Q\left(\sqrt{\frac{E_d}{2\eta}}\right) = Q\left(\sqrt{\frac{A^2T}{\eta}}\right) = Q\left(\sqrt{\frac{2E_b}{\eta}}\right)$$

where $E_b = A^2T/2$ is the average signal energy per bit.

7.31. An on-off binary system uses the pulse waveforms

$$s_i(t) = \begin{cases} s_1(t) = A\sin\dfrac{\pi t}{T} & 0 \le t \le T \\ s_2(t) = 0 & 0 \le t \le T \end{cases}$$

Let $A = 0.2$ mV and $T = 2\ \mu$s. Additive white noise with a power spectral density $\eta/2 = 10^{-15}$ W/Hz is added to the signal. Determine the probability of error when $P(s_1) = P(s_2) = \frac{1}{2}$.

From Eq. (*7.92*)

$$P_e = Q\left(\sqrt{\frac{E_d}{2\eta}}\right)$$

where

$$E_d = \int_0^T [s_1(t) - s_2(t)]^2\,dt$$

$$= \int_0^T A^2 \sin^2 \frac{\pi t}{T}\,dt = \frac{A^2 T}{2}$$

and

$$\frac{E_d}{2\eta} = \frac{A^2 T}{4\eta} = \frac{(2\times 10^{-4})^2(2\times 10^{-6})}{4(2\times 10^{-15})} = 10$$

From Table C-1

$$P_e = Q(\sqrt{10}) = 7.83(10^{-4})$$

7.32. A binary system uses the pulse waveforms

$$s_i(t) = \begin{cases} s_1(t) = A_1 \sin \dfrac{\pi t}{T} & 0 \le t \le T \\ s_2(t) = -s_1(t) & 0 \le t \le T \end{cases}$$

Determine the probability of error under the same conditions given in Prob. 7.31 except for the value of A_1 which is chosen such that the system transmits the same average power as in Prob. 7.31.

In Prob. 7.31 the average power transmitted is

$$P_{\text{av}} = \frac{1}{2}\left(\frac{A^2}{2}\right) + \frac{1}{2}(0^2) = \frac{A^2}{4}$$

With the present case, the average power transmitted is

$$P_{\text{av}} = \frac{1}{2}\left(\frac{A_1^2}{2}\right) + \frac{1}{2}\left(\frac{A_1^2}{2}\right) = \frac{A_1^2}{2}$$

Thus, for the same average power

$$A_1 = \frac{A}{\sqrt{2}} = \frac{2(10^{-4})}{\sqrt{2}} = \sqrt{2}\,(10^{-4})\text{ V} = 0.14\text{ mV}$$

From Eq. (7.92)

$$P_e = Q\left(\sqrt{\frac{E_d}{2\eta}}\right)$$

where

$$E_d = \int_0^T [s_1(t) - s_2(t)]^2\,dt$$

$$= \int_0^T 4A_1^2 \sin^2 \frac{\pi t}{T}\,dt = 2A_1^2 T$$

and

$$\frac{E_d}{2\eta} = \frac{A_1^2 T}{\eta} = \frac{(\sqrt{2}\times 10^{-4})^2(2\times 10^{-6})}{2\times 10^{-15}} = 20$$

From Table C-1

$$P_e = Q(\sqrt{20}) = 3.88(10^{-6})$$

7.33. Derive Eq. (*7.102*), that is,

$$P_e = Q\left(\sqrt{\frac{A^2T}{2\eta}}\right) = Q\left(\sqrt{\frac{E_b}{\eta}}\right)$$

where $E_b = A^2T/2$ is the average signal energy per bit.

From Eq. (*7.101*)

$$s_i(t) = \begin{cases} s_1(t) = A\cos\omega_1 t & 0 \le t \le T \\ s_2(t) = A\cos\omega_2 t & 0 \le t \le T \end{cases}$$

Then
$$s_1(t) - s_2(t) = A\cos\omega_1 t - A\cos\omega_2 t$$

and
$$\begin{aligned} E_d &= \int_0^T [s_1(t) - s_2(t)]^2\,dt \\ &= \int_0^T A^2(\cos\omega_1 t - \cos\omega_2 t)^2\,dt \\ &= A^2\int_0^T \left(\cos^2\omega_1 t + \cos^2\omega_2 t - 2\cos\omega_1 t\cos\omega_2 t\right)dt \\ &= A^2T\left[1 + \frac{\sin 2\omega_1 T}{4\omega_1 T} + \frac{\sin 2\omega_2 T}{4\omega_2 T} - \frac{\sin(\omega_1 - \omega_2)T}{2(\omega_1 - \omega_2)T} - \frac{\sin 2(\omega_1 + \omega_2)T}{2(\omega_1 + \omega_2)T}\right] \end{aligned}$$

If we assume $\omega_1 T \gg 1$, $\omega_2 T \gg 1$, and $(\omega_1 - \omega_2)T \gg 1$, then

$$E_d \approx A^2T$$

Thus,
$$P_e = Q\left(\sqrt{\frac{E_d}{2\eta}}\right) = Q\left(\sqrt{\frac{A^2T}{2\eta}}\right) = Q\left(\sqrt{\frac{E_b}{\eta}}\right)$$

Supplementary Problems

7.34. Rewrite Eqs. (*7.19*) and (*7.26*) in terms of $\gamma_p = S_p/(\eta B)$, where S_p is the peak envelope power of the DSB or AM signal.

Ans. DSB: $\left(\frac{S}{N}\right)_o = \frac{1}{2}S_X\gamma_p$; AM: $\left(\frac{S}{N}\right)_o = \frac{1}{8}S_X\gamma_p$

7.35. An AM receiver operates with a tone modulation, and the modulation index $\mu = 0.3$. The message signal is $20\cos 1000\pi t$.

(*a*) Compute the output SNR relative to the baseband performance.

(*b*) Determine the improvement (in decibels) in the output SNR that results if μ is increased from 0.3 to 0.7.

Ans. (*a*) $\left(\frac{S}{N}\right)_o = 0.043\gamma$, (*b*) 6.6 dB

7.36. An AM system with envelope detection is operating at threshold. Find the power gain in decibels needed at the transmitter to produce $(S/N)_o = 30$ dB for tone modulation with $\mu = 1$.

Ans. ≈ 22 dB

7.37. An AM system with envelope detection has $(S/N)_o = 30$ dB and tone modulation with $\mu = 1$ with $B = 8$ kHz. If all bandwidths are increased accordingly while other parameters remain fixed, what is the largest usable value of B?

Ans. 1.2 MHz

7.38. Find the detector gain α_d for an SSB system.

Ans. $\alpha_d = 1$

7.39. Find the output SNR in a PM system for tone modulation.

Ans. $\left(\dfrac{S}{N}\right)_o = \dfrac{1}{2}\beta^2\gamma$

7.40. Find the detection gain α_d for an FM system with $\beta = 2$.

Ans. $\alpha_d = 36$.

7.41. Show that for tone modulation, FM is superior to PM by a factor of 3 from the SNR point of view.

Hint: Use the result of Prob. 7.39 and Eq. (*7.123*).

7.42. For a modulating signal $X(t) = \cos^3 \omega_m t$, show that PM is superior to FM by a factor of 2.25 from the output SNR point of view.

Hint: Use Eqs. (*7.56*) and (*7.62*).

7.43. Consider a communication system with the following characteristics:

$$S_x = E[X^2(t)] = \frac{1}{2} \qquad B = 10 \text{ kHz} \qquad \frac{\eta}{2} = 10^{-12} \text{ W/Hz}$$

Transmission loss = 70 dB

Calculate the required transmission power S_T needed to achieve $(S/N)_o = 40$ dB when the modulation is (*a*) SSB, (*b*) AM with $\mu = 1$ and $\mu = 0.5$, (*c*) PM with $k_p = \pi$, (*d*) FM with $D = 1$ and $D = 5$.

Ans. (*a*) SSB, $S_T = 1$ kW; (*b*) AM; $\mu = 1$, $S_T = 3$ kW; $\mu = 0.5$, $S_T = 9$ kW; (*c*) PM, $S_T = 202.6$ W; (*d*) FM; $D = 1$, $S_T = 667$ W; $D = 5$, $S_T = 26.7$ W

7.44. A binary communication system transmits signals $s_i(t)$ $(i = 1, 2)$. The receiver test statistic $z(T)$ is

$$z(T) = a_i + n_o$$

where $a_1 = +1$ and $a_2 = -1$ and n_o is uniformly distributed, yielding the conditional density functions $f(z|s_i)$ given by

$$f(z|s_1) = \begin{cases} \frac{1}{2} & -0.1 \le z \le 1.9 \\ 0 & \text{otherwise} \end{cases}$$

$$f(z|s_2) = \begin{cases} \frac{1}{2} & -1.9 \le z \le 0.1 \\ 0 & \text{otherwise} \end{cases}$$

Find the probability of error P_e for the case of equally likely signals, using an optimum decision threshold.

Ans. $P_e = 0.05$

7.45. A binary communication system transmits signals $s_i(t)$ $(i = 1, 2)$ with equal probability. The receiver test statistic $z(T)$ is

$$z(T) = a_i(T) + n_o(T)$$

where $a_1(T) = +1$ and $a_2(T) = -1$ and $n_o(T)$ is a zero-mean gaussian random variable with variance 0.1.

(*a*) Determine the optimum decision rule.

(*b*) Calculate the probability of error.

Ans. (*a*) $z \underset{H_2}{\overset{H_1}{\gtrless}} 0$ (*b*) $P_e = 7.8(10^{-4})$

7.46. A binary communication system transmits signals $s_i(t)$ $(i = 1, 2)$ with the probabilities $P(s_1) = 0.75$ and $P(s_2) = 0.25$. The receiver test statistic $z(T)$ is

$$z(T) = a_i(T) + n_o(T)$$

where $a_1(T) = 1$ and $a_2(T) = 0$ and $n_o(T)$ is a zero-mean gaussian random variable with variance 0.1.

(*a*) Determine the optimum decision rule.

(*b*) Calculate the probability of error.

Hint: Use Eq. (*7.140*) of Prob. 7.19 and Eq. (*7.69*).

Ans. (*a*) $Z \underset{H_2}{\overset{H_1}{\gtrless}} 0.61$, (*b*) $P_e = 0.0883$

7.47. Compute the matched filter output over $(0, T)$ to the pulse waveform

$$s(t) = \begin{cases} e^{-t} & 0 \le t \le T \\ 0 & \text{otherwise} \end{cases}$$

Ans. $e^{-T} \sinh t$

7.48. Derive Eqs. (*7.94*) and (*7.98*).

Hint: Use Eqs. (*7.91*) and (*7.92*).

7.49. It is required to transmit 2.08 Mb/s with an error probability of $P_e \le 10^{-6}$. The channel noise power spectrum is $\eta/2 = 10^{-11}$ W/Hz. Determine the signal power required at the receiver input, using polar signaling.

Ans. 0.47 mW

7.50. In a PSK system, the received waveforms $s_1(t) = A \cos \omega_c t$ and $s_2(t) = -A \cos \omega_c t$ are coherently detected with a matched filter. The value of A is 20 mV, and the bit rate is 1 Mb/s. Assume that the noise power spectral density $\eta/2 = 10^{-11}$ W/Hz. Find the probability of error P_e.

Ans. $P_e = 3.9(10^{-6})$

Chapter 8

Information and Coding

8.1 INTRODUCTION

The purpose of a communication system is, in the broadest sense, the transmission of information from one point to another with high efficiency and reliability. In the preceding chapters we have examined several ways of accomplishing this goal by use of electric signals.

Information theory provides a quantitative measure of the information contained in message signals and allows us to determine the capacity of a communication system to transfer this information from source to destination. Through the use of coding, a major topic of information theory, redundancy can be reduced from message signals so that channels can be used with improved efficiency. In addition, systematic redundancy can be introduced to the transmitted signal so that channels can be used with improved reliability.

In this chapter we briefly explore some basic ideas involved in information and coding theory.

8.2 MEASURE OF INFORMATION

A. Information Sources:

An information source is an object that produces an event, the outcome of which is selected at random according to a probability distribution. A practical source in a communication system is a device that produces messages, and it can be either analog or discrete. In this chapter we deal mainly with the discrete sources since analog sources can be transformed to discrete sources through the use of sampling and quantization techniques, described in Chap. 4. A discrete information source is a source that has only a finite set of symbols as possible outputs. The set of source symbols is called the *source alphabet*, and the elements of the set are called *symbols* or *letters*.

Information sources can be classified as having memory or being memoryless. A source with memory is one for which a current symbol depends on the previous symbols. A memoryless source is one for which each symbol produced is independent of the previous symbols.

A *discrete memoryless source* (DMS) can be characterized by the list of the symbols, the probability assignment to these symbols, and the specification of the rate of generating these symbols by the source.

B. Information Content of a Discrete Memoryless Source:

The amount of information contained in an event is closely related to its uncertainty. Messages containing knowledge of high probability of occurrence convey relatively little information. We note that if an event is certain (that is, the event occurs with probability 1), it conveys zero information. Thus a mathematical measure of information should be a function of the probability of the outcome and should satisfy the following axioms:

1. Information should be proportional to the uncertainty of an outcome.
2. Information contained in independent outcomes should add.

1. Information Content of a Symbol:

Consider a DMS, denoted by X, with alphabet $\{x_1, x_2, \ldots, x_m\}$. The *information content* of a symbol x_i, denoted by $I(x_i)$, is defined by

$$I(x_i) = \log_b \frac{1}{P(x_i)} = -\log_b P(x_i) \tag{8.1}$$

where $P(x_i)$ is the probability of occurrence of symbol x_i. Note that $I(x_i)$ satisfies the following properties:

$$I(x_i) = 0 \qquad \text{for } P(x_i) = 1 \tag{8.2}$$

$$I(x_i) \geq 0 \tag{8.3}$$

$$I(x_i) > I(x_j) \qquad \text{if } P(x_i) < P(x_j) \tag{8.4}$$

$$I(x_i x_j) = I(x_i) + I(x_j) \qquad \text{if } x_i \text{ and } x_j \text{ are independent} \tag{8.5}$$

The unit of $I(x_i)$ is the bit (*bi*nary uni*t*) if $b = 2$, hartley if $b = 10$, and nat (*na*tural uni*t*) if $b = e$. It is standard to use $b = 2$. Here the unit bit (abbreviated "b") is a measure of information content and is not to be confused with the term *bit* meaning "binary digit." The conversion of these units to other units can be achieved by the following relationships.

$$\log_2 a = \frac{\ln a}{\ln 2} = \frac{\log a}{\log 2} \tag{8.6}$$

2. *Average Information or Entropy:*

In a practical communication system, we usually transmit long sequences of symbols from an information source. Thus we are more interested in the average information that a source produces than the information content of a single symbol.

The mean value of $I(x_i)$ over the alphabet of source X with m different symbols is given by

$$H(X) = E[I(x_i)] = \sum_{i=1}^{m} P(x_i) I(x_i)$$

$$= -\sum_{i=1}^{m} P(x_i) \log_2 P(x_i) \qquad \text{b/symbol} \tag{8.7}$$

The quantity $H(X)$ is called the *entropy* of source X. It is a measure of the *average information content per source symbol*. The source entropy $H(X)$ can be considered as the average amount of uncertainty within source X that is resolved by use of the alphabet.

Note that for a binary source X that generates independent symbols 0 and 1 with equal probability, the source entropy $H(X)$ is

$$H(X) = -\tfrac{1}{2} \log_2 \tfrac{1}{2} - \tfrac{1}{2} \log_2 \tfrac{1}{2} = 1 \text{ b/symbol} \tag{8.8}$$

The source entropy $H(X)$ satisfies the following relation:

$$0 \leq H(X) \leq \log_2 m \tag{8.9}$$

where m is the size (number of symbols) of the alphabet of source X (Prob. 8.4). The lower bound corresponds to no uncertainty which occurs when one symbol has probability $P(x_i) = 1$ while $P(x_j) = 0$ for $j \neq i$, so X emits the same symbol x_i all the time. The upper bound corresponds to the maximum uncertainty which occurs when $P(x_i) = 1/m$ for all i, that is, when all symbols are equally likely to be emitted by X.

3. *Information Rate:*

If the time rate at which source X emits symbols is r (symbols/s), the *information rate R* of the source is given by

$$R = rH(X) \qquad \text{b/s} \tag{8.10}$$

8.3 DISCRETE MEMORYLESS CHANNELS

A. Channel Representation:

A communication channel is the path or medium through which the symbols flow to the receiver. A *discrete memoryless channel* (DMC) is a statistical model with an input X and an output Y (Fig. 8-1) During each unit of time (signaling interval), the channel accepts an input symbol from X, and in response it generates an output symbol from Y. The channel is "discrete" when the alphabets of X and Y are both finite. It is "memoryless" when the current output depends on only the current input and not on any of the previous inputs.

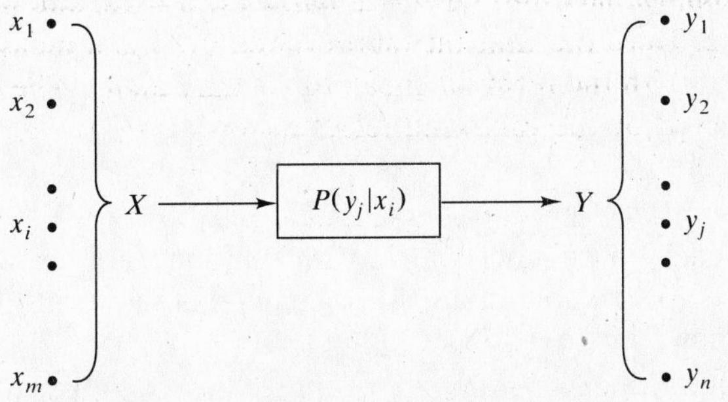

Fig. 8-1 Discrete memoryless channel

A diagram of a DMC with m inputs and n outputs is illustrated in Fig. 8-1. The input X consists of input symbols $x_1, x_2, \ldots, x_m$. The a priori probabilities of these source symbols $P(x_i)$ are assumed to be known. The output Y consists of output symbols $y_1, y_2, \ldots, y_n$. Each possible input-to-output path is indicated along with a conditional probability $P(y_j|x_i)$, where $P(y_j|x_i)$ is the conditional probability of obtaining output y_j given that the input is x_i, and is called a *channel transition probability*.

B. Channel Matrix:

A channel is completely specified by the complete set of transition probabilities. Accordingly, the channel of Fig. 8-1 is often specified by the matrix of transition probabilities $[P(Y|X)]$, given by

$$[P(Y|X)] = \begin{bmatrix} P(y_1|x_1) & P(y_2|x_1) & \cdots & P(y_n|x_1) \\ P(y_1|x_2) & P(y_2|x_2) & \cdots & P(y_n|x_2) \\ \cdots & \cdots & \cdots & \cdots \\ P(y_1|x_m) & P(y_2|x_m) & \cdots & P(y_n|x_m) \end{bmatrix} \tag{8.11}$$

The matrix $[P(Y|X)]$ is called the *channel matrix*. Since each input to the channel results in some output, each row of the channel matrix must sum to unity, that is,

$$\sum_{j=1}^{n} P(y_j|x_i) = 1 \quad \text{for all } i \tag{8.12}$$

Now, if the input probabilities $P(X)$ are represented by the row matrix

$$[P(X)] = [P(x_1) \quad P(x_2) \quad \cdots \quad P(x_m)] \tag{8.13}$$

and the output probabilities $P(Y)$ are represented by the row matrix

$$[P(Y)] = [P(y_1) \quad P(y_2) \quad \cdots \quad P(y_n)] \tag{8.14}$$

then

$$[P(Y)] = [P(X)][P(Y|X)] \tag{8.15}$$

If $P(X)$ is represented as a diagonal matrix

$$[P(X)]_d = \begin{bmatrix} P(x_1) & 0 & \cdots & 0 \\ 0 & P(x_2) & \cdots & 0 \\ \cdots & \cdots & \cdots & \cdots \\ 0 & 0 & \cdots & P(x_m) \end{bmatrix} \tag{8.16}$$

then

$$[P(X,Y)] = [P(X)]_d[P(Y|X)] \tag{8.17}$$

where the (i, j) element of matrix $[P(X,Y)]$ has the form $P(x_i, y_j)$. The matrix $[P(X,Y)]$ is known as the *joint probability matrix*, and the element $P(x_i, y_j)$ is the joint probability of transmitting x_i and receiving y_j.

C. Special Channels:

1. *Lossless Channel:*

A channel described by a channel matrix with only one nonzero element in each column is called a *lossless channel*. An example of a lossless channel is shown in Fig. 8-2, and the corresponding channel matrix is shown in Eq. (*8.18*).

$$[P(Y|X)] = \begin{bmatrix} \frac{3}{4} & \frac{1}{4} & 0 & 0 & 0 \\ 0 & 0 & \frac{1}{3} & \frac{2}{3} & 0 \\ 0 & 0 & 0 & 0 & 1 \end{bmatrix} \tag{8.18}$$

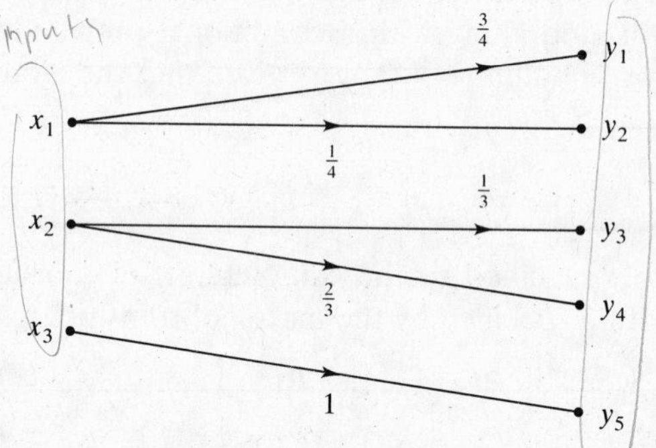

Fig. 8-2 Lossless channel

It can be shown that in the lossless channel no source information is lost in transmission. [See Eq. (*8.35*) and Prob. 8.10.]

2. *Deterministic Channel:*

A channel described by a channel matrix with only one nonzero element in each row is called a *deterministic channel*. An example of a deterministic channel is shown in Fig. 8-3, and the corresponding channel matrix is shown in Eq. (*8.19*).

$$[P(Y|X)] = \begin{bmatrix} 1 & 0 & 0 \\ 1 & 0 & 0 \\ 0 & 1 & 0 \\ 0 & 1 & 0 \\ 0 & 0 & 1 \end{bmatrix} \tag{8.19}$$

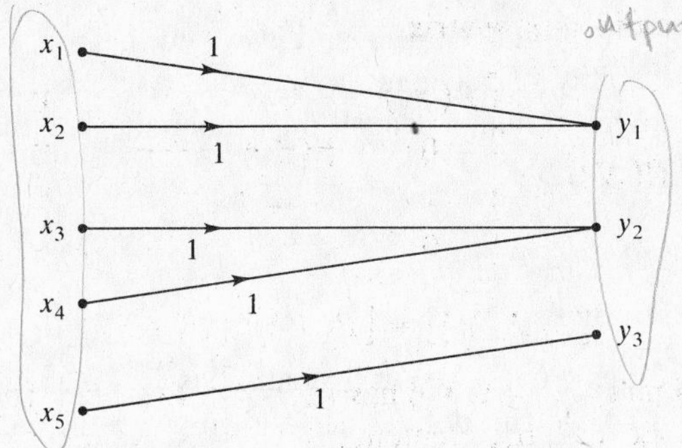

Fig. 8-3 Deterministic channel

Note that since each row has only one nonzero element, this element must be unity by Eq. (8.12). Thus, when a given source symbol is sent in the deterministic channel, it is clear which output symbol will be received.

3. ***Noiseless Channel:***

A channel is called *noiseless* if it is both lossless and deterministic. A noiseless channel is shown in Fig. 8-4. The channel matrix has only one element in each row and in each column, and this element is unity. Note that the input and output alphabets are of the same size; that is, $m = n$ for the noiseless channel.

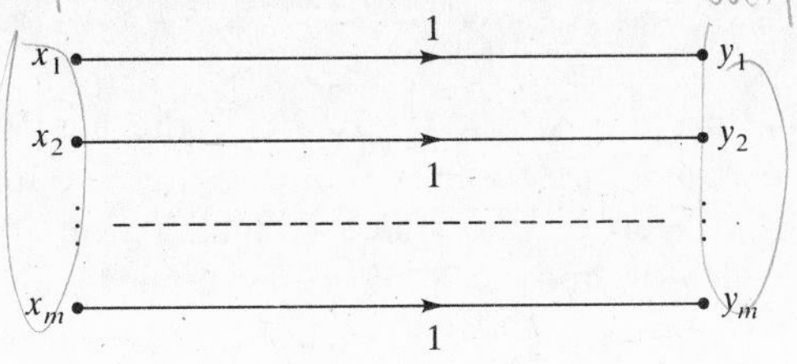

Fig. 8-4 Noiseless channel

4. ***Binary Symmetric Channel:***

The *binary symmetric channel* (BSC) is defined by the channel diagram shown in Fig. 8-5, and its channel matrix is given by

$$[P(Y|X)] = \begin{bmatrix} 1-p & p \\ p & 1-p \end{bmatrix} \tag{8.20}$$

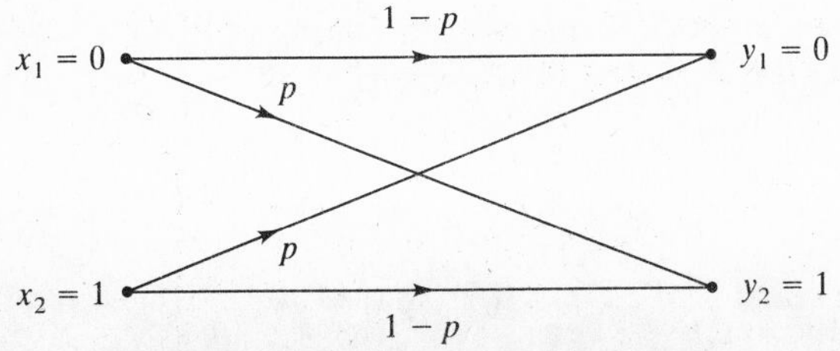

Fig. 8-5 Binary symmetric channel

The channel has two inputs ($x_1 = 0$, $x_2 = 1$) and two outputs ($y_1 = 0$, $y_2 = 1$). The channel is symmetric because the probability of receiving a 1 if a 0 is sent is the same as the probability of receiving a 0 if a 1 is sent. This common transition probability is denoted by p. (See Prob. 8.35.)

8.4 MUTUAL INFORMATION

A. Conditional and Joint Entropies:

Using the input probabilities $P(x_i)$, output probabilities $P(y_j)$, transition probabilities $P(y_j | x_i)$, and joint probabilities $P(x_i, y_j)$, we can define the following various entropy functions for a channel with m inputs and n outputs:

$$H(X) = -\sum_{i=1}^{m} P(x_i) \log_2 P(x_i) \tag{8.21}$$

$$H(Y) = -\sum_{j=1}^{n} P(y_j) \log_2 P(y_j) \tag{8.22}$$

$$H(X|Y) = -\sum_{j=1}^{n} \sum_{i=1}^{m} P(x_i, y_j) \log_2 P(x_i | y_j) \tag{8.23}$$

$$H(Y|X) = -\sum_{j=1}^{n} \sum_{i=1}^{m} P(x_i, y_j) \log_2 P(y_j | x_i) \tag{8.24}$$

$$H(X,Y) = -\sum_{j=1}^{n} \sum_{i=1}^{m} P(x_i, y_j) \log_2 P(x_i, y_j) \tag{8.25}$$

These entropies can be interpreted as follows: $H(X)$ is the average uncertainty of the channel input, and $H(Y)$ is the average uncertainty of the channel output. The conditional entropy $H(X|Y)$ is a measure of the average uncertainty remaining about the channel input after the channel output has been observed. And $H(X|Y)$ is sometimes called the *equivocation* of X with respect to Y. The conditional entropy $H(Y|X)$ is the average uncertainty of the channel output given that X was transmitted. The joint entropy $H(X,Y)$ is the average uncertainty of the communication channel as a whole.

Two useful relationships among the above various entropies are

$$H(X,Y) = H(X|Y) + H(Y) \tag{8.26}$$

$$H(X,Y) = H(Y|X) + H(X) \tag{8.27}$$

B. Mutual Information:

The *mutual information* $I(X;Y)$ of a channel is defined by

$$I(X;Y) = H(X) - H(X|Y) \quad \text{b/symbol} \tag{8.28}$$

Since $H(X)$ represents the uncertainty about the channel input before the channel output is observed and $H(X|Y)$ represents the uncertainty about the channel input after the channel output is observed, the mutual information $I(X;Y)$ represents the uncertainty about the channel input that is resolved by observing the channel output.

Properties of $I(X;Y)$:

1. $$I(X;Y) = I(Y;X) \tag{8.29}$$

2. $$I(X;Y) \geq 0 \tag{8.30}$$

3. $$I(X;Y) = H(Y) - H(Y|X) \tag{8.31}$$

4. $$I(X;Y) = H(X) + H(Y) - H(X,Y) \tag{8.32}$$

8.5 CHANNEL CAPACITY

A. Channel Capacity per Symbol C_s:

The *channel capacity per symbol* of a DMC is defined as

$$C_s = \max_{\{P(x_i)\}} I(X;Y) \quad \text{b/symbol} \tag{8.33}$$

where the maximization is over all possible input probability distributions $\{P(x_i)\}$ on X. Note that the channel capacity C_s is a function of only the channel transition probabilities which define the channel.

B. Channel Capacity per Second C:

If r symbols are being transmitted per second, then the maximum rate of transmission of information per second is rC_s. This is the *channel capacity per second* and is denoted by C (b/s):

$$C = rC_s \quad \text{b/s} \tag{8.34}$$

C. Capacities of Special Channels:

1. Lossless Channel:

For a lossless channel, $H(X|Y) = 0$ (Prob. 8.10) and

$$I(X;Y) = H(X) \tag{8.35}$$

Thus, the mutual information (information transfer) is equal to the input (source) entropy, and no source information is lost in transmission. Consequently the channel capacity per symbol is

$$C_s = \max_{\{P(x_i)\}} H(X) = \log_2 m \tag{8.36}$$

where m is the number of symbols in X.

2. Deterministic Channel:

For a deterministic channel, $H(Y|X) = 0$ for all input distributions $P(x_i)$, and

$$I(X;Y) = H(Y) \tag{8.37}$$

Thus, the information transfer is equal to the output entropy. The channel capacity per symbol is

$$C_s = \max_{\{P(x_i)\}} H(Y) = \log_2 n \tag{8.38}$$

where n is the number of symbols in Y.

3. Noiseless Channel:

Since a noiseless channel is both lossless and deterministic, we have

$$I(X;Y) = H(X) = H(Y) \tag{8.39}$$

and the channel capacity per symbol is

$$C_s = \log_2 m = \log_2 n \tag{8.40}$$

4. Binary Symmetric Channel:

For the BSC of Fig. 8-5, the mutual information is (Prob. 8.16)

$$I(X;Y) = H(Y) + p\log_2 p + (1-p)\log_2(1-p) \tag{8.41}$$

and the channel capacity per symbol is

$$C_s = 1 + p \log_2 p + (1-p)\log_2(1-p) \tag{8.42}$$

8.6 ADDITIVE WHITE GAUSSIAN NOISE CHANNEL

In a continuous channel an information source produces a continuous signal $x(t)$. The set of possible signals is considered as an ensemble of waveforms generated by some ergodic random process. It is further assumed that $x(t)$ has a finite bandwidth so that $x(t)$ is completely characterized by its periodic sample values. Thus, at any sampling instant, the collection of possible sample values constitutes a continuous random variable X described by its probability density function $f_X(x)$.

A. Differential Entropy:

The average amount of information per sample value of $x(t)$ is measured by

$$H(X) = -\int_{-\infty}^{\infty} f_X(x)\log_2 f_X(x)\,dx \qquad \text{b/sample} \tag{8.43}$$

The entropy $H(X)$ defined by Eq. (*8.43*) is known as the *differential entropy* of X.

The average mutual information in a continuous channel is defined (by analogy with the discrete case) as

$$I(X;Y) = H(X) - H(X|Y)$$

or

$$I(X;Y) = H(Y) - H(Y|X)$$

where

$$H(Y) = -\int_{-\infty}^{\infty} f_Y(y)\log_2 f_Y(y)\,dy \tag{8.44}$$

$$H(X|Y) = -\int_{-\infty}^{\infty}\int_{-\infty}^{\infty} f_{XY}(x,y)\log_2 f_X(x|y)\,dx\,dy \tag{8.45a}$$

$$H(Y|X) = -\int_{-\infty}^{\infty}\int_{-\infty}^{\infty} f_{XY}(x,y)\log_2 f_Y(y|x)\,dx\,dy \tag{8.45b}$$

B. Additive White Gaussian Noise Channel:

In an additive white gaussian noise (AWGN) channel, the channel output Y is given by

$$Y = X + n \tag{8.46}$$

where X is the channel input and n is an additive band-limited white gaussian noise with zero mean and variance σ^2.

The capacity C_s of an AWGN channel is given by

$$C_s = \max_{\{f_X(x)\}} I(X;Y) = \frac{1}{2}\log_2\left(1+\frac{S}{N}\right) \qquad \text{b/sample} \tag{8.47}$$

where S/N is the signal-to-noise ratio at the channel output. If the channel bandwidth B Hz is fixed, then the output $y(t)$ is also a band-limited signal completely characterized by its periodic sample values taken at the Nyquist rate $2B$ samples/s. Then the capacity C (b/s) of the AWGN channel is given by

$$C = 2BC_s = B\log_2\left(1+\frac{S}{N}\right) \qquad \text{b/s} \tag{8.48}$$

Equation (*8.48*) is known as the *Shannon-Hartley law*.

The Shannon-Hartley law underscores the fundamental role of bandwidth and signal-to-noise ratio in communication. It also shows that we can exchange increased bandwidth for decreased signal power (Prob. 8.24) for a system with given capacity C.

8.7 SOURCE CODING

A conversion of the output of a DMS into a sequence of binary symbols (binary code word) is called *source coding*. The device which performs this conversion is called the *source encoder* (Fig. 8-6).

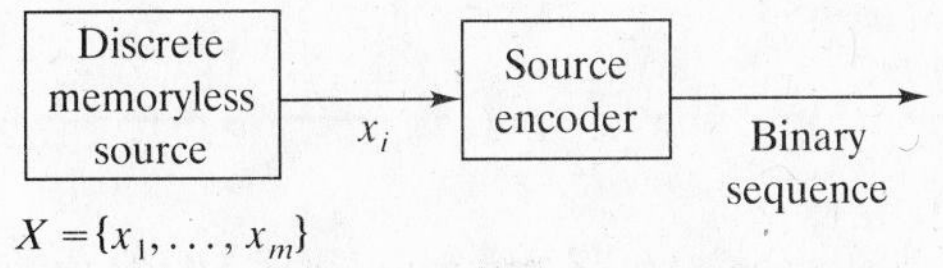

Fig. 8-6 Source coding

An objective of source coding is to minimize the average bit rate required for representation of the source by reducing the redundancy of the information source.

A. Code Length and Code Efficiency:

Let X be a DMS with finite entropy $H(X)$ and an alphabet $\{x_1, \ldots, x_m\}$ with corresponding probabilities of occurrence $P(x_i)$ $(i = 1, \ldots, m)$. Let the binary code word assigned to symbol x_i by the encoder have length n_i, measured in bits. The length of a code word is the number of binary digits in the code word. The average code word length L per source symbol is given by

$$L = \sum_{i=1}^{m} P(x_i) n_i \tag{8.49}$$

The parameter L represents the average number of bits per source symbol used in the source coding process.

The *code efficiency* η is defined as

$$\eta = \frac{L_{\min}}{L} \tag{8.50}$$

where $L_{\min}$ is the minimum possible value of L. When η approaches unity, the code is said to be *efficient*.

The *code redundancy* γ is defined as

$$\gamma = 1 - \eta \tag{8.51}$$

B. Source Coding Theorem:

The source coding theorem states that for a DMS X with entropy $H(X)$, the average code word length L per symbol is bounded as

$$L \geq H(X) \tag{8.52}$$

and further, L can be made as close to $H(X)$ as desired for some suitably chosen code.

Thus, with $L_{\min} = H(X)$, the code efficiency can be rewritten as

$$\eta = \frac{H(X)}{L} \tag{8.53}$$

C. Classification of Codes:

Classification of codes is best illustrated by an example. Consider Table 8-1 where a source of size 4 has been encoded in binary codes with symbols 0 and 1.

Table 8-1 Binary Codes

x_i	Code 1	Code 2	Code 3	Code 4	Code 5	Code 6
x_1	00	00	0	0	0	1
x_2	01	01	1	10	01	01
x_3	00	10	00	110	011	001
x_4	11	11	11	111	0111	0001

1. *Fixed-Length Codes:*

A *fixed-length* code is one whose code word length is fixed. Code 1 and code 2 of Table 8-1 are fixed-length codes with length 2.

2. *Variable-Length Codes:*

A *variable-length* code is one whose code word length is not fixed. All codes of Table 8-1 except codes 1 and 2 are variable-length codes.

3. *Distinct Codes:*

A code is *distinct* if each code word is distinguishable from other code words. All codes of Table 8-1 except code 1 are distinct codes—notice the codes for x_1 and x_3.

4. *Prefix-Free Codes:*

A code in which no code word can be formed by adding code symbols to another code word is called a *prefix-free code*. Thus, in a prefix-free code no code word is a *prefix* of another. Codes 2, 4, and 6 of Table 8-1 are prefix-free codes.

5. *Uniquely Decodable Codes:*

A distinct code is *uniquely decodable* if the original source sequence can be reconstructed perfectly from the encoded binary sequence. Note that code 3 of Table 8-1 is not a uniquely decodable code. For example, the binary sequence 1001 may correspond to the source sequences $x_2x_3x_2$ or $x_2x_1x_1x_2$. A sufficient condition to ensure that a code is uniquely decodable is that no code word is a prefix of another. Thus, the prefix-free codes 2, 4, and 6 are uniquely decodable codes. Note that the prefix-free condition is not a necessary condition for unique decodability. For example, code 5 of Table 8-1 does not satisfy the prefix-free condition, and yet it is uniquely decodable since the bit 0 indicates the beginning of each code word of the code.

6. *Instantaneous Codes:*

A uniquely decodable code is called an *instantaneous code* if the end of any code word is recognizable without examining subsequent code symbols. The instantaneous codes have the property previously mentioned that no code word is a prefix of another code word. For this reason, prefix-free codes are sometimes called instantaneous codes.

7. *Optimal Codes:*

A code is said to be *optimal* if it is instantaneous and has minimum average length L for a given source with a given probability assignment for the source symbols.

D. Kraft Inequality:

Let X be a DMS with alphabet $\{x_i\}$ $(i = 1, 2, \ldots, m)$. Assume that the length of the assigned binary code word corresponding to x_i is n_i.

A necessary and sufficient condition for the existence of an instantaneous binary code is

$$K = \sum_{i=1}^{m} 2^{-n_i} \leq 1 \tag{8.54}$$

which is known as the *Kraft inequality*.

Note that the Kraft inequality assures us of the existence of an instantaneously decodable code with code word lengths that satisfy the inequality. But it does not show us how to obtain these code words, nor does it say that any code which satisfies the inequality is automatically uniquely decodable (Prob. 8.27).

8.8 ENTROPY CODING

The design of a variable-length code such that its average code word length approaches the entropy of the DMS is often referred to as *entropy coding*. In this section we present two examples of entropy coding.

A. Shannon-Fano Coding:

An efficient code can be obtained by the following simple procedure, known as *Shannon-Fano algorithm*:

1. List the source symbols in order of decreasing probability.
2. Partition the set into two sets that are as close to equiprobable as possible, and assign 0 to the upper set and 1 to the lower set.
3. Continue this process, each time partitioning the sets with as nearly equal probabilities as possible until further partitioning is not possible.

An example of Shannon-Fano encoding is shown in Table 8-2. Note that in Shannon-Fano encoding the ambiguity may arise in the choice of approximately equiprobable sets. (See Prob. 8.33.)

Table 8-2 Shannon-Fano Encoding

x_i	$P(x_i)$	Step 1	Step 2	Step 3	Step 4	Code 5
x_1	0.30	0	0			00
x_2	0.25	0	1			01
x_3	0.20	1	0			10
x_4	0.12	1	1	0		110
x_5	0.08	1	1	1	0	1110
x_6	0.05	1	1	1	1	1111

$$H(X) = 2.36 \text{ b/symbol}$$
$$L = 2.38 \text{ b/symbol}$$
$$\eta = H(X)/L = 0.99$$

B. Huffman Encoding:

In general, Huffman encoding results in an optimum code. Thus it is the code that has the highest efficiency (Prob. 8.34). The Huffman encoding procedure is as follows:

1. List the source symbols in order of decreasing probability.
2. Combine the probabilities of the two symbols having the lowest probabilities, and reorder the resultant probabilities; this step is called reduction 1. The same procedure is repeated until there are two ordered probabilities remaining.
3. Start encoding with the last reduction, which consists of exactly two ordered probabilities. Assign 0 as the first digit in the code words for all the source symbols associated with the first probability; assign 1 to the second probability.
4. Now go back and assign 0 and 1 to the second digit for the two probabilities that were combined in the previous reduction step, retaining all assignments made in step 3.
5. Keep regressing this way until the first column is reached.

An example of Huffman encoding is shown in Table 8-3.

Table 8-3 Huffman Encoding

x_i	$P(x_i)$	Code							
x_1	0.30	00	0.30	00	0.30	00	0.45	1	0.55	0
x_2	0.25	01	0.25	01	0.25	01	0.30	00	0.45	1
x_3	0.20	11	0.20	11	0.25	10	0.25	01		
x_4	0.12	101	0.13	100	0.20	11				
x_5	0.08	1000	0.12	101						
x_6	0.05	1001								

$$H(X) = 2.36 \text{ b/symbol}$$
$$L = 2.38 \text{ b/symbol}$$
$$\eta = 0.99$$

8.9 CHANNEL CODING

In this section, we treat the subject of designing codes for the reliable transmission of digital information over a noisy channel.

A. Channel Coding:

A basic block diagram for the channel coding is shown in Fig. 8-7. The binary message sequence at the input of the channel encoder may be the output of a source encoder or the output of a source

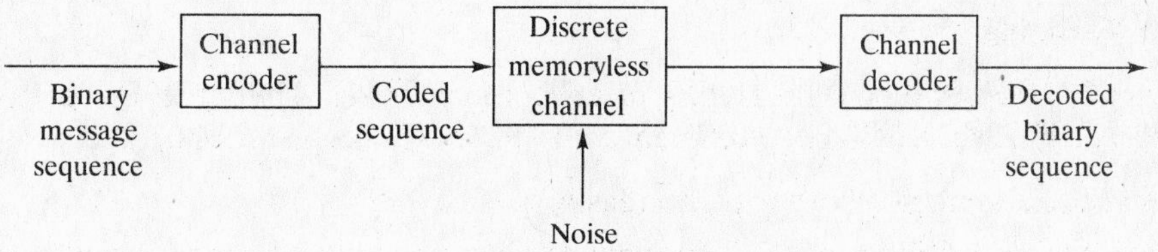

Fig. 8-7 Channel coding

directly. The channel encoder introduces systematic redundancy into the data stream by adding bits to the message bits in such a way as to facilitate the detection and/or correction of bit errors in the original binary message sequence at the receiver.

B. Channel Coding Theorem:

The channel coding theorem for a DMC is stated as follows:

Given a DMS X with entropy $H(X)$ b/symbol and a DMC with capacity C_s b/symbol, if $H(X) \leq C_s$, there exists a coding scheme for which the source output can be transmitted over the channel with an arbitrarily small probability of error.

Conversely, if $H(X) > C_s$, it is not possible to transmit information over the channel with an arbitrarily small probability of error.

Note that the channel coding theorem only asserts the existence of codes; it does not tell us how to construct these codes.

C. Block Codes:

In block codes, the binary message or data sequence is divided into sequential blocks each k b long, and each k-b block is converted to an n-b block, where $n > k$. The resultant block code is called an (n, k) block code. Thus, in an (n, k) block code, a code word can be represented as

$$\underbrace{c_1 c_2 \cdots c_k}_{\text{Data bits}} \quad \underbrace{c_{k+1} \cdots c_n}_{\text{Parity-check bits}}$$

The $n - k$ bits in the second portion are referred to as the *parity-check bits*.

Thus, in this case, the channel encoder performs a mapping

$$T: U \longrightarrow V \tag{8.55}$$

where U is a set of binary data words of length k and V is a set of binary code words of length n with $n > k$. Each of the 2^k data words is mapped to a unique code word. The ratio k/n is called the *code rate*.

8.10 ERROR CONTROL CODING

Codes can either correct or merely detect errors, depending on the amount of redundancy contained in the parity-check bits. Codes that can correct errors are known as *error-correcting* codes. There are many different error-correcting codes, such as linear parity-check codes, cyclic codes, and convolutional codes. We treat only the parity-check codes in the following.

A. Linear Parity-Check Codes:

In an (n, k) block code, it is convenient to represent a binary code word in matrix form as a row vector whose elements are the code symbols. Thus we define a code vector **c** and a data vector **d** as follows:

$$\mathbf{c} = \begin{bmatrix} c_1 & c_2 & \cdots & c_n \end{bmatrix}$$
$$\mathbf{d} = \begin{bmatrix} d_1 & d_2 & \cdots & d_k \end{bmatrix}$$

If the first k bits of a code word are the data bits, then the code is called a *systematic* code. In a systematic parity-check code, the first k bits are the data bits, and the last $m = n - k$ bits are the parity-check bits formed by linear combination of data bits

$$\begin{aligned} c_1 &= d_1 \\ c_2 &= d_2 \\ &\vdots \\ c_{k+1} &= p_{11} d_1 \oplus p_{12} d_2 \oplus \cdots \oplus p_{1k} d_k \\ c_{k+2} &= p_{21} d_1 \oplus p_{22} d_2 \oplus \cdots \oplus p_{2k} d_k \\ &\vdots \\ c_{k+m} &= p_{m1} d_1 \oplus p_{m2} d_2 \oplus \cdots \oplus p_{mk} d_k \end{aligned} \tag{8.56}$$

where $\oplus$ denotes modulo 2 addition, defined by

$$0 \oplus 0 = 1 \oplus 1 = 0$$
$$0 \oplus 1 = 1 \oplus 0 = 1$$

Equation (*8.56*) can be written in matrix form as

$$\mathbf{c} = \mathbf{d}G = \begin{bmatrix} d_1 & d_2 & \cdots & d_k \end{bmatrix} \begin{bmatrix} 1 & 0 & \cdots & 0 & p_{11} & p_{21} & \cdots & p_{m1} \\ 0 & 1 & \cdots & 0 & p_{12} & p_{22} & \cdots & p_{m2} \\ \cdots & \cdots & \cdots & \cdots & \cdots & \cdots & \cdots & \cdots \\ 0 & 0 & \cdots & 1 & p_{1k} & p_{2k} & \cdots & p_{mk} \end{bmatrix} \tag{8.57}$$

where
$$G = \begin{bmatrix} I_k & P^T \end{bmatrix} \tag{8.58}$$

where I_k is the kth-order identity matrix and P^T is the transpose of the coefficient matrix P given by

$$P = \begin{bmatrix} p_{11} & p_{12} & \cdots & p_{1k} \\ p_{21} & p_{22} & \cdots & p_{2k} \\ \cdots & \cdots & \cdots & \cdots \\ p_{m1} & p_{m2} & \cdots & p_{mk} \end{bmatrix} \tag{8.59}$$

The $k \times n$ matrix G is called the *generator matrix*. The code (the full set of code words) generated by Eq. (*8.56*) or Eq. (*8.57*) is called the *linear parity-check code*.

B. Parity-Check Matrix:

Let H denote an $m \times n$ matrix defined by

$$H = \begin{bmatrix} P & I_m \end{bmatrix} \tag{8.60}$$

where $m = n - k$. Then

$$H^T = \begin{bmatrix} P^T \\ I_m \end{bmatrix} \tag{8.61}$$

and
$$GH^T = \begin{bmatrix} I_k & P^T \end{bmatrix} \begin{bmatrix} P^T \\ I_m \end{bmatrix} = P^T \oplus P^T = 0 \tag{8.62}$$

where 0 denotes the $k \times m$ zero matrix. Postmultiplying both sides of Eq. (8.57) by H^T and then using Eq. (8.62), we obtain

$$\mathbf{c}H^T = \mathbf{d}GH^T = \mathbf{0} \tag{8.63}$$

The matrix H is called the *parity-check matrix* of the code, and Eq. (8.63) is called the *parity-check equation*.

C. Syndrome Decoding:

Let $\mathbf{r}$ denote the $1 \times n$ received vector that results from sending the code vector $\mathbf{c}$ over a noisy channel. Consider first the case of a single error in the ith position. Then

$$\mathbf{r} = \mathbf{c} \oplus \mathbf{e} \tag{8.64}$$

where
$$\mathbf{e} = [0 \cdots 0 \ \ \underset{\substack{\uparrow \\ i\text{th position}}}{1} \ \ 0 \cdots 0] \tag{8.65}$$

The vector $\mathbf{e}$ is called the *error vector*.

Note that the sum (modulo 2) of two code vectors, say, $\mathbf{a}$ and $\mathbf{b}$, is defined as

$$\mathbf{a} \oplus \mathbf{b} = \begin{bmatrix} a_1 \oplus b_1 & a_2 \oplus b_2 & \cdots & a_n \oplus b_n \end{bmatrix}$$

where
$$\mathbf{a} = \begin{bmatrix} a_1 & a_2 & \cdots & a_n \end{bmatrix}$$
$$\mathbf{b} = \begin{bmatrix} b_1 & b_2 & \cdots & b_n \end{bmatrix}$$

Next, we evaluate $\mathbf{r}H^T$ and obtain

$$\mathbf{r}H^T = (\mathbf{c} \oplus \mathbf{e})H^T = \mathbf{0} \oplus \mathbf{e}H^T = \mathbf{e}H^T = \mathbf{s} \tag{8.66}$$

The $1 \times m$ vector $\mathbf{s}$ is called the *syndrome* of $\mathbf{r}$. Thus, using $\mathbf{s}$ and noting that $\mathbf{e}H^T$ is the ith row of H^T we can identify the error position by comparing $\mathbf{s}$ to the rows of H^T. Decoding by this simple comparison method is called *syndrome decoding*. Note that not all error patterns can be correctly decoded by syndrome decoding. The zero syndrome indicates that $\mathbf{r}$ is a code vector and is presumably correct.

With syndrome decoding, an (n, k) linear block code can correct up to t errors per code word if n and k satisfy the following *Hamming bound*

$$2^{n-k} \geq \sum_{i=0}^{t} \binom{n}{i} \tag{8.67}$$

where
$$\binom{n}{i} = \frac{n!}{(n-i)!\,i!}$$

A block code for which the equality holds for Eq. (8.67) is known as the *perfect code*. Single error-correcting perfect codes are called *Hamming codes*.

Note that the Hamming bound is necessary but not sufficient for the construction of a t error-correcting parity-check code.

8.11 ERROR DETECTION AND CORRECTION CAPABILITIES OF LINEAR BLOCK CODES

A. Hamming Distance:

The *Hamming distance* $d(\mathbf{c}_i, \mathbf{c}_j)$ between two code vectors $\mathbf{c}_i$ and $\mathbf{c}_j$ (having the same number of elements) is defined as the number of positions in which their elements differ.

The *Hamming weight* $w(\mathbf{c}_i)$ of a code vector $\mathbf{c}_i$ is defined as the number of 1s in $\mathbf{c}_i$. Thus, the Hamming weight of $\mathbf{c}_i$ is the Hamming distance between $\mathbf{c}_i$ and $\mathbf{0}$, that is,

$$w(\mathbf{c}_i) = d(\mathbf{c}_i, \mathbf{0}) \tag{8.68}$$

where $\mathbf{0}$ is the zero code vector whose elements are all zeros. Similarly, the Hamming distance can be written in terms of Hamming weight as

$$d(\mathbf{c}_i, \mathbf{c}_j) = w(\mathbf{c}_i \oplus \mathbf{c}_j) \tag{8.69}$$

B. Minimum Distance:

The minimum distance d_{min} of a linear block code is defined as the smallest Hamming distance between any pair of code vectors in the code.

From the closure property of linear block codes [that is, the sum (or difference) (modulo 2) of two code vectors is also a code vector], we can derive the following theorem (Prob. 8.48).

THEOREM 8.1

The minimum distance of a linear block code is the smallest Hamming weight of the nonzero code vector in the code.

From Eq. (*8.63*), a linear block code can also be defined as the set of all code vectors for which

$$\mathbf{c}H^T = \mathbf{0} \tag{8.70}$$

where H^T is the transpose of the parity-check matrix H.

The minimum distance d_{min} of a linear block code is closely related to the structure of the parity-check matrix H of the code. This is stated by the following theorem (Prob. 8.49).

THEOREM 8.2

The minimum distance d_{min} of a linear block code is equal to the minimum number of rows of H^T that sum to $\mathbf{0}$.

C. Error Detection and Correction Capabilities:

The minimum distance d_{min} of a linear block code is an important parameter of the code. It determines the error detection and correction capabilities of the code. This is stated in the following theorem.

THEOREM 8.3

An (n, k) linear block code of minimum distance d_{min} can correct up to t errors if and only if

$$d_{\text{min}} \geq 2t + 1 \tag{8.71}$$

Equation (*8.71*) can be illustrated geometrically by Fig. 8-8. In Fig. 8-8 two Hamming spheres, each of radius t, are constructed around the points that represent code vectors $\mathbf{c}_i$ and $\mathbf{c}_j$. Figure 8-8(a) depicts the case where two spheres are disjoint, that is, $d(\mathbf{c}_i, \mathbf{c}_j) \geq 2t + 1$. For this case, if the code vector $\mathbf{c}_i$ is transmitted, the received vector is $\mathbf{r}$, and $d(\mathbf{c}_i, \mathbf{r}) \leq t$, then it is clear that the decoder will

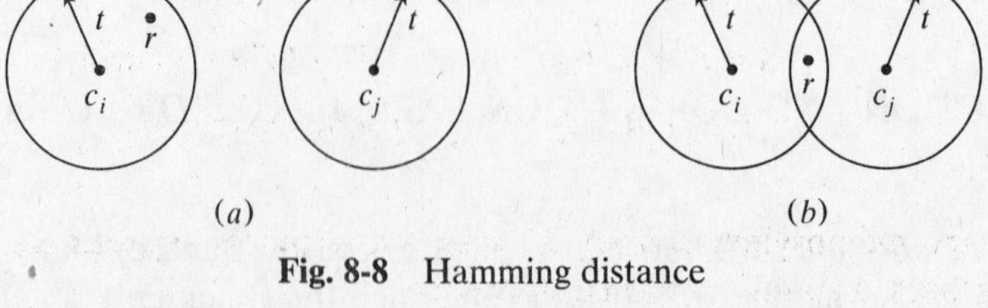

(a) (b)

Fig. 8-8 Hamming distance

choose $\mathbf{c}_i$ since it is the code vector closest to the received vector $\mathbf{r}$. On the other hand, Fig. 8-8(*b*) depicts the case where the two spheres intersect, that is, $d(\mathbf{c}_i, \mathbf{c}_j) < 2t$. In this case we see that if $\mathbf{c}_i$ is transmitted, there exists a received vector $\mathbf{r}$ such that $d(\mathbf{c}_i, \mathbf{r}) \le t$, and yet $\mathbf{r}$ is as close to $\mathbf{c}_j$ as it is to $\mathbf{c}_i$. Thus the decoder may choose $\mathbf{c}_j$, which is incorrect.

Solved Problems

MEASURE OF INFORMATION

8.1. Verify Eq. (*8.5*), that is,

$$I(x_i x_j) = I(x_i) + I(x_j) \qquad \text{if } x_i \text{ and } x_j \text{ are independent}$$

If x_i and x_j are independent, then by Eq. (*5.25*)

$$P(x_i x_j) = P(x_i)P(x_j)$$

By Eq. (*8.1*)

$$I(x_i x_j) = \log \frac{1}{P(x_i x_j)} = \log \frac{1}{P(x_i)P(x_j)}$$

$$= \log \frac{1}{P(x_i)} + \log \frac{1}{P(x_j)}$$

$$= I(x_i) + I(x_j)$$

8.2. A DMS X has four symbols x_1, x_2, x_3, x_4 with probabilities $P(x_1) = 0.4$, $P(x_2) = 0.3$, $P(x_3) = 0.2$, and $P(x_4) = 0.1$.

(*a*) Calculate $H(X)$.

(*b*) Find the amount of information contained in the messages $x_1 x_2 x_1 x_3$ and $x_4 x_3 x_3 x_2$, and compare with the $H(X)$ obtained in part (*a*).

(*a*)
$$H(X) = -\sum_{i=1}^{4} P(x_i) \log_2 [P(x_i)]$$
$$= -0.4 \log_2 0.4 - 0.3 \log_2 0.3 / - 0.2 \log_2 0.2 - 0.1 \log_2 0.1$$
$$= 1.85 \text{ b/symbol}$$

(*b*)
$$P(x_1 x_2 x_1 x_3) = (0.4)(0.3)(0.4)(0.2) = 0.0096$$
$$I(x_1 x_2 x_1 x_3) = -\log_2 0.0096 = 6.70 \text{ b/symbol}$$

Thus,
$$I(x_1 x_2 x_1 x_3) < 7.4[= 4H(X)] \text{ b/symbol}$$

$$P(x_4 x_3 x_3 x_2) = (0.1)(0.2)^2(0.3) = 0.0012$$
$$I(x_4 x_3 x_3 x_2) = -\log_2 0.0012 = 9.70 \text{ b/symbol}$$

Thus,
$$I(x_4 x_3 x_3 x_2) > 7.4[= 4H(X)] \text{ b/symbol}$$

8.3. Consider a binary memoryless source X with two symbols x_1 and x_2. Show that $H(X)$ is maximum when both x_1 and x_2 are equiprobable.

Let $P(x_1) = \alpha$. Then $P(x_2) = 1 - \alpha$.

$$H(X) = -\alpha \log_2 \alpha - (1-\alpha)\log_2(1-\alpha) \tag{8.72}$$

$$\frac{dH(X)}{d\alpha} = \frac{d}{d\alpha}\left[-\alpha \log_2 \alpha - (1-\alpha)\log_2(1-\alpha)\right]$$

Using the relation

$$\frac{d}{dx}\log_b y = \frac{1}{y}\log_b e \frac{dy}{dx}$$

we obtain

$$\frac{dH(X)}{d\alpha} = -\log_2 \alpha + \log_2(1-\alpha) = \log_2 \frac{1-\alpha}{\alpha}$$

The maximum value of $H(X)$ requires that

$$\frac{dH(X)}{d\alpha} = 0$$

that is,

$$\frac{1-\alpha}{\alpha} = 1 \longrightarrow \alpha = \frac{1}{2}$$

Note that $H(X) = 0$ when $\alpha = 0$ or 1. When $P(x_1) = P(x_2) = \frac{1}{2}$, $H(X)$ is maximum and is given by

$$H(X) = \frac{1}{2}\log_2 2 + \frac{1}{2}\log_2 2 = 1 \text{ b/symbol} \tag{8.73}$$

8.4. Verify Eq. (*8.9*), that is,

$$0 \le H(X) \le \log_2 m$$

where m is the size of the alphabet of X.

Proof of the lower bound: Since $0 \le P(x_i) \le 1$,

$$\frac{1}{P(x_i)} \ge 1 \quad \text{and} \quad \log_2 \frac{1}{P(x_i)} \ge 0$$

Then it follows that

$$P(x_i)\log_2 \frac{1}{P(x_i)} \ge 0$$

Thus,

$$H(X) = \sum_{i=1}^{m} P(x_i)\log_2 \frac{1}{P(x_i)} \ge 0 \tag{8.74}$$

Next we note that

$$P(x_i)\log_2 \frac{1}{P(x_i)} = 0$$

if and only if $P(x_i) = 0$ or 1. Since

$$\sum_{i=1}^{m} P(x_i) = 1$$

when $P(x_i) = 1$, then $P(x_j) = 0$ for $j \ne i$. Thus, only in this case, $H(X) = 0$.

Proof of the upper bound: Consider two probability distributions $\{P(x_i) = P_i\}$ and $\{Q(x_i) = Q_i\}$ on the alphabet $\{x_i\}$, $i = 1, 2, \dots, m$, such that

$$\sum_{i=1}^{m} P_i = 1 \quad \text{and} \quad \sum_{i=1}^{m} Q_i = 1 \tag{8.75}$$

Using Eq. (*8.6*), we have

$$\sum_{i=1}^{m} P_i \log_2 \frac{Q_i}{P_i} = \frac{1}{\ln 2} \sum_{i=1}^{m} P_i \ln \frac{Q_i}{P_i}$$

Next, using the inequality

$$\ln \alpha \le \alpha - 1 \qquad \alpha \ge 0 \tag{8.76}$$

and noting that the equality holds only if $\alpha = 1$, we get

$$\sum_{i=1}^{m} P_i \ln \frac{Q_i}{P_i} \le \sum_{i=1}^{m} P_i \left(\frac{Q_i}{P_i} - 1 \right) = \sum_{i=1}^{m} (Q_i - P_i)$$

$$= \sum_{i=1}^{m} Q_i - \sum_{i=1}^{m} P_i = 0 \tag{8.77}$$

by using Eq. (*8.75*). Thus,

$$\sum_{i=1}^{m} P_i \log_2 \frac{Q_i}{P_i} \le 0 \tag{8.78}$$

where the equality holds only if $Q_i = P_i$ for all i. Setting

$$Q_i = \frac{1}{m} \qquad i = 1, 2, \ldots, m \tag{8.79}$$

we obtain

$$\sum_{i=1}^{m} P_i \log_2 \frac{1}{P_i m} = -\sum_{i=1}^{m} P_i \log_2 P_i - \sum_{i=1}^{m} P_i \log_2 m$$

$$= H(X) - \log_2 m \sum_{i=1}^{m} P_i$$

$$= H(X) - \log_2 m \le 0 \tag{8.80}$$

Hence

$$H(X) \le \log_2 m$$

and the equality holds only if the symbols in X are equiprobable, as in Eq. (*8.79*).

8.5. A high-resolution black-and-white TV picture consists of about 2×10^6 picture elements and 16 different brightness levels. Pictures are repeated at the rate of 32 per second. All picture elements are assumed to be independent, and all levels have equal likelihood of occurrence. Calculate the average rate of information conveyed by this TV picture source.

$$H(X) = -\sum_{i=1}^{16} \frac{1}{16} \log_2 \frac{1}{16} = 4 \text{ b/element}$$

$$r = 2(10^6)(32) = 64(10^6) \text{ elements/s}$$

Hence, by Eq. (*8.10*)

$$R = rH(X) = 64(10^6)(4) = 256(10^6) \text{ b/s} = 256 \text{ Mb/s}$$

8.6. Consider a telegraph source having two symbols, dot and dash. The dot duration is 0.2 s. The dash duration is 3 times the dot duration. The probability of the dot's occurring is twice that of the dash, and the time between symbols is 0.2 s. Calculate the information rate of the telegraph source.

$$P(\text{dot}) = 2P(\text{dash})$$

$$P(\text{dot}) + P(\text{dash}) = 3P(\text{dash}) = 1$$

Thus,

$$P(\text{dash}) = \frac{1}{3} \qquad \text{and} \qquad P(\text{dot}) = \frac{2}{3}$$

By Eq. (8.7)

$$H(X) = -P(\text{dot})\log_2 P(\text{dot}) - P(\text{dash})\log_2 P(\text{dash})$$
$$= 0.667(0.585) + 0.333(1.585) = 0.92 \text{ b/symbol}$$
$$t_{\text{dot}} = 0.2 \text{ s} \qquad t_{\text{dash}} = 0.6 \text{ s} \qquad t_{\text{space}} = 0.2 \text{ s}$$

Thus the average time per symbol is

$$T_s = P(\text{dot})t_{\text{dot}} + P(\text{dash})t_{\text{dash}} + t_{\text{space}} = 0.5333 \text{ s/symbol}$$

and the average symbol rate is

$$r = \frac{1}{T_s} = 1.875 \qquad \text{symbols/s}$$

Thus, the average information rate of the telegraph source is

$$R = rH(X) = 1.875(0.92) = 1.725 \text{ b/s}$$

DISCRETE MEMORYLESS CHANNELS

8.7. Consider a binary channel shown in Fig. 8-9. (See Prob. 5.8.)

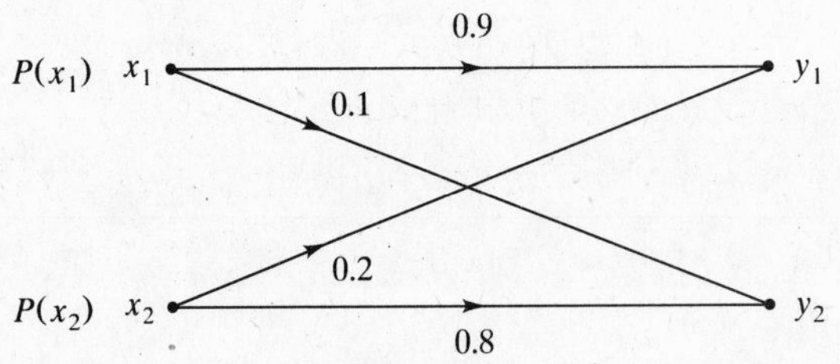

Fig. 8-9

(a) Find the channel matrix of the channel.

(b) Find $P(y_1)$ and $P(y_2)$ when $P(x_1) = P(x_2) = 0.5$.

(c) Find the joint probabilities $P(x_1, y_2)$ and $P(x_2, y_1)$ when $P(x_1) = P(x_2) = 0.5$.

(a) Using Eq. (8.11), we see the channel matrix is given by

$$[P(Y|X)] = \begin{bmatrix} P(y_1|x_1) & P(y_2|x_1) \\ P(y_1|x_2) & P(y_2|x_2) \end{bmatrix} = \begin{bmatrix} 0.9 & 0.1 \\ 0.2 & 0.8 \end{bmatrix}$$

(b) Using Eq. (8.13), (8.14), and (8.15), we obtain

$$[P(Y)] = [P(X)][P(Y|X)]$$
$$= [0.5 \quad 0.5]\begin{bmatrix} 0.9 & 0.1 \\ 0.2 & 0.8 \end{bmatrix}$$
$$= [0.55 \quad 0.45] = [P(y_1) \quad P(y_2)]$$

Hence $P(y_1) = 0.55$ and $P(y_2) = 0.45$.

(c) Using Eqs. (8.16) and (8.17), we obtain

$$[P(X,Y)] = [P(X)]_d[P(Y|X)]$$
$$= \begin{bmatrix} 0.5 & 0 \\ 0 & 0.5 \end{bmatrix}\begin{bmatrix} 0.9 & 0.1 \\ 0.2 & 0.8 \end{bmatrix}$$
$$= \begin{bmatrix} 0.45 & 0.05 \\ 0.1 & 0.4 \end{bmatrix} = \begin{bmatrix} P(x_1, y_1) & P(x_1, y_2) \\ P(x_2, y_1) & P(x_2, y_2) \end{bmatrix}$$

Hence $P(x_1, y_2) = 0.05$ and $P(x_2, y_1) = 0.1$.

8.8. Two binary channels of Prob. 8.7 are connected in cascade, as shown in Fig. 8-10.

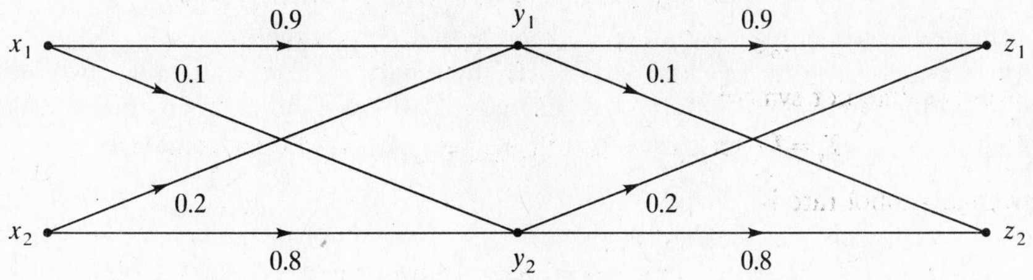

Fig. 8-10

(*a*) Find the overall channel matrix of the resultant channel, and draw the resultant equivalent channel diagram.

(*b*) Find $P(z_1)$ and $P(z_2)$ when $P(x_1) = P(x_2) = 0.5$.

(*a*) By Eq. (*8.15*)

$$[P(Y)] = [P(X)][P(Y|X)]$$
$$[P(Z)] = [P(Y)][P(Z|Y)] = [P(X)][P(Y|X)][P(Z|Y)] = [P(X)][P(Z|X)]$$

Thus from Fig. 8-10

$$[P(Z|X)] = [P(Y|X)][P(Z|Y)] = \begin{bmatrix} 0.9 & 0.1 \\ 0.2 & 0.8 \end{bmatrix} \begin{bmatrix} 0.9 & 0.1 \\ 0.2 & 0.8 \end{bmatrix} = \begin{bmatrix} 0.83 & 0.17 \\ 0.34 & 0.66 \end{bmatrix}$$

The resultant equivalent channel diagram is shown in Fig. 8-11.

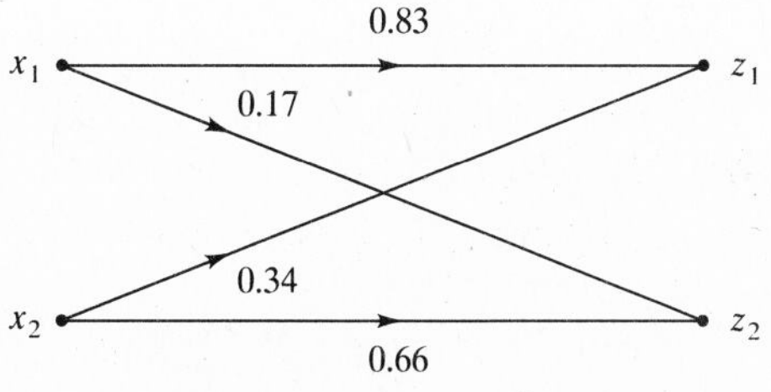

Fig. 8-11

(*b*)

$$[P(Z)] = [P(X)][P(Z|X)] = [0.5 \quad 0.5] \begin{bmatrix} 0.83 & 0.17 \\ 0.34 & 0.66 \end{bmatrix} = [0.585 \quad 0.415]$$

Hence $P(z_1) = 0.585$ and $P(z_2) = 0.415$.

8.9. A channel has the following channel matrix:

$$[P(Y|X)] = \begin{bmatrix} 1-p & p & 0 \\ 0 & p & 1-p \end{bmatrix} \tag{8.81}$$

(*a*) Draw the channel diagram.

(*b*) If the source has equally likely outputs, compute the probabilities associated with the channel outputs for $p = 0.2$.

(*a*) The channel diagram is shown in Fig. 8-12. Note that the channel represented by Eq. (*8.81*) (see Fig. 8-12) is known as the *binary erasure channel*. The binary erasure channel has two inputs $x_1 = 0$ and $x_2 = 1$ and three outputs $y_1 = 0$, $y_2 = e$, and $y_3 = 1$, where e indicates an erasure; that is, the output is in doubt, and it should be erased.

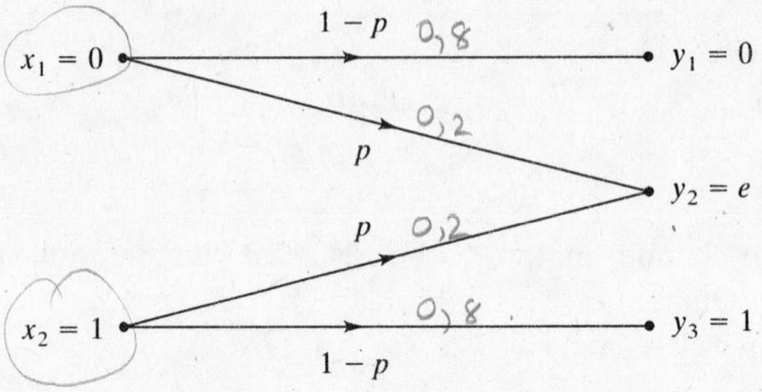

Fig. 8-12 Binary erasure channel

(*b*) By Eq. (*8.15*)

$$[P(Y)] = [0.5 \quad 0.5]\begin{bmatrix} 0.8 & 0.2 & 0 \\ 0 & 0.2 & 0.8 \end{bmatrix}$$
$$= [0.4 \quad 0.2 \quad 0.4]$$

Thus $P(y_1) = 0.4$, $P(y_2) = 0.2$, and $P(y_3) = 0.4$.

MUTUAL INFORMATION

8.10. For a lossless channel show that

$$H(X|Y) = 0 \tag{8.82}$$

When we observe the output y_j in a lossless channel (Fig. 8-2), it is clear which x_i was transmitted, that is,

$$P(x_i|y_j) = 0 \quad \text{or} \quad 1 \tag{8.83}$$

Now by Eq. (*8.23*)

$$H(X|Y) = -\sum_{j=1}^{n}\sum_{i=1}^{m} P(x_i, y_j)\log_2 P(x_i|y_j)$$
$$= -\sum_{j=1}^{n} P(y_j)\sum_{i=1}^{m} P(x_i|y_j)\log_2 P(x_i|y_j) \tag{8.84}$$

Note that all the terms in the inner summation are zero because they are in the form of $1 \times \log_2 1$ or $0 \times \log_2 0$. Hence, we conclude that for a lossless channel

$$H(X|Y) = 0$$

8.11. Consider a noiseless channel with m input symbols and m output symbols (Fig. 8-4). Show that

$$H(X) = H(Y) \tag{8.85}$$

and

$$H(Y|X) = 0 \tag{8.86}$$

For a noiseless channel the transition probabilities are

$$P(y_j \mid x_i) = \begin{cases} 1 & i = j \\ 0 & i \neq j \end{cases} \tag{8.87}$$

Hence,

$$P(x_i, y_j) = P(y_j \mid x_i) P(x_i) = \begin{cases} P(x_i) & i = j \\ 0 & i \neq j \end{cases} \tag{8.88}$$

and

$$P(y_j) = \sum_{i=1}^{m} P(x_i, y_j) = P(x_j) \tag{8.89}$$

Thus, by Eqs. (8.7) and (8.89)

$$\begin{aligned} H(Y) &= -\sum_{j=1}^{m} P(y_j) \log_2 P(y_j) \\ &= -\sum_{i=1}^{m} P(x_i) \log_2 P(x_i) = H(X) \end{aligned}$$

Next, by Eqs. (8.24), (8.87), and (8.88)

$$\begin{aligned} H(Y \mid X) &= -\sum_{j=1}^{m} \sum_{i=1}^{m} P(x_i, y_j) \log_2 P(y_j \mid x_i) \\ &= -\sum_{i=1}^{m} P(x_i) \sum_{j=1}^{m} \log_2 P(y_j \mid x_i) \\ &= -\sum_{i=1}^{m} P(x_i) \log_2 1 = 0 \end{aligned}$$

8.12. Verify Eq. (8.26), that is,

$$H(X,Y) = H(X \mid Y) + H(Y)$$

From Eqs. (5.23) and (5.30)

$$P(x_i, y_j) = P(x_i \mid y_j) P(y_j)$$

and

$$\sum_{i=1}^{m} P(x_i, y_j) = P(y_j)$$

So by Eq. (8.25) and using Eqs. (8.22) and (8.23), we have

$$\begin{aligned} H(X,Y) &= -\sum_{j=1}^{n} \sum_{i=1}^{m} P(x_i, y_j) \log P(x_i, y_j) \\ &= -\sum_{j=1}^{n} \sum_{i=1}^{m} P(x_i, y_j) \log \left[P(x_i \mid y_j) P(y_j) \right] \\ &= -\sum_{j=1}^{n} \sum_{i=1}^{m} P(x_i, y_j) \log P(x_i \mid y_j) \\ &\quad - \sum_{j=1}^{n} \left[\sum_{i=1}^{m} P(x_i, y_j) \right] \log P(y_j) \\ &= H(X \mid Y) - \sum_{j=1}^{n} P(y_j) \log P(y_j) \\ &= H(X \mid Y) + H(Y) \end{aligned}$$

8.13. Show that the mutual information $I(X;Y)$ of the channel described by Eq. (*8.11*), with the input probabilities $P(x_i)$, $i = 1, 2, \ldots, m$, and the output probabilities $P(y_j)$, $j = 1, 2, \ldots, n$, can be expressed as

$$I(X;Y) = \sum_{i=1}^{m} \sum_{j=1}^{n} P(x_i, y_j) \log_2 \frac{P(x_i \mid y_j)}{P(x_i)} \tag{8.90}$$

By Eq. (*8.28*)

$$I(X;Y) = H(X) - H(X \mid Y)$$

Using Eqs. (*8.21*) and (*8.23*), we obtain

$$\begin{aligned} I(X;Y) &= \sum_{i=1}^{m} P(x_i) \log_2 \frac{1}{P(x_i)} + \sum_{j=1}^{n} \sum_{i=1}^{m} P(x_i, y_j) \log_2 P(x_i \mid y_j) \\ &= \sum_{i=1}^{m} \left[\sum_{j=1}^{n} P(x_i, y_j) \right] \log_2 \frac{1}{P(x_i)} + \sum_{j=1}^{n} \sum_{i=1}^{m} P(x_i, y_j) \log_2 P(x_i \mid y_j) \\ &= \sum_{i=1}^{m} \sum_{j=1}^{n} P(x_i, y_j) \left[\log_2 \frac{1}{P(x_i)} + \log_2 P(x_i \mid y_j) \right] \\ &= \sum_{i=1}^{m} \sum_{j=1}^{n} P(x_i, y_j) \log_2 \frac{P(x_i \mid y_j)}{P(x_i)} \end{aligned}$$

8.14. Verify Eq. (*8.29*), that is,

$$I(X;Y) = I(Y;X)$$

Using Eq. (*8.90*), we can express $I(Y;X)$ as

$$I(Y;X) = \sum_{i=1}^{m} \sum_{j=1}^{n} P(y_j, x_i) \log_2 \frac{P(y_j \mid x_i)}{P(y_j)} \tag{8.91}$$

Now, by Eqs. (*5.8*) and (*5.24*)

$$P(y_j, x_i) = P(x_i, y_j)$$

and

$$\frac{P(y_j \mid x_i)}{P(y_j)} = \frac{P(x_i \mid y_j)}{P(x_i)}$$

Thus, comparing Eqs. (*8.91*) with (*8.90*), we conclude that

$$I(X;Y) = I(Y;X)$$

8.15. Verify Eq. (*8.30*), that is,

$$I(X;Y) \geq 0$$

From Eq. (*8.90*) and using the relation $\log(a/b) = -\log(b/a)$, we have

$$-I(X;Y) = \sum_{i=1}^{m} \sum_{j=1}^{n} P(x_i, y_j) \log_2 \frac{P(x_i)}{P(x_i \mid y_j)} \tag{8.92}$$

Using Bayes' rule [Eq. (*5.24*)], we have

$$\frac{P(x_i)}{P(x_i \mid y_j)} = \frac{P(x_i) P(y_j)}{P(x_i, y_j)}$$

Then by using Eq. (8.6), Eq. (8.92) can be rewritten as

$$-I(X;Y) = \frac{1}{\ln 2} \sum_{i=1}^{m} \sum_{j=1}^{n} P(x_i, y_j) \ln \frac{P(x_i)P(y_j)}{P(x_i, y_j)} \tag{8.93}$$

Using the inequality (8.76), that is,

$$\ln \alpha \le \alpha - 1$$

we have

$$-I(X;Y) \le \frac{1}{\ln 2} \sum_{i=1}^{m} \sum_{j=1}^{n} P(x_i, y_j) \left[\frac{P(x_i)P(y_j)}{P(x_i, y_j)} - 1 \right]$$

or

$$-I(X;Y) \le \frac{1}{\ln 2} \left[\sum_{i=1}^{m} \sum_{j=1}^{n} P(x_i)P(y_j) - \sum_{i=1}^{m} \sum_{j=1}^{n} P(x_i, y_j) \right] \tag{8.94}$$

Since

$$\sum_{i=1}^{m} \sum_{j=1}^{n} P(x_i)P(y_j) = \sum_{i=1}^{m} P(x_i) \sum_{j=1}^{n} P(y_j) = (1)(1) = 1$$

$$\sum_{i=1}^{m} \sum_{j=1}^{n} P(x_i, y_j) = \sum_{i=1}^{m} \left[\sum_{j=1}^{n} P(x_i, y_j) \right] = \sum_{i=1}^{m} P(x_i) = 1$$

Equation (8.94) reduces to

$$-I(X;Y) \le 0$$

or

$$I(X;Y) \ge 0$$

8.16. Consider a BSC (Fig. 8-5) with $P(x_1) = \alpha$.

(a) Show that the mutual information $I(X;Y)$ is given by

$$I(X;Y) = H(Y) + p \log_2 p + (1-p) \log_2 (1-p) \tag{8.95}$$

(b) Calculate $I(X;Y)$ for $\alpha = 0.5$ and $p = 0.1$.

(c) Repeat (b) for $\alpha = 0.5$ and $p = 0.5$, and comment on the result.

Figure 8-13 shows the diagram of the BSC with associated input probabilities.

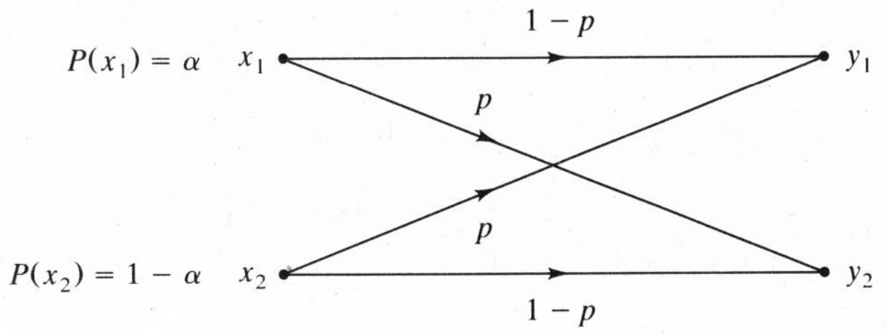

Fig. 8-13 BSC

(a) Using Eqs. (8.16), (8.17), and (8.20), we have

$$[P(X,Y)] = \begin{bmatrix} \alpha & 0 \\ 0 & 1-\alpha \end{bmatrix} \begin{bmatrix} 1-p & p \\ p & 1-p \end{bmatrix}$$

$$= \begin{bmatrix} \alpha(1-p) & \alpha p \\ (1-\alpha)p & (1-\alpha)(1-p) \end{bmatrix} = \begin{bmatrix} P(x_1, y_1) & P(x_1, y_2) \\ P(x_2, y_1) & P(x_2, y_2) \end{bmatrix}$$

Then by Eq. (*8.24*)

$$\begin{aligned} H(Y|X) &= -P(x_1,y_1)\log_2 P(y_1|x_1) - P(x_1,y_2)\log_2 P(y_2|x_1) \\ &\quad - P(x_2,y_1)\log_2 P(y_1|x_2) - P(x_2,y_2)\log_2 P(y_2|x_2) \\ &= -\alpha(1-p)\log_2(1-p) - \alpha p\log_2 p \\ &\quad -(1-\alpha)p\log_2 p - (1-\alpha)(1-p)\log_2(1-p) \\ &= -p\log_2 p - (1-p)\log_2(1-p) \end{aligned} \tag{8.96}$$

Hence, by Eq. (*8.31*)

$$\begin{aligned} I(X;Y) &= H(Y) - H(Y|X) \\ &= H(Y) + p\log_2 p + (1-p)\log_2(1-p) \end{aligned}$$

(*b*) When $\alpha = 0.5$ and $p = 0.1$, by Eq. (*8.15*)

$$[P(Y)] = [0.5 \quad 0.5]\begin{bmatrix} 0.9 & 0.1 \\ 0.1 & 0.9 \end{bmatrix} = [0.5 \quad 0.5]$$

Thus, $P(y_1) = P(y_2) = 0.5$.
By Eq. (*8.22*)

$$\begin{aligned} H(Y) &= -P(y_1)\log_2 P(y_1) - P(y_2)\log_2 P(y_2) \\ &= -0.5\log_2 0.5 - 0.5\log_2 0.5 = 1 \end{aligned}$$

$$\begin{aligned} p\log_2 p + (1-p)\log_2(1-p) &= 0.1\log_2 0.1 + 0.9\log_2 0.9 \\ &= -0.469 \end{aligned}$$

Thus, $$I(X;Y) = 1 - 0.469 = 0.531$$

(*c*) When $\alpha = 0.5$ and $p = 0.5$,

$$[P(Y)] = [0.5 \quad 0.5]\begin{bmatrix} 0.5 & 0.5 \\ 0.5 & 0.5 \end{bmatrix} = [0.5 \quad 0.5]$$

$$H(Y) = 1$$

$$\begin{aligned} p\log_2 p + (1-p)\log_2(1-p) &= 0.5\log_2 0.5 + 0.5\log_2 0.5 \\ &= -1 \end{aligned}$$

Thus, $$I(X;Y) = 1 - 1 = 0$$

Note that in this case ($p = 0.5$) no information is being transmitted at all. An equally acceptable decision could be made by dispensing with the channel entirely and "flipping a coin" at the receiver. When $I(X;Y) = 0$, the channel is said to be *useless*.

CHANNEL CAPACITY

8.17. Verify Eq. (*8.36*), that is,

$$C_s = \log_2 m$$

where C_s is the channel capacity of a lossless channel and m is the number of symbols in X.

For a lossless channel [Eq. (*8.82*), Prob. 8.10]

$$H(X|Y) = 0$$

Then by Eq. (*8.28*)

$$I(X;Y) = H(X) - H(X|Y) = H(X) \tag{8.97}$$

Hence, by Eqs. (*8.33*) and (*8.9*)

$$C_s = \max_{\{P(X)\}} I(X;Y) = \max_{\{P(x_i)\}} H(X) = \log_2 m$$

8.18. Verify Eq. (*8.42*), that is,

$$C_s = 1 + p\log_2 p + (1-p)\log_2(1-p)$$

where C_s is the channel capacity of a BSC (Fig. 8-12).

By Eq. (*8.95*) (Prob. 8.16) the mutual information $I(X;Y)$ of a BSC is given by

$$I(X;Y) = H(Y) + p\log_2 p + (1-p)\log_2(1-p)$$

which is maximum when $H(Y)$ is maximum. Since the channel output is binary, $H(Y)$ is maximum when each output has a probability of 0.5 and is achieved for equally likely inputs [Eq. (*8.9*)]. For this case $H(Y) = 1$, and the channel capacity is

$$C_s = \max_{\{P(X)\}} I(X;Y) = 1 + p\log_2 p + (1-p)\log_2(1-p)$$

8.19. Find the channel capacity of the binary erasure channel of Fig. 8-14 (Prob. 8.9).

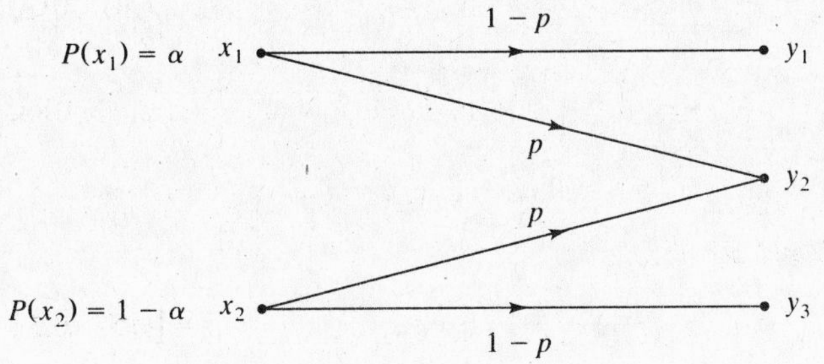

Fig. 8-14 Binary erasure channel

Let $P(x_1) = \alpha$. Then $P(x_2) = 1 - \alpha$. By Eq. (*8.81*)

$$[P(Y|X)] = \begin{bmatrix} 1-p & p & 0 \\ 0 & p & 1-p \end{bmatrix} = \begin{bmatrix} P(y_1|x_1) & P(y_2|x_1) & P(y_3|x_1) \\ p(y_1|x_2) & P(y_2|x_2) & P(y_3|x_2) \end{bmatrix}$$

By Eq. (*8.15*)

$$[P(Y)] = [\alpha \quad 1-\alpha]\begin{bmatrix} 1-p & p & 0 \\ 0 & p & 1-p \end{bmatrix}$$

$$= [\alpha(1-p) \quad p \quad (1-\alpha)(1-p)]$$

$$= [P(y_1) \quad P(y_2) \quad P(y_3)]$$

By Eq. (*8.17*)

$$[P(X,Y)] = \begin{bmatrix} \alpha & 0 \\ 0 & 1-\alpha \end{bmatrix}\begin{bmatrix} 1-p & p & 0 \\ 0 & p & 1-p \end{bmatrix}$$

$$= \begin{bmatrix} \alpha(1-p) & \alpha p & 0 \\ 0 & (1-\alpha)p & (1-\alpha)(1-p) \end{bmatrix}$$

$$= \begin{bmatrix} P(x_1,y_1) & P(x_1,y_2) & P(x_1,y_3) \\ P(x_2,y_1) & P(x_2,y_2) & P(x_2,y_3) \end{bmatrix}$$

In addition, from Eqs. (*8.22*) and (*8.24*) we can calculate

$$H(Y) = -\sum_{j=1}^{3} P(y_j)\log_2 P(y_j)$$
$$= -\alpha(1-p)\log_2 \alpha(1-p) - p\log_2 p - (1-\alpha)(1-p)\log_2[(1-\alpha)(1-p)]$$
$$= (1-p)[-\alpha\log_2\alpha - (1-\alpha)\log_2(1-\alpha)]$$
$$-p\log_2 p - (1-p)\log_2(1-p) \tag{8.98}$$

$$H(Y|X) = -\sum_{j=1}^{3}\sum_{i=1}^{2} P(x_i, y_j)\log_2 P(y_j|x_i)$$
$$= -\alpha(1-p)\log_2(1-p) - \alpha p\log_2 p$$
$$-(1-\alpha)p\log_2 p - (1-\alpha)(1-p)\log_2(1-p)$$
$$= -p\log_2 p - (1-p)\log_2(1-p) \tag{8.99}$$

Thus, by Eqs. (*8.31*) and (*8.72*)

$$I(X;Y) = H(Y) - H(Y|X)$$
$$= (1-p)[-\alpha\log_2\alpha - (1-\alpha)\log_2(1-\alpha)]$$
$$= (1-p)H(X) \tag{8.100}$$

And by Eqs. (*8.33*) and (*8.73*)

$$C_s = \max_{\{P(X)\}} I(X;Y) = \max_{\{P(x_i)\}} (1-p)H(X) = (1-p)\max_{\{P(x_i)\}} H(X) = 1-p \tag{8.101}$$

ADDITIVE WHITE GAUSSIAN NOISE CHANNEL

8.20. Find the differential entropy $H(X)$ of the uniformly distributed random variable X with probability density function

$$f_X(x) = \begin{cases} \dfrac{1}{a} & 0 \le x \le a \\ 0 & \text{otherwise} \end{cases}$$

for (*a*) $a = 1$, (*b*) $a = 2$, and (*c*) $a = \frac{1}{2}$.

By Eq. (*8.43*)

$$H(X) = -\int_{-\infty}^{\infty} f_X(x)\log_2 f_X(x)\,dx$$
$$= -\int_0^a \frac{1}{a}\log_2\frac{1}{a}\,dx = \log_2 a \tag{8.102}$$

(*a*) $a = 1,\ H(X) = \log_2 1 = 0$

(*b*) $a = 2,\ H(X) = \log_2 2 = 1$

(*c*) $a = \frac{1}{2},\ H(X) = \log_2\frac{1}{2} = -\log_2 2 = -1$

Note that the differential entropy $H(X)$ is not an absolute measure of information.

8.21. With the differential entropy of a random variable X defined by Eq. (*8.43*), that is,

$$H(X) = -\int_{-\infty}^{\infty} f_X(x)\log_2 f_X(x)\,dx$$

find the probability density function $f_X(x)$ for which $H(X)$ is maximum.

From Eqs. (*5.34b*) and (*5.74*), $f_X(x)$ must satisfy the following two conditions:

$$\int_{-\infty}^{\infty} f_X(x)\, dx = 1 \tag{8.103}$$

$$\int_{-\infty}^{\infty} (x-\mu)^2 f_X(x)\, dx = \sigma^2 \tag{8.104}$$

where μ is the mean of X and σ^2 is its variance. Since the problem is the maximization of $H(X)$ under constraints of Eqs. (*8.103*) and (*8.104*), we use the method of *Lagrange multipliers*.

First, we form the function

$$G[f_X(x), \lambda_1, \lambda_2] = H(X) + \lambda_1\left[\int_{-\infty}^{\infty} f_X(x)\, dx - 1\right] + \lambda_2\left[\int_{-\infty}^{\infty} (x-\mu)^2 f_X(x)\, dx - \sigma^2\right]$$

$$= \int_{-\infty}^{\infty}\left[-f_X(x)\log_2 f_X(x) + \lambda_1 f_X(x) + \lambda_2 (x-\mu)^2 f_X(x)\right] dx - \lambda_1 - \lambda_2\sigma^2 \tag{8.105}$$

where parameters λ_1 and λ_2 are the Lagrange multipliers. Then the maximization of $H(X)$ requires that

$$\frac{\partial G}{\partial f_X(x)} = -\log_2 f_X(x) - \log_2 e + \lambda_1 + \lambda_2 (x-\mu)^2 = 0 \tag{8.106}$$

Thus

$$\log_2 f_X(x) = -\log_2 e + \lambda_1 + \lambda_2 (x-\mu)^2$$

or

$$\ln f_X(x) = -1 + \frac{\lambda_1}{\log_2 e} + \frac{\lambda_2}{\log_2 e}(x-\mu)^2$$

Hence, we obtain

$$f_X(x) = \exp\left[-1 + \frac{\lambda_1}{\log_2 e} + \frac{\lambda_2}{\log_2 e}(x-\mu)^2\right] \tag{8.107}$$

In view of the constraints of Eqs. (*8.103*) and (*8.104*), it is required that $\lambda_2 < 0$. Let

$$\exp\left(-1 + \frac{\lambda_1}{\log_2 e}\right) = a \qquad \text{and} \qquad \frac{\lambda_2}{\log_2 e} = -b^2$$

Then Eq. (*8.107*) can be rewritten as

$$f_X(x) = a e^{-b^2 (x-\mu)^2} \tag{8.108}$$

Substituting Eq. (*8.108*) into Eqs. (*8.103*) and (*8.104*), we obtain

$$a\int_{-\infty}^{\infty} e^{-b^2(x-\mu)^2}\, dx = a\frac{\sqrt{\pi}}{b} = 1 \tag{8.109}$$

$$a\int_{-\infty}^{\infty} (x-\mu)^2 e^{-b^2(x-\mu)^2}\, dx = a\frac{\sqrt{\pi}}{2b^3} = \sigma^2 \tag{8.110}$$

Solving Eqs. (*8.109*) and (*8.110*) for a and b^2, we obtain

$$a = \frac{1}{\sqrt{2\pi}\,\sigma} \qquad \text{and} \qquad b^2 = \frac{1}{2\sigma^2}$$

Substituting these values in Eq. (*8.108*), we see that the desired $f_X(x)$ is given by

$$f_X(x) = \frac{1}{\sqrt{2\pi}\,\sigma} e^{-(x-\mu)^2/(2\sigma^2)} \tag{8.111}$$

which is the probability density function of a gaussian random variable X of mean μ and variance σ^2 [Eq. (*5.101*)].

8.22. Show that the channel capacity of an ideal AWGN channel with infinite bandwidth is given by

$$C_\infty = \frac{1}{\ln 2}\frac{S}{\eta} \approx 1.44\frac{S}{\eta} \quad \text{b/s} \tag{8.112}$$

where S is the average signal power and $\eta/2$ is the power spectral density of white gaussian noise.

From Eq. (7.6) the noise power N is given by $N = \eta B$. Thus, by Eq. (8.48)

$$C = B\log_2\left(1 + \frac{S}{\eta B}\right)$$

Let $S/(\eta B) = \lambda$. Then

$$C = \frac{S}{\eta\lambda}\log_2(1+\lambda) = \frac{1}{\ln 2}\frac{S}{\eta}\frac{\ln(1+\lambda)}{\lambda} \tag{8.113}$$

Now

$$C_\infty = \lim_{B\to\infty} B\log_2\left(1 + \frac{S}{\eta B}\right)$$

$$= \frac{1}{\ln 2}\frac{S}{\eta}\lim_{\lambda\to 0}\frac{\ln(1+\lambda)}{\lambda}$$

Since $\lim_{\lambda\to 0}[\ln(1+\lambda)]/\lambda = 1$, we obtain

$$C_\infty = \frac{1}{\ln 2}\frac{S}{\eta} \approx 1.44\frac{S}{\eta} \quad \text{b/s}$$

Note that Eq. (8.112) can be used to estimate upper limits on the performance of any practical communication system whose transmission channel can be approximated by the AWGN channel.

8.23. Consider an AWGN channel with 4-kHz bandwidth and the noise power spectral density $\eta/2 = 10^{-12}$ W/Hz. The signal power required at the receiver is 0.1 mW. Calculate the capacity of this channel.

$$B = 4000 \text{ Hz} \qquad S = 0.1(10^{-3}) \text{ W}$$

$$N = \eta B = 2(10^{-12})(4000) = 8(10^{-9}) \text{ W}$$

Thus

$$\frac{S}{N} = \frac{0.1(10^{-3})}{8(10^{-9})} = 1.25(10^4)$$

And by Eq. (8.48)

$$C = B\log_2\left(1 + \frac{S}{N}\right)$$

$$= 4000\log_2\left[1 + 1.25(10^4)\right] = 54.44(10^3) \text{ b/s}$$

8.24. An analog signal having 4-kHz bandwidth is sampled at 1.25 times the Nyquist rate, and each sample is quantized into one of 256 equally likely levels. Assume that the successive samples are statistically independent.

(*a*) What is the information rate of this source?

(*b*) Can the output of this source be transmitted without error over an AWGN channel with a bandwidth of 10 kHz and an S/N ratio of 20 dB?

(*c*) Find the S/N ratio required for error-free transmission for part (*b*).

(*d*) Find the bandwidth required for an AWGN channel for error-free transmission of the output of this source if the S/N ratio is 20 dB.

(*a*)
$$f_M = 4(10^3)\text{ Hz}$$
$$\text{Nyquist rate} = 2f_M = 8(10^3)\text{ samples/s}$$
$$r = 8(10^3)(1.25) = 10^4\text{ samples/s}$$
$$H(X) = \log_2 256 = 8\text{ b/sample}$$

By Eq. (*8.10*) the information rate R of the source is

$$R = rH(X) = 10^4(8)\text{ b/s} = 80\text{ kb/s}$$

(*b*) By Eq. (*8.48*)

$$C = B\log_2\left(1 + \frac{S}{N}\right) = 10^4\log_2(1 + 10^2) = 66.6(10^3)\text{ b/s}$$

Since $R > C$, error-free transmission is not possible.

(*c*) The required S/N ratio can be found by

$$C = 10^4\log_2\left(1 + \frac{S}{N}\right) \geq 8(10^4)$$

or
$$\log_2\left(1 + \frac{S}{N}\right) \geq 8$$

or
$$1 + \frac{S}{N} \geq 2^8 = 256 \longrightarrow \frac{S}{N} \geq 255 \quad (= 24.1\text{ dB})$$

Thus the required S/N ratio must be greater than or equal to 24.1 dB for error-free transmission.

(*d*) The required bandwidth B can be found by

$$C = B\log_2(1 + 100) \geq 8(10^4)$$

or
$$B \geq \frac{8(10^4)}{\log_2(1 + 100)} = 1.2(10^4)\text{Hz} = 12\text{ kHz}$$

and the required bandwidth of the channel must be greater than or equal to 12 kHz.

SOURCE CODING

8.25. Consider a DMS X with two symbols x_1 and x_2 and $P(x_1) = 0.9$, $P(x_2) = 0.1$. Symbols x_1 and x_2 are encoded as follows (Table 8-4):

Table 8-4

x_i	$P(x_i)$	Code
x_1	0.9	0
x_2	0.1	1

Find the efficiency η and the redundancy γ of this code.

By Eq. (*8.49*) the average code length L per symbol is

$$L = \sum_{i=1}^{2} P(x_i)n_i = (0.9)(1) + (0.1)(1) = 1\text{ b}$$

By Eq. (8.7)

$$H(X) = -\sum_{i=1}^{2} P(x_i)\log_2 P(x_i)$$
$$= -0.9\log_2 0.9 - 0.1\log_2 0.1 = 0.469 \text{ b/symbol}$$

Thus by Eq. (8.53) the code efficiency η is

$$\eta = \frac{H(X)}{L} = 0.469 = 46.9\%$$

By Eq. (8.51) the code redundancy γ is

$$\gamma = 1 - \eta = 0.531 = 53.1\%$$

8.26. The second-order extension of the DMS X of Prob. 8.25, denoted by X^2, is formed by taking the source symbols two at a time. The coding of this extension is shown in Table 8-5. Find the efficiency η and the redundancy γ of this extension code.

Table 8-5

a_i	$P(a_i)$	Code
$a_1 = x_1x_1$	0.81	0
$a_2 = x_1x_2$	0.09	10
$a_3 = x_2x_1$	0.09	110
$a_4 = x_2x_2$	0.01	111

$$L = \sum_{i=1}^{4} P(a_i)n_i = 0.81(1) + 0.09(2) + 0.09(3) + 0.01(3)$$
$$= 1.29 \text{ b/symbol}$$

The entropy of the second-order extension of X, $H(X^2)$, is given by

$$H(X^2) = -\sum_{i=1}^{4} P(a_i)\log_2 P(a_i)$$
$$= -0.81\log_2 0.81 - 0.09\log_2 0.09 - 0.09\log_2 0.09 - 0.01\log_2 0.01$$
$$= 0.938 \text{ b/symbol}$$

Therefore the code efficiency η is

$$\eta = \frac{H(X^2)}{L} = \frac{0.938}{1.29} = 0.727 = 72.7\%$$

and the code redundancy γ is

$$\gamma = 1 - \eta = 0.273 = 27.3\%$$

Note that $H(X^2) = 2H(X)$.

8.27. Consider a DMS X with symbols x_i, $i = 1, 2, 3, 4$. Table 8-6 lists four possible binary codes.

Table 8-6

x_i	Code A	Code B	Code C	Code D
x_1	00	0	0	0
x_2	01	10	11	100
x_3	10	11	100	110
x_4	11	110	110	111

(*a*) Show that all codes except code B satisfy the Kraft inequality.

(*b*) Show that codes A and D are uniquely decodable but codes B and C are not uniquely decodable.

(*a*) From Eq. (*8.54*) we obtain the following:

For code A:

$$n_1 = n_2 = n_3 = n_4 = 2$$

$$K = \sum_{i=1}^{4} 2^{-n_i} = \frac{1}{4} + \frac{1}{4} + \frac{1}{4} + \frac{1}{4} = 1$$

For code B:

$$n_1 = 1 \qquad n_2 = n_3 = 2 \qquad n_4 = 3$$

$$K = \sum_{i=1}^{4} 2^{-n_i} = \frac{1}{2} + \frac{1}{4} + \frac{1}{4} + \frac{1}{8} = 1\frac{1}{8} > 1$$

For code C:

$$n_1 = 1 \qquad n_2 = 2 \qquad n_3 = n_4 = 3$$

$$K = \sum_{i=1}^{4} 2^{-n_i} = \frac{1}{2} + \frac{1}{4} + \frac{1}{8} + \frac{1}{8} = 1$$

For code D:

$$n_1 = 1 \qquad n_2 = n_3 = n_4 = 3$$

$$K = \sum_{i=1}^{4} 2^{-n_i} = \frac{1}{2} + \frac{1}{8} + \frac{1}{8} + \frac{1}{8} = \frac{7}{8} < 1$$

All codes except code B satisfy the Kraft inequality.

(*b*) Codes A and D are prefix-free codes. They are therefore uniquely decodable. Code B does not satisfy the Kraft inequality, and it is not uniquely decodable. Although code C does satisfy the Kraft inequality, it is not uniquely decodable. This can be seen by the following example: Given the binary sequence 0110110. This sequence may correspond to the source sequences $x_1x_2x_1x_4$ or $x_1x_4x_4$.

8.28. Verify Eq. (*8.52*), that is,

$$L \geq H(X)$$

where L is the average code word length per symbol and $H(X)$ is the source entropy.

From Eq. (*8.78*) (Prob. 8.4), we have

$$\sum_{i=1}^{m} P_i \log_2 \frac{Q_i}{P_i} \leq 0$$

where the equality holds only if $Q_i = P_i$. Let

$$Q_i = \frac{2^{-n_i}}{K} \tag{8.114}$$

where

$$K = \sum_{i=1}^{m} 2^{-n_i} \tag{8.115}$$

which is defined in Eq. (*8.54*). Then

$$\sum_{i=1}^{m} Q_i = \frac{1}{K} \sum_{i=1}^{m} 2^{-n_i} = 1 \tag{8.116}$$

and

$$\sum_{i=1}^{m} P_i \log_2 \frac{2^{-n_i}}{KP_i} = \sum_{i=1}^{m} P_i \left(\log_2 \frac{1}{P_i} - n_i - \log_2 K \right)$$

$$= -\sum_{i=1}^{m} P_i \log_2 P_i - \sum_{i=1}^{m} P_i n_i - (\log_2 K) \sum_{i=1}^{m} P_i$$

$$= H(X) - L - \log_2 K \leq 0 \tag{8.117}$$

From the Kraft inequality (*8.54*) we have

$$\log_2 K \le 0 \tag{8.118}$$

Thus,

$$H(X) - L \le \log_2 K \le 0 \tag{8.119}$$

or

$$L \ge H(X)$$

The equality holds when $K = 1$ and $P_i = Q_i$.

8.29. Let X be a DMS with symbols x_i and corresponding probabilities $P(x_i) = P_i$, $i = 1, 2, \ldots, m$. Show that for the optimum source encoding we require that

$$K = \sum_{i=1}^{m} 2^{-n_i} = 1 \tag{8.120}$$

and

$$n_i = \log_2 \frac{1}{P_i} = I_i \tag{8.121}$$

where n_i is the length of the code word corresponding to x_i and I_i is the information content of x_i.

From the result of Prob. 8.28, the optimum source encoding with $L = H(X)$ requires $K = 1$ and $P_i = Q_i$. Thus, by Eqs. (*8.115*) and (*8.114*)

$$K = \sum_{i=1}^{m} 2^{-n_i} = 1 \tag{8.122}$$

and

$$P_i = Q_i = 2^{-n_i} \tag{8.123}$$

Hence

$$n_i = -\log_2 P_i = \log_2 \frac{1}{P_i} = I_i$$

Note that Eq. (*8.121*) implies the following commonsense principle: Symbols that occur with high probability should be assigned shorter code words than symbols that occur with low probability.

8.30. Consider a DMS X with symbols x_i and corresponding probabilities $P(x_i) = P_i$, $i = 1, 2, \ldots, m$. Let n_i be the length of the code word for x_i such that

$$\log_2 \frac{1}{P_i} \le n_i \le \log_2 \frac{1}{P_i} + 1 \tag{8.124}$$

Show that this relationship satisfies the Kraft inequality (*8.54*), and find the bound on K in Eq. (*8.54*).

Equation (*8.124*) can be rewritten as

$$-\log_2 P_i \le n_i \le -\log_2 P_i + 1 \tag{8.125}$$

or

$$\log_2 P_i \ge -n_i \ge \log_2 P_i - 1$$

Then

$$2^{\log_2 P_i} \ge 2^{-n_i} \ge 2^{\log_2 P_i} 2^{-1}$$

or

$$P_i \ge 2^{-n_i} \ge \tfrac{1}{2} P_i \tag{8.126}$$

Thus,

$$\sum_{i=1}^{m} P_i \ge \sum_{i=1}^{m} 2^{-n_i} \ge \tfrac{1}{2} \sum_{i=1}^{m} P_i \tag{8.127}$$

or

$$1 \ge \sum_{i=1}^{m} 2^{-n_i} \ge \tfrac{1}{2} \tag{8.128}$$

which indicates that the Kraft inequality (*8.54*) is satisfied, and the bound on K is

$$\tfrac{1}{2} \le K \le 1 \tag{8.129}$$

8.31. Consider a DMS X with symbols x_i and corresponding probabilities $P(x_i) = P_i$, $i = 1, 2, \ldots, m$. Show that a code constructed in agreement with Eq. (*8.124*) will satisfy the following relation:

$$H(X) \le L \le H(X) + 1 \tag{8.130}$$

where $H(X)$ is the source entropy and L is the average code word length.

Multiplying Eq. (*8.125*) by P_i and summing over i yields

$$-\sum_{i=1}^{m} P_i \log_2 P_i \le \sum_{i=1}^{m} n_i P_i \le \sum_{i=1}^{m} P_i(-\log_2 P_i + 1) \tag{8.131}$$

Now
$$\sum_{i=1}^{m} P_i(-\log_2 P_i + 1) = -\sum_{i=1}^{m} P_i \log_2 P_i + \sum_{i=1}^{m} P_i$$
$$= H(X) + 1$$

Thus, Eq. (*8.131*) reduces to

$$H(X) \le L \le H(X) + 1$$

ENTROPY CODING

8.32. A DMS X has four symbols x_1, x_2, x_3, and x_4 with $P(x_1) = \frac{1}{2}$, $P(x_2) = \frac{1}{4}$, and $P(x_3) = P(x_4) = \frac{1}{8}$. Construct a Shannon-Fano code for X; show that this code has the optimum property that $n_i = I(x_i)$ and that the code efficiency is 100 percent.

The Shannon-Fano code is constructed as follows (see Table 8-7):

Table 8-7

x_i	$P(x_i)$	Step 1	Step 2	Step 3	Code
x_1	$\frac{1}{2}$	0			0
x_2	$\frac{1}{4}$	1	0		10
x_3	$\frac{1}{8}$	1	1	0	110
x_4	$\frac{1}{8}$	1	1	1	111

$$I(x_1) = -\log_2 \tfrac{1}{2} = 1 = n_1 \qquad I(x_2) = -\log_2 \tfrac{1}{4} = 2 = n_2$$
$$I(x_3) = -\log_2 \tfrac{1}{8} = 3 = n_3 \qquad I(x_4) = -\log_2 \tfrac{1}{8} = 3 = n_4$$

$$H(X) = \sum_{i=1}^{4} P(x_i) I(x_i) = \tfrac{1}{2}(1) + \tfrac{1}{4}(2) + \tfrac{1}{8}(3) + \tfrac{1}{8}(3) = 1.75$$

$$L = \sum_{i=1}^{4} P(x_i) n_i = \tfrac{1}{2}(1) + \tfrac{1}{4}(2) + \tfrac{1}{8}(3) + \tfrac{1}{8}(3) = 1.75$$

$$\eta = \frac{H(X)}{L} = 1 = 100\%$$

8.33. A DMS X has five equally likely symbols.

(*a*) Construct a Shannon-Fano code for X, and calculate the efficiency of the code.

(*b*) Construct another Shannon-Fano code, and compare the results.

(*c*) Repeat for the Huffman code and compare the results.

(*a*) A Shannon-Fano code [by choosing two approximately equiprobable (0.4 versus 0.6) sets] is constructed as follows (see Table 8-8):

Table 8-8

x_i	$P(x_i)$	Step 1	Step 2	Step 3	Code
x_1	0.2	0	0		00
x_2	0.2	0	1		01
x_3	0.2	1	0		10
x_4	0.2	1	1	0	110
x_5	0.2	1	1	1	111

$$H(X) = -\sum_{i=1}^{5} P(x_i)\log_2 P(x_i) = 5(-0.2\log_2 0.2) = 2.32$$

$$L = \sum_{i=1}^{5} P(x_i)n_i = 0.2(2+2+2+3+3) = 2.4$$

The efficiency η is

$$\eta = \frac{H(X)}{L} = \frac{2.32}{2.4} = 0.967 = 96.7\%$$

(*b*) Another Shannon-Fano code [by choosing another two approximately equiprobable (0.6 versus 0.4) sets] is constructed as follows (see Table 8-9):

Table 8-9

x_i	$P(x_i)$	Step 1	Step 2	Step 3	Code
x_1	0.2	0	0		00
x_2	0.2	0	1	0	010
x_3	0.2	0	1	1	011
x_4	0.2	1	0		10
x_5	0.2	1	1		11

$$L = \sum_{i=1}^{5} P(x_i)n_i = 0.2(2+3+3+2+2) = 2.4$$

Since the average code word length is the same as that for the code of part (*a*), the efficiency is the same.

(*c*) The Huffman code is constructed as follows (see Table 8-10):

Table 8-10

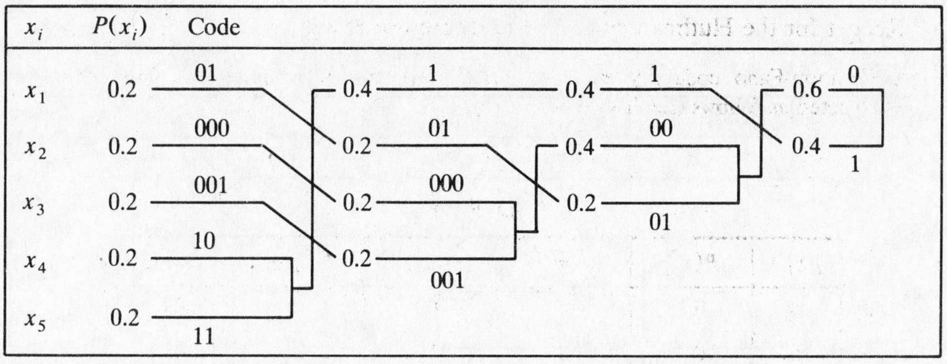

$$L = \sum_{i=1}^{5} P(x_i)n_i = 0.2(2 + 3 + 3 + 2 + 2) = 2.4$$

Since the average code word length is the same as that for the Shannon-Fano code, the efficiency is also the same.

8.34. A DMS X has five symbols x_1, x_2, x_3, x_4, and x_5 with $P(x_1) = 0.4$, $P(x_2) = 0.19$, $P(x_3) = 0.16$, $P(x_4) = 0.15$, and $P(x_5) = 0.1$.

(*a*) Construct a Shannon-Fano code for X, and calculate the efficiency of the code.

(*b*) Repeat for the Huffman code and compare the results.

(*a*) The Shannon-Fano code is constructed as follows (see Table 8-11):

Table 8-11

x_i	$P(x_i)$	Step 1	Step 2	Step 3	Code
x_1	0.4	0	0		00
x_2	0.19	0	1		01
x_3	0.16	1	0		10
x_4	0.15	1	1	0	110
x_5	0.1	1	1	1	111

$$\begin{aligned} H(X) &= -\sum_{i=1}^{5} P(x_i)\log_2 P(x_i) \\ &= -0.4\log_2 0.4 - 0.19\log_2 0.19 - 0.16\log_2 0.16 \\ &\quad -0.15\log_2 0.15 - 0.1\log_2 0.1 \\ &= 2.15 \end{aligned}$$

$$\begin{aligned} L &= \sum_{i=1}^{5} P(x_i)n_i \\ &= 0.4(2) + 0.19(2) + 0.16(2) + 0.15(3) + 0.1(3) = 2.25 \end{aligned}$$

$$\eta = \frac{H(X)}{L} = \frac{2.15}{2.25} = 0.956 = 95.6\%$$

(*b*) The Huffman code is constructed as follows (see Table 8-12):

Table 8-12

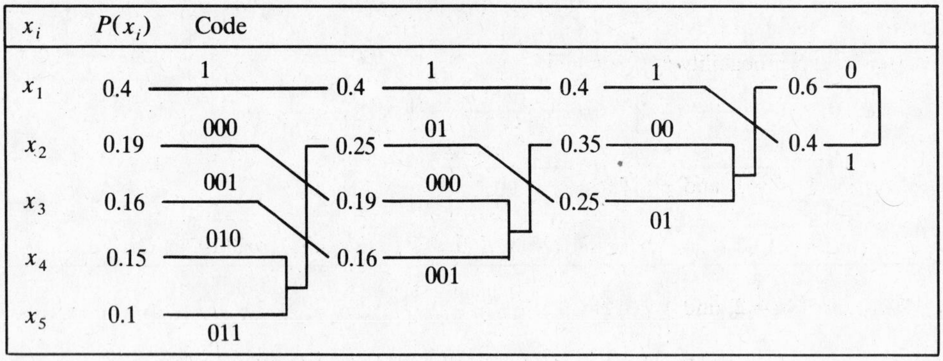

$$L = \sum_{i=1}^{5} P(x_i)n_i$$

$$= 0.4(1) + (0.19 + 0.16 + 0.15 + 0.1)(3) = 2.2$$

$$\eta = \frac{H(X)}{L} = \frac{2.15}{2.2} = 0.977 = 97.7\%$$

The average code word length of the Huffman code is shorter than that of the Shannon-Fano code, and thus the efficiency is higher than that of the Shannon-Fano code.

CHANNEL CODING

8.35. Consider a BSC of Fig. 8-13 (Prob. 8.16). Show that the probability of error P_e is

$$P_e = p \tag{8.132}$$

From Eq. (*8.20*)

$$[P(Y|X)] = \begin{bmatrix} 1-p & p \\ p & 1-p \end{bmatrix} = \begin{bmatrix} P(y_1|x_1) & P(y_2|x_1) \\ P(y_1|x_2) & P(y_2|x_2) \end{bmatrix}$$

The average probability of error P_e is

$$P_e = P(y_2|x_1)P(x_1) + P(y_1|x_2)P(x_2)$$

$$= p\alpha + p(1-\alpha) = p$$

8.36. In the BSC of Prob. 8.35, consider a simple coding scheme that involves the use of a repetition code in which each binary symbol is repeated several times. Let each symbol (0 or 1) be repeated n times, where $n = 2m + 1$ is an odd integer. For decoding, a *majority rule* is employed. That is, if in a block of n received bits the number of 0s exceeds the number of 1s, then the decoder decides in favor of a 0. Otherwise, it decides in favor of a 1. Hence, an error occurs when $m + 1$ or more bits out of $n = 2m + 1$ bits are received incorrectly.

(*a*) Show that the probability of error P_e is given by

$$P_e = \sum_{i=m+1}^{n} \binom{n}{i} p^i (1-p)^{n-i} \tag{8.133}$$

(*b*) Calculate P_e when $p = 0.01$ for $n = 3$, 5, and 7.

(*a*) The probability of i b being received in error is

$$\binom{n}{i} p^i (1-p)^{n-i}$$

Hence, the probability of error P_e is

$$P_e = \sum_{i=m+1}^{n} \binom{n}{i} p^i (1-p)^{n-i}$$

(*b*) For $n = 3$, $m = 1$, and $p = 0.01$,

$$P_e = \binom{3}{2}(0.01)^2(0.99) + \binom{3}{3}(0.01)^3 \approx 3(10^{-4})$$

For $n = 5$, $m = 2$, and $p = 0.01$,

$$P_e = \binom{5}{3}(0.01)^3(0.99)^2 + \binom{5}{4}(0.01)^4(0.99) + \binom{5}{5}(0.01)^5$$
$$= 9.85(10^{-6}) \approx 10^{-5}$$

For $n = 7$, $m = 3$, and $p = 0.01$,

$$P_e = \binom{7}{4}(0.01)^4(0.99)^3 + \binom{7}{5}(0.01)^5(0.99)^2$$
$$+ \binom{7}{6}(0.01)^6(0.99) + \binom{7}{7}(0.01)^7$$
$$= 3.416(10^{-7})$$

8.37. In a *single-parity-check* code, a single parity bit is appended to a block of k data bits $(d_1 d_2 \cdots d_k)$. The single parity bit c_1 is chosen so that the code word satisfies the *even parity rule*:

$$d_1 \oplus d_2 \oplus \cdots \oplus d_k \oplus c_1 = 0$$

(*a*) For $k = 3$, set up all possible code words in the (4, 3) code.

(*b*) Which error patterns can the (4, 3) code detect?

(*c*) Compute the probability of an undetected symbol error, assuming that all symbol errors are independent and that the probability of symbol error is $p = 0.01$.

(*a*) There are $2^k = 2^3 = 8$ possible code words, which are shown in Table 8-13.

Table 8-13

Data code	Parity bit	Code word
000	0	0000
001	1	0011
010	1	0101
011	0	0110
100	1	1001
101	0	1010
110	0	1100
111	1	1111

(*b*) The (4, 3) code is capable of detecting all single-error and triple-error patterns.

(*c*) The probability of an undetected symbol error P_{ue} is equal to the probability that two or four errors occur anywhere in a code word.

$$P_{ue} = \binom{4}{2} p^2(1-p)^2 + \binom{4}{4} p^4$$

$$= 6p^2(1-p)^2 + p^4$$

$$= 6(0.01)^2(0.99)^2 + (0.01)^4 \approx 5.88(10^{-4})$$

ERROR CONTROL CODING

8.38. For a (6, 3) systematic linear block code, the three parity-check bits c_4, c_5, and c_6 are formed from the following equations:

$$c_4 = d_1 \oplus d_3$$

$$c_5 = d_1 \oplus d_2 \oplus d_3$$

$$c_6 = d_1 \oplus d_2$$

(*a*) Write down the generator matrix G.

(*b*) Construct all possible code words.

(*c*) Suppose that the received word is 010111. Decode this received word by finding the location of the error and the transmitted data bits.

(*a*) From the given equation we obtain [Eqs. (*8.56*) and (*8.59*)]

$$P = \begin{bmatrix} 1 & 0 & 1 \\ 1 & 1 & 1 \\ 1 & 1 & 0 \end{bmatrix}$$

Then by Eq. (*8.58*)

$$G = \begin{bmatrix} I_3 & P^T \end{bmatrix} = \begin{bmatrix} 1 & 0 & 0 & 1 & 1 & 1 \\ 0 & 1 & 0 & 0 & 1 & 1 \\ 0 & 0 & 1 & 1 & 1 & 0 \end{bmatrix}$$

(*b*) Since $k = 3$, we have $2^3 = 8$ data words. Thus, if $\mathbf{d} = [101]$, then using Eq. (*8.59*), we obtain

$$\mathbf{c} = \mathbf{d}G = \begin{bmatrix} 1 & 0 & 1 \end{bmatrix} \begin{bmatrix} 1 & 0 & 0 & 1 & 1 & 1 \\ 0 & 1 & 0 & 0 & 1 & 1 \\ 0 & 0 & 1 & 1 & 1 & 0 \end{bmatrix} = \begin{bmatrix} 1 & 0 & 1 & 0 & 0 & 1 \end{bmatrix}$$

In a similar manner, the other code words can be constructed. They are listed in Table 8-14.

Table 8-14

d	**c**	**d**	**c**
000	000000	100	100111
001	001110	101	101001
010	010011	110	110100
011	011101	111	111010

(*c*) By Eq. (*8.61*)

$$H^T = \begin{bmatrix} P^T \\ I_3 \end{bmatrix} = \begin{bmatrix} 1 & 1 & 1 \\ 0 & 1 & 1 \\ 1 & 1 & 0 \\ 1 & 0 & 0 \\ 0 & 1 & 0 \\ 0 & 0 & 1 \end{bmatrix}$$

Now $$\mathbf{r} = [0 \quad 1 \quad 0 \quad 1 \quad 1 \quad 1]$$

By Eq. (*8.66*) the syndrome **s** of **r** is

$$\mathbf{s} = \mathbf{r}H^T = [0 \quad 1 \quad 0 \quad 1 \quad 1 \quad 1] \begin{bmatrix} 1 & 1 & 1 \\ 0 & 1 & 1 \\ 1 & 1 & 0 \\ 1 & 0 & 0 \\ 0 & 1 & 0 \\ 0 & 0 & 1 \end{bmatrix} = [1 \quad 0 \quad 0]$$

Since **s** is equal to the fourth row of H^T, an error is at the fourth bit, the correct code word is 010011, and the data bits are 010.

8.39. A parity-check code has the parity-check matrix

$$H = \begin{bmatrix} 1 & 0 & 1 & 1 & 0 & 0 \\ 1 & 1 & 0 & 0 & 1 & 0 \\ 0 & 1 & 1 & 0 & 0 & 1 \end{bmatrix}$$

(*a*) Determine the generator matrix *G*.

(*b*) Find the code word that begins 101. . . .

(*c*) Suppose that the received word is 110110. Decode this received word.

(*a*) Since H is a 6×3 matrix, $n = 6$ and $k = 3$. Using Eq. (*8.61*), we obtain

$$P^T = \begin{bmatrix} 1 & 1 & 0 \\ 0 & 1 & 1 \\ 1 & 0 & 1 \end{bmatrix}$$

Then by Eq. (*8.58*) the generator matrix G is

$$G = \begin{bmatrix} I_3 & P^T \end{bmatrix} = \begin{bmatrix} 1 & 0 & 0 & 1 & 1 & 0 \\ 0 & 1 & 0 & 0 & 1 & 1 \\ 0 & 0 & 1 & 1 & 0 & 1 \end{bmatrix}$$

(*b*)

$$\mathbf{c} = \mathbf{d}G = [1 \quad 0 \quad 1] \begin{bmatrix} 1 & 0 & 0 & 1 & 1 & 0 \\ 0 & 1 & 0 & 0 & 1 & 1 \\ 0 & 0 & 1 & 1 & 0 & 1 \end{bmatrix} = [1 \quad 0 \quad 1 \quad 0 \quad 1 \quad 1]$$

(*c*)

$$\mathbf{r} = [1 \quad 1 \quad 0 \quad 1 \quad 1 \quad 0]$$

$$\mathbf{s} = \mathbf{r}H^T = [1 \quad 1 \quad 0 \quad 1 \quad 1 \quad 0] \begin{bmatrix} 1 & 1 & 0 \\ 0 & 1 & 1 \\ 1 & 0 & 1 \\ 1 & 0 & 0 \\ 0 & 1 & 0 \\ 0 & 0 & 1 \end{bmatrix} = [0 \quad 1 \quad 1]$$

Since **s** is equal to the second row of H^T, an error is at the second bit, the correct code word is 100110, and the data bits are 100.

8.40. The repetition code of Prob. 8.36 is an $(n, 1)$ block code. There are only two code words in the repetition code, an all-0 code word and an all-1 code word. Consider the case of a repetition code with $n = 5$.

(*a*) Construct the generator matrix G for this $(5, 1)$ block code.

(*b*) Using G find all code words.

(*c*) Find the parity-check matrix H for this code.

(*d*) Show that $GH^T = \mathbf{0}$.

(*a*) We have 4 parity bits that are the same as the data bit. With $k = 1$, the identity matrix $I_k = 1$, and by Eqs. (*8.56*) and (*8.57*) matrix P^T is given by

$$P^T = [1 \quad 1 \quad 1 \quad 1]$$

Then by Eq. (*8.58*) the generator matrix G is

$$G = [1 \quad 1 \quad 1 \quad 1 \quad 1]$$

(*b*) For $d_1 = 0$,

$$\mathbf{c}_1 = [0][1 \quad 1 \quad 1 \quad 1 \quad 1] = [0 \quad 0 \quad 0 \quad 0 \quad 0]$$

For $d_1 = 1$,

$$\mathbf{c}_2 = [1][1 \quad 1 \quad 1 \quad 1 \quad 1] = [1 \quad 1 \quad 1 \quad 1 \quad 1]$$

(*c*) By Eq. (*8.60*) the parity-check matrix H is

$$H = \begin{bmatrix} P & I_4 \end{bmatrix} = \begin{bmatrix} 1 & 1 & 0 & 0 & 0 \\ 1 & 0 & 1 & 0 & 0 \\ 1 & 0 & 0 & 1 & 0 \\ 1 & 0 & 0 & 0 & 1 \end{bmatrix}$$

(*d*)

$$GH^T = [1 \quad 1 \quad 1 \quad 1 \quad 1] \begin{bmatrix} 1 & 1 & 1 & 1 \\ 1 & 0 & 0 & 0 \\ 0 & 1 & 0 & 0 \\ 0 & 0 & 1 & 0 \\ 0 & 0 & 0 & 1 \end{bmatrix} = [0 \quad 0 \quad 0 \quad 0] = \mathbf{0}$$

8.41. Consider the $(5, 1)$ repetition code of Prob. 8.40.

(*a*) Evaluate the syndrome $\mathbf{s}$ for all five possible single-error patterns.

(*b*) Repeat (*a*) for all ten possible double-error patterns.

(*c*) Show that the $(5, 1)$ repetition code is capable of correcting up to two errors.

(*a*) From Prob. 8.40

$$H^T = \begin{bmatrix} 1 & 1 & 1 & 1 \\ 1 & 0 & 0 & 0 \\ 0 & 1 & 0 & 0 \\ 0 & 0 & 1 & 0 \\ 0 & 0 & 0 & 1 \end{bmatrix}$$

From Eq. (*8.66*) syndrome $\mathbf{s}$ is given by

$$\mathbf{e}H^T = \mathbf{s}$$

where $\mathbf{e}$ is an error vector. Let $\mathbf{e} = [1 \ 0 \ 0 \ 0 \ 0]$. Then

$$\mathbf{s} = [1 \quad 0 \quad 0 \quad 0 \quad 0] \begin{bmatrix} 1 & 1 & 1 & 1 \\ 1 & 0 & 0 & 0 \\ 0 & 1 & 0 & 0 \\ 0 & 0 & 1 & 0 \\ 0 & 0 & 0 & 1 \end{bmatrix} = [1 \quad 1 \quad 1 \quad 1]$$

In a similar manner, the other syndromes can be evaluated. They are listed in Table 8-15.

Table 8-15

e	s
10000	1111
01000	1000
00100	0100
00010	0010
00001	0001

(*b*) Let $\mathbf{e} = [1\ 1\ 0\ 0\ 0]$. Then

$$\mathbf{s} = [1\quad 1\quad 0\quad 0\quad 0]\begin{bmatrix} 1 & 1 & 1 & 1 \\ 1 & 0 & 0 & 0 \\ 0 & 1 & 0 & 0 \\ 0 & 0 & 1 & 0 \\ 0 & 0 & 0 & 1 \end{bmatrix} = [0\quad 1\quad 1\quad 1]$$

Note that s is equal to the sum (modulo 2) of the first and second rows of H^T (Prob. 8.43).

In a similar manner, the other syndromes can be evaluated. They are listed in Table 8-16.

Table 8-16

e	s
11000	0111
10100	1011
10010	1101
10001	1110
01100	1100
01010	1010
01001	1001
00110	0110
00101	0101
00011	0011

(*c*) Since the syndromes for all single-error and double-error patterns are distinct, the (5, 1) repetition code is capable of correcting up to two errors.

8.42. Show that all error vectors that differ by a code vector have the same syndrome.

For k data bits, there are 2^k distinct code vectors, denoted as $\mathbf{c}_i$, $i = 0, 1, \ldots, 2^k - 1$. Thus, for any error vector $\mathbf{e}$, we define the 2^k distinct vectors $\mathbf{e}_i$ as

$$\mathbf{e}_i = \mathbf{e} \oplus \mathbf{c}_i \qquad i = 0, 1, \ldots, 2^k - 1 \tag{8.134}$$

Postmultiplying both sides of Eq. (*8.134*) by H^T and using Eq. (*8.62*), we obtain

$$\begin{aligned} \mathbf{e}_i H^T &= (\mathbf{e} \oplus \mathbf{c}_i)H^T = \mathbf{e}H^T \oplus \mathbf{c}_i H^T \\ &= \mathbf{e}H^T \oplus \mathbf{0} = \mathbf{e}H^T = \mathbf{s} \end{aligned} \tag{8.135}$$

8.43. Show that the syndrome **s** is the sum (modulo 2) of those rows of matrix H^T corresponding to the error locations in the error pattern.

Let matrix H^T be expressed in terms of its rows as

$$H^T = \begin{bmatrix} \mathbf{h}_1 \\ \mathbf{h}_2 \\ \vdots \\ \mathbf{h}_n \end{bmatrix} \tag{8.136}$$

Substituting Eq. (8.136) into Eq. (8.66), we can express the syndrome **s** as

$$\mathbf{s} = \mathbf{e}H^T = [e_1 \quad e_2 \quad \cdots \quad e_n] \begin{bmatrix} \mathbf{h}_1 \\ \mathbf{h}_2 \\ \vdots \\ \mathbf{h}_n \end{bmatrix} = \sum_{i=1}^{n} e_i \mathbf{h}_i \tag{8.137}$$

where e_i is the ith element of the error vector **e**, that is,

$$e_i = \begin{cases} 1 & \text{if an error occurred in } i\text{th location} \\ 0 & \text{if no error occurred in } i\text{th location} \end{cases}$$

Hence, Eq. (8.137) indicates that the syndrome **s** equals the sum of those rows of matrix H^T that correspond to the error locations in the error pattern.

8.44. Verify Eq. (8.67). That is, if an (n, k) linear block code can correct up to t errors per code word, the number of check bits $n - k$ in the code word must satisfy the Hamming bound given by

$$2^{n-k} \geq \sum_{i=0}^{t} \binom{n}{i}$$

There are total of 2^{n-k} syndromes including the all-0 syndrome. Each syndrome corresponds to a specific error pattern. The number of possible i-tuple errors in an n-b code word is equal to the number of ways of choosing i bits out of n, namely, $\binom{n}{i}$

Accordingly, the total number of all possible error patterns equals

$$\sum_{i=0}^{t} \binom{n}{i}$$

where t is the maximum number of errors in an error pattern. Therefore, if an (n, k) linear block code is capable of correcting up to t errors, the total number of syndromes must not be less than the total number of all possible error patterns. Thus, we must have

$$2^{n-k} \geq \sum_{i=0}^{t} \binom{n}{i}$$

8.45. Consider a single-error-correcting code for 11 data bits.

(a) How many check bits are required?

(b) Find a parity-check matrix H for this code.

(a) By Eq. (8.67)

$$2^{n-k} \geq \sum_{i=0}^{1} \binom{n}{i} = \binom{n}{0} + \binom{n}{1} = 1 + n$$

Let $n - k = m$. Since $k = 11$, we have $n = m + 11$, and

$$2^m \geq 12 + m \longrightarrow m \geq 4$$

Thus, at least 4 check bits are required.

(*b*) By Eq. (*8.61*)

$$H^T = \begin{bmatrix} P^T \\ I_m \end{bmatrix} = \begin{bmatrix} P^T \\ I_4 \end{bmatrix}$$

For a single-error-correcting condition, it is required that the first 11 rows of matrix H^T be unique. They must also differ from the last 4 rows containing a single 1 in each row and cannot include an all-0 row.

With this requirement, a parity-check matrix H (transpose of H^T) for the (15, 11) code is given by

$$H = \begin{bmatrix} 1 & 1 & 1 & 0 & 0 & 0 & 1 & 1 & 1 & 0 & 1 & 1 & 0 & 0 & 0 \\ 1 & 0 & 0 & 1 & 1 & 0 & 1 & 1 & 0 & 1 & 1 & 0 & 1 & 0 & 0 \\ 0 & 1 & 0 & 1 & 0 & 1 & 1 & 0 & 1 & 1 & 1 & 0 & 0 & 1 & 0 \\ 0 & 0 & 1 & 0 & 1 & 1 & 0 & 1 & 1 & 1 & 1 & 0 & 0 & 0 & 1 \end{bmatrix}$$

ERROR DETECTION AND CORRECTION CAPABILITIES OF LINEAR BLOCK CODES

8.46. Consider the following code vectors:

$$\mathbf{c}_1 = [1 \quad 0 \quad 0 \quad 1 \quad 0]$$
$$\mathbf{c}_2 = [0 \quad 1 \quad 1 \quad 0 \quad 1]$$
$$\mathbf{c}_3 = [1 \quad 1 \quad 0 \quad 0 \quad 1]$$

(*a*) Find $d(\mathbf{c}_1, \mathbf{c}_2)$, $d(\mathbf{c}_1, \mathbf{c}_3)$, and $d(\mathbf{c}_2, \mathbf{c}_3)$.

(*b*) Show that

$$d(\mathbf{c}_1, \mathbf{c}_2) + d(\mathbf{c}_2, \mathbf{c}_3) \geq d(\mathbf{c}_1, \mathbf{c}_3)$$

(*a*) From Eq. (*8.69*) we obtain

$$d(\mathbf{c}_1, \mathbf{c}_2) = w(\mathbf{c}_1 \oplus \mathbf{c}_2) = w[1 \quad 1 \quad 1 \quad 1 \quad 1] = 5$$
$$d(\mathbf{c}_1, \mathbf{c}_3) = w(\mathbf{c}_1 \oplus \mathbf{c}_3) = w[0 \quad 1 \quad 0 \quad 1 \quad 1] = 3$$
$$d(\mathbf{c}_2, \mathbf{c}_3) = w(\mathbf{c}_2 \oplus \mathbf{c}_3) = w[1 \quad 0 \quad 1 \quad 0 \quad 0] = 2$$

(*b*)
$$d(\mathbf{c}_1, \mathbf{c}_2) + d(\mathbf{c}_2, \mathbf{c}_3) = 5 + 2 \geq 3 = d(\mathbf{c}_1, \mathbf{c}_3)$$

8.47. Show that if $\mathbf{c}_i$ and $\mathbf{c}_j$ are two code vectors in an (n, k) linear block code, then their sum is also a code vector.

Since all code vectors $\mathbf{c}$ must satisfy Eq. (*8.70*), we have

$$\mathbf{c}_i H^T = \mathbf{0} \qquad \text{and} \qquad \mathbf{c}_j H^T = \mathbf{0}$$

Then
$$(\mathbf{c}_i \oplus \mathbf{c}_j) H^T = \mathbf{c}_i H^T \oplus \mathbf{c}_j H^T = \mathbf{0} + \mathbf{0} = \mathbf{0} \tag{8.138}$$

which indicates that $\mathbf{c}_i \oplus \mathbf{c}_j$ is also a code vector.

8.48. Prove Theorem 8.1, that is, the minimum distance of a linear block code is the smallest Hamming weight of the nonzero code vectors in the code.

From Eq. (*8.69*)

$$d(\mathbf{c}_i, \mathbf{c}_j) = w(\mathbf{c}_i \oplus \mathbf{c}_j)$$

Then
$$d_{\min} = \min_{\mathbf{c}_i \neq \mathbf{c}_j} d(\mathbf{c}_i, \mathbf{c}_j) = \min_{\mathbf{c}_i \neq \mathbf{c}_j} w(\mathbf{c}_i \oplus \mathbf{c}_j) \tag{8.139}$$

Hence by using the result of Prob. 8.47, Eq. (*8.139*) becomes

$$d_{\min} = \min_{\mathbf{c} \neq \mathbf{0}} w(\mathbf{c}) \tag{8.140}$$

8.49. Prove Theorem 8.2, that is, the minimum distance of a linear block code is equal to the minimum number of rows of H^T that sum to **0**.

By Eq. (*8.70*)

$$\mathbf{c}H^T = \mathbf{0}$$

The product $\mathbf{c}H^T$ is a linear combination of rows of H^T. (See Prob. 8.43.) Hence, the minimum number of rows of H^T that can be added to produce **0** is

$$\min_{\mathbf{c}\neq\mathbf{0}} w(\mathbf{c})$$

which equals $d_{\min}$ by Eq. (*8.140*).

8.50. Consider the $(6, 3)$ code of Prob. 8.38.

(*a*) Show that $d_{\min} = 3$ and that it can correct a single error.

(*b*) Using the minimum distance decoding rule, repeat part (*c*) of Prob. 8.38.

(*a*) From part (*b*) of Prob. 8.38, the code vectors and their Hamming weights are listed in Table 8-17.

Table 8-17

$\mathbf{c}_i$	Hamming weight
$\mathbf{c}_1 = [0\ \ 0\ \ 0\ \ 0\ \ 0\ \ 0]$	0
$\mathbf{c}_2 = [0\ \ 0\ \ 1\ \ 1\ \ 1\ \ 0]$	3
$\mathbf{c}_3 = [0\ \ 1\ \ 0\ \ 0\ \ 1\ \ 1]$	3
$\mathbf{c}_4 = [0\ \ 1\ \ 1\ \ 1\ \ 0\ \ 1]$	4
$\mathbf{c}_5 = [1\ \ 0\ \ 0\ \ 1\ \ 1\ \ 1]$	4
$\mathbf{c}_6 = [1\ \ 0\ \ 1\ \ 0\ \ 0\ \ 1]$	3
$\mathbf{c}_7 = [1\ \ 1\ \ 0\ \ 1\ \ 0\ \ 0]$	3
$\mathbf{c}_8 = [1\ \ 1\ \ 1\ \ 0\ \ 1\ \ 0]$	4

Since $d_{\min}$ is the smallest Hamming weight of the nonzero code vectors in the code, $d_{\min} = 3$. Then by Eq. (*8.71*)

$$d_{\min} = 3 \geq 2t + 1$$

which is satisfied by $t = 1$. Hence, the code can correct a single error.

(*b*) The received vector is $\mathbf{r} = [0\ \ 1\ \ 0\ \ 1\ \ 1\ \ 1]$. Then by Eq. (*8.69*)

$$d(\mathbf{r}, \mathbf{c}_1) = 4 \qquad d(\mathbf{r}, \mathbf{c}_5) = 2$$

$$d(\mathbf{r}, \mathbf{c}_2) = 3 \qquad d(\mathbf{r}, \mathbf{c}_6) = 5$$

$$d(\mathbf{r}, \mathbf{c}_3) = 1 \qquad d(\mathbf{r}, \mathbf{c}_7) = 3$$

$$d(\mathbf{r}, \mathbf{c}_4) = 2 \qquad d(\mathbf{r}, \mathbf{c}_8) = 4$$

The minimum distance between the code vectors and the received vector is

$$d(\mathbf{r}, \mathbf{c}_3) = 1$$

Thus, we conclude that the code vector $\mathbf{c}_3 = [0\ \ 1\ \ 0\ \ 0\ \ 1\ \ 1]$ was sent and the data bits are 010 [which is the same result obtained in part (*c*) of Prob. 8.38].

8.51. Consider a $(7,4)$ linear block code with the parity-check matrix H given by

$$H = \begin{bmatrix} 1 & 0 & 1 & 1 & 1 & 0 & 0 \\ 1 & 1 & 0 & 1 & 0 & 1 & 0 \\ 0 & 1 & 1 & 1 & 0 & 0 & 1 \end{bmatrix}$$

(*a*) Construct code words for this $(7,4)$ code.

(*b*) Show that this code is a Hamming code.

(*c*) Illustrate the relation between the minimum distance and the structure of the parity-check matrix H by considering the code word 0101100.

(*a*) By Eqs. (*8.60*) and (*8.58*) the generating matrix G for this code is

$$G = \begin{bmatrix} 1 & 0 & 0 & 0 & 1 & 1 & 0 \\ 0 & 1 & 0 & 0 & 0 & 1 & 1 \\ 0 & 0 & 1 & 0 & 1 & 0 & 1 \\ 0 & 0 & 0 & 1 & 1 & 1 & 1 \end{bmatrix}$$

With $k = 4$, there are $2^k = 16$ distinct data words, which are listed in Table 8-18. For a given data word, the corresponding code word is obtained by using Eq. (*8.57*). The resultant code words are listed in Table 8-18.

Table 8-18

Data word	Code word	Hamming weight
0000	0000000	0
0001	0001111	4
0010	0010101	3
0011	0011010	3
0100	0100011	3
0101	0101100	3
0110	0110110	4
0111	0111001	4
1000	1000110	3
1001	1001001	3
1010	1010011	4
1011	1011100	4
1100	1100101	4
1101	1101010	4
1110	1110000	3
1111	1111111	7

(*b*) Table 8-18 also lists the Hamming weights of all code words. Since the smallest of the Hamming weights for the nonzero code words is 3, we have $d_{\min} = 3$. Thus, by Eq. (*8.71*) the code can correct a single error.

Next, $n = 7$ and $k = 4$, and we have

$$2^{n-k} = 2^3 = 8$$

$$\sum_{i=0}^{1} \binom{7}{i} = \binom{7}{0} + \binom{7}{1} = 1 + 7 = 8$$

Thus, the equality holds for the Hamming bound of Eq. (*8.67*), and the code is a Hamming code.

(*c*) With code vector $\mathbf{c} = [0\ 1\ 0\ 1\ 1\ 0\ 0]$, the matrix multiplication defined by Eq. (*8.70*) indicates that the second, fourth, and fifth rows of matrix H^T yield

$$[0\quad 1\quad 1] \oplus [1\quad 1\quad 1] \oplus [1\quad 0\quad 0] = [0\quad 0\quad 0]$$

Similar calculations for the remaining 14 nonzero code vectors indicate that the smallest number of rows in H^T that sums to $\mathbf{0}$ is 3, which is equal to $d_{\min}$ (Theorem 8.2).

Supplementary Problems

8.52. Consider a source X that produces five symbols with probabilities $\frac{1}{2}$, $\frac{1}{4}$, $\frac{1}{8}$, $\frac{1}{16}$, and $\frac{1}{16}$. Determine the source entropy $H(X)$.

Ans. 1.875 b/symbol

8.53. Calculate the average information content in the English language, assuming that each of the 26 characters in the alphabet occurs with equal probability.

Ans. 4.7 b/character

8.54. Two BSCs are connected in cascade, as shown in Fig. 8-15.

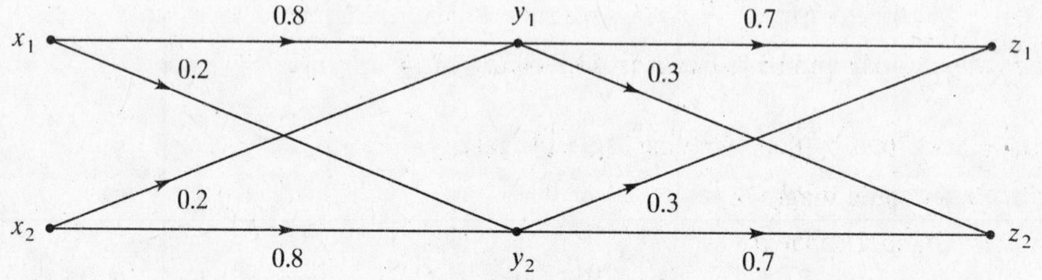

Fig. 8-15

(*a*) Find the channel matrix of the resultant channel.

(*b*) Find $P(z_1)$ and $P(z_2)$ if $P(x_1) = 0.6$ and $P(x_2) = 0.4$.

Ans.

(*a*) $\begin{bmatrix} 0.62 & 0.38 \\ 0.38 & 0.62 \end{bmatrix}$ (b) $P(z_1) = 0.524$, $P(z_2) = 0.476$

8.55. Consider the DMC shown in Fig. 8-16.

(*a*) Find the output probabilities if $P(x_1) = \frac{1}{2}$ and $P(x_2) = P(x_3) = \frac{1}{4}$.

(*b*) Find the output entropy $H(Y)$.

Ans. (*a*) $P(y_1) = 7/24$, $P(y_2) = 17/48$, and $P(y_3) = 17/48$

(*b*) 1.58 b/symbol

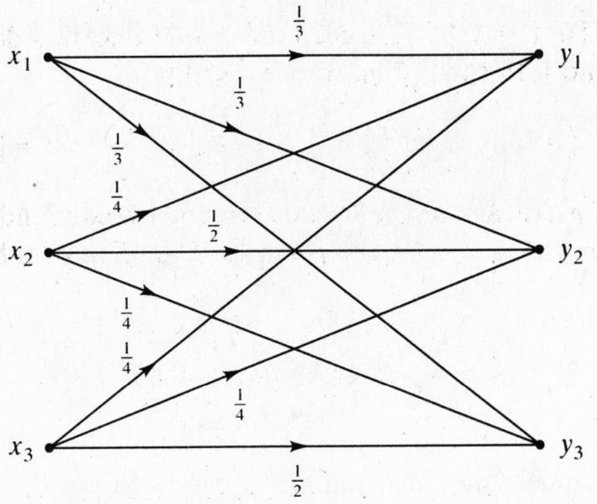

Fig. 8-16

8.56. Verify Eq. (*8.32*), that is,

$$I(X;Y) = H(X) + H(Y) - H(X,Y)$$

Hint: Use Eqs. (*8.28*) and (*8.26*).

8.57. Show that $H(X,Y) \le H(X) + H(Y)$ with equality if and only if X and Y are independent.

Hint: Use Eqs. (*8.30*) and (*8.32*).

8.58. Show that for a deterministic channel

$$H(Y|X) = 0$$

Hint: Use Eq. (*8.24*), and note that for a deterministic channel $P(y_j|x_i)$ are either 0 or 1.

8.59. Consider a channel with an input X and an output Y. Show that if X and Y are statistically independent, then $H(X|Y) = H(X)$ and $I(X;Y) = 0$.

Hint: Use Eqs. (*5.48*) and (*5.49*) in Eqs. (*8.24*) and (*8.28*).

8.60. A channel is described by the following channel matrix.

(*a*) Draw the channel diagram.

(*b*) Find the channel capacity.

$$\begin{bmatrix} \frac{1}{2} & \frac{1}{2} & 0 \\ 0 & 0 & 1 \end{bmatrix}$$

Ans. (*a*) See Fig. 8-17.

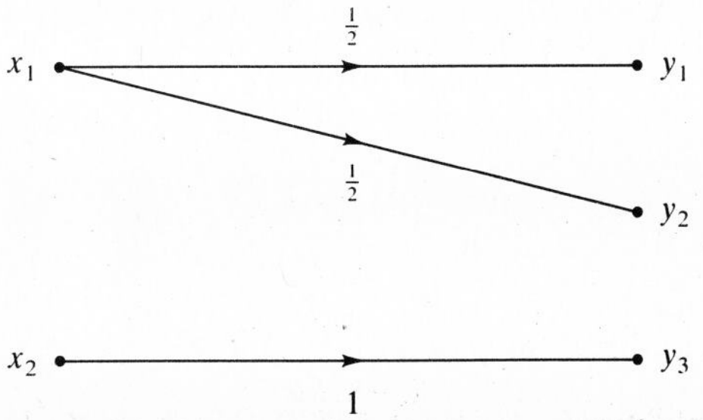

Fig. 8-17

(*b*) 1 b/symbol

8.61. Let X be a random variable with probability density function $f_X(x)$, and let $Y = aX + b$, where a and b are constants. Find $H(Y)$ in terms of $H(X)$.

Ans. $H(Y) = H(X) + \log_2 a$

8.62. Find the differential entropy $H(X)$ of a gaussian random variable X with zero mean and variance σ_X^2.

Ans. $H(X) = \frac{1}{2}\log_2(2\pi e \sigma_X^2)$

8.63. Consider an AWGN channel described by Eq. (*8.46*), that is,

$$Y = X + n$$

where X and Y are the channel input and output, respectively, and n is an additive white gaussian noise with zero mean and variance σ_n^2. Find the average mutual information $I(X;Y)$ when the channel input X is gaussian with zero mean and variance σ_X^2.

Ans. $I(X;Y) = \dfrac{1}{2}\log_2\left(1 + \dfrac{\sigma_X^2}{\sigma_n^2}\right)$

8.64. Calculate the capacity of an AWGN channel with a bandwidth of 1 MHz and an S/N ratio of 40 dB.

Ans. 13.29 Mb/s

8.65. Consider a DMS X with m equiprobable symbols x_i, $i = 1, 2, \ldots, m$.

(*a*) Show that the use of a fixed-length code for the representation of x_i is most efficient.

(*b*) Let n_o be the fixed code word length. Show that if $n_o = \log_2 m$, then the code efficiency is 100 percent.

Hint: Use Eqs. (*8.49*) and (*8.52*).

8.66. Construct a Huffman code for the DMS X of Prob. 8.32, and show that the code is an optimum code.

Ans.

Symbols:	x_1	x_2	x_3	x_4
Code:	0	10	110	111

8.67. A DMS X has five symbols x_1, x_2, x_3, x_4, and x_5 with respective probabilities 0.2, 0.15, 0.05, 0.1, and 0.5.

(*a*) Construct a Shannon-Fano code for X, and calculate the code efficiency.

(*b*) Repeat (*a*) for the Huffman code.

Ans.

(*a*)

Symbols:	x_1	x_2	x_3	x_4	x_5
Code:	10	110	1111	1110	0

Code efficiency $\eta = 98.6$ percent.

(*b*)

Symbols:	x_1	x_2	x_3	x_4	x_5
Code:	11	100	1011	1010	0

Code efficiency $\eta = 98.6$ percent.

8.68. Show that the Kraft inequality is satisfied by the codes of Prob. 8.33.

Hint: Use Eq. (*8.54*).

8.69. Consider a (6, 3) linear block code with the parity-check matrix H given by

$$H = \begin{bmatrix} 1 & 0 & 1 & 1 & 0 & 0 \\ 0 & 1 & 1 & 0 & 1 & 0 \\ 1 & 1 & 1 & 0 & 0 & 1 \end{bmatrix}$$

(*a*) Find the generator matrix G.

(*b*) Find the code word for the data bit 101.

Ans.

(*a*)
$$G = \begin{bmatrix} 1 & 0 & 0 & 1 & 0 & 1 \\ 0 & 1 & 0 & 0 & 1 & 1 \\ 0 & 0 & 1 & 1 & 1 & 1 \end{bmatrix}$$

(*b*) 101010

8.70. For the (7, 4) Hamming code of Prob. 8.51, decode the received word 0111100.

Ans. 0101

8.71. Consider an (n, k) linear block code with generator matrix G and parity-check matrix H. The $(n, n-k)$ code generated by H is called the *dual code* of the (n, k) code. Show that matrix G is the parity-check matrix for the dual code.

Hint: Take the transpose of Eq. (*8.62*).

8.72. Show that all code vectors of an (n, k) linear block code are orthogonal to the code vectors of its dual code. (The vector $\mathbf{x}$ is orthogonal to $\mathbf{c}$ if $\mathbf{x}\mathbf{c}^T = \mathbf{0}$, where $\mathbf{c}^T$ is the transpose of $\mathbf{c}$.)

Hint: Use Eq. (*8.62*).

8.73. Find the dual code of the (7, 4) Hamming code of Prob. 8.51, and find $d_{\min}$ of this dual code.

Ans.

0000000	0111001	1101010	1010011
1011100	1100101	0110110	0001111

$d_{\min} = 4$

8.74. A code consists of code words 1101000, 0111001, 0011010, 1001011, 1011100, and 0001101. If 1101011 is received, what is the decoded code word?

Ans. 1001011

8.75. Show that for all (n, k) linear block codes

$$d_{\min} \leq n - k + 1$$

Hint: Apply Theorem 8.2 to show that the rank of matrix H is $d_{\min} - 1$.

Appendix A

Fourier Transform

DEFINITIONS

$$X(\omega) = \mathscr{F}[x(t)] = \int_{-\infty}^{\infty} x(t)e^{-j\omega t}\,dt$$

$$x(t) = \mathscr{F}^{-1}[X(\omega)] = \frac{1}{2\pi}\int_{-\infty}^{\infty} X(\omega)e^{j\omega t}\,d\omega$$

Parseval's theorems:

$$\int_{-\infty}^{\infty} x_1(t)x_2(t)\,dt = \frac{1}{2\pi}\int_{-\infty}^{\infty} X_1(\omega)X_2^*(\omega)\,d\omega$$

$$\int_{-\infty}^{\infty} |x(t)|^2\,dt = \frac{1}{2\pi}\int_{-\infty}^{\infty} |X(\omega)|^2\,d\omega$$

Table A-1 Properties of the Fourier Transform

Property	$x(t)$	$X(\omega)$
Linearity	$a_1x_1(t) + a_2x_2(t)$	$a_1X_1(\omega) + a_2X_2(\omega)$
Time shifting	$x(t - t_0)$	$X(\omega)e^{-j\omega t_0}$
Scaling	$x(at)$	$\frac{1}{\lvert a\rvert}X\left(\frac{\omega}{a}\right)$
Time reversal	$x(-t)$	$X(-\omega)$
Duality	$X(t)$	$2\pi x(-\omega)$
Frequency shifting	$x(t)e^{j\omega_0 t}$	$X(\omega - \omega_0)$
Modulation	$x(t)\cos\omega_0 t$	$\frac{1}{2}[X(\omega - \omega_0) + X(\omega + \omega_0)]$
Time differentiation	$x'(t)$	$j\omega X(\omega)$
Frequency differentiation	$-jtx(t)$	$X'(\omega)$
Integration	$\int_{-\infty}^{t} x(\tau)\,d\tau$	$\frac{1}{j\omega}X(\omega) + \pi X(0)\delta(\omega)$
Convolution	$x_1(t) * x_2(t)$	$X_1(\omega)X_2(\omega)$
Multiplication	$x_1(t)x_2(t)$	$\frac{1}{2\pi}X_1(\omega) * X_2(\omega)$

Table A-2 Some Fourier Transform Pairs

$x(t)$	$X(\omega)$
$\delta(t)$	1
$\delta(t-t_0)$	$e^{-j\omega t_0}$
1	$2\pi\delta(\omega)$
$u(t)$	$\pi\delta(\omega)+\dfrac{1}{j\omega}$
$\operatorname{sgn}(t)$	$\dfrac{2}{j\omega}$
$\dfrac{1}{\pi t}$	$-j\operatorname{sgn}(\omega)$
$e^{j\omega_0 t}$	$2\pi\delta(\omega-\omega_0)$
$\cos\omega_0 t$	$\pi[\delta(\omega-\omega_0)+\delta(\omega+\omega_0)]$
$\sin\omega_0 t$	$-j\pi[\delta(\omega-\omega_0)-\delta(\omega+\omega_0)]$
$e^{-at}u(t) \quad a>0$	$\dfrac{1}{j\omega+a}$
$te^{-at}u(t) \quad a>0$	$\dfrac{1}{(j\omega+a)^2}$
$e^{-a\|t\|} \quad a>0$	$\dfrac{2a}{\omega^2+a^2}$
$e^{-t^2/(2\sigma^2)}$	$\sigma\sqrt{2\pi}\,e^{-\sigma^2\omega^2/2}$
$p_a(t)=\begin{cases}1 & \|t\|<a\\ 0 & \|t\|>a\end{cases}$	$2a\dfrac{\sin\omega a}{\omega a}$
$\dfrac{\sin at}{\pi t}$	$p_a(\omega)=\begin{cases}1 & \|\omega\|<a\\ 0 & \|\omega\|>a\end{cases}$
$x(t)=\begin{cases}1-\dfrac{\|t\|}{a} & \|t\|<a\\ 0 & \|t\|>a\end{cases}$	$a\left[\dfrac{\sin(\omega a/2)}{\omega a/2}\right]^2$
$\displaystyle\sum_{n=-\infty}^{\infty}\delta(t-nT)$	$\displaystyle\omega_0\sum_{n=-\infty}^{\infty}\delta(\omega-n\omega_0) \qquad \omega_0=\frac{2\pi}{T}$
$\displaystyle\hat{x}(t)=\frac{1}{\pi}\int_{-\infty}^{\infty}\frac{x(\tau)}{t-\tau}\,d\tau$	$-j\operatorname{sgn}(\omega)\,X(\omega)$

Appendix B

Bessel Functions $J_n(\beta)$

Bessel functions of the first kind of order n and argument β:

GENERATING FUNCTION AND DEFINITION

$$e^{j\beta \sin \omega_m t} = \sum_{n=-\infty}^{\infty} J_n(\beta) e^{jn\omega_m t}$$

$$J_n(\beta) = \frac{1}{2\pi} \int_{-\pi}^{\pi} e^{j(\beta \sin \lambda - n\lambda)}\, d\lambda$$

$$J_n(\beta) = \sum_{k=0}^{\infty} \frac{(-1)^k (\beta/2)^{2k+n}}{k!(k+n)!}$$

PROPERTIES OF $J_n(\beta)$:

1. $J_{-n}(\beta) = (-1)^n J_n(\beta)$
2. $J_{n-1}(\beta) + J_{n+1}(\beta) = \dfrac{2n}{\beta} J_n(\beta)$
3. $\displaystyle\sum_{n=-\infty}^{\infty} J_n^2(\beta) = 1$

Table B-1 Selected Values of $J_n(\beta)$

$n\backslash\beta$	0.1	0.2	0.5	1	2	5	8	10
0	0.997	0.990	0.938	0.765	0.224	−0.178	0.172	−0.246
1	0.050	0.100	0.242	0.440	0.577	−0.328	0.235	0.043
2	0.001	0.005	0.031	0.115	0.353	0.047	−0.113	0.255
3			0.003	0.020	0.129	0.365	−0.291	0.058
4				0.002	0.034	0.391	−0.105	−0.220
5					0.007	0.261	0.286	−0.234
6					0.001	0.131	0.338	−0.014
7						0.053	0.321	0.217
8						0.018	0.224	0.318
9						0.006	0.126	0.292
10						0.001	0.061	0.208
11							0.026	0.123
12							0.010	0.063
13							0.003	0.029
14							0.001	0.012
15								0.005
16								0.002

Appendix C

The Complementary Error Function $Q(z)$

$$Q(z) = \frac{1}{\sqrt{2\pi}} \int_z^{\infty} e^{-\lambda^2/2}\, d\lambda$$

$$Q(0) = \tfrac{1}{2} \qquad Q(-z) = 1 - Q(z) \qquad z \geq 0$$

$$Q(z) = \tfrac{1}{2} - \operatorname{erf}(z)$$

$$\operatorname{erf}(z) = \frac{1}{\sqrt{2\pi}} \int_0^{z} e^{-\lambda^2/2}\, d\lambda$$

$$Q(z) \approx \frac{1}{\sqrt{2\pi}\, z} e^{-z^2/2} \qquad z \gg 1\ (z > 4)$$

Table C-1 $Q(z)$

z	$Q(z)$	z	$Q(z)$	z	$Q(z)$	z	$Q(z)$
0.00	0.5000	1.00	0.1587	2.00	0.0228	3.00	0.00135
0.05	0.4801	1.05	0.1469	2.05	0.0202	3.05	0.00114
0.10	0.4602	1.10	0.1357	2.10	0.0179	3.10	0.00097
0.15	0.4404	1.15	0.1251	2.15	0.0158	3.15	0.00082
0.20	0.4207	1.20	0.1151	2.20	0.0139	3.20	0.00069
0.25	0.4013	1.25	0.1056	2.25	0.0122	3.25	0.00058
0.30	0.3821	1.30	0.0968	2.30	0.0107	3.30	0.00048
0.35	0.3632	1.35	0.0885	2.35	0.0094	3.35	0.00040
0.40	0.3446	1.40	0.0808	2.40	0.0082	3.40	0.00034
0.45	0.3264	1.45	0.0735	2.45	0.0071	3.45	0.00028
0.50	0.3085	1.50	0.0668	2.50	0.0062	3.50	0.00023
0.55	0.2912	1.55	0.0606	2.55	0.0054	3.55	0.00019
0.60	0.2743	1.60	0.0548	2.60	0.0047	3.60	0.00016
0.65	0.2578	1.65	0.0495	2.65	0.0040	3.65	0.00013
0.70	0.2420	1.70	0.0446	2.70	0.0035	3.70	0.00011
0.75	0.2266	1.75	0.0401	2.75	0.0030	3.75	0.00009
0.80	0.2169	1.80	0.0359	2.80	0.0026	3.80	0.00007
0.85	0.1977	1.85	0.0322	2.85	0.0022	3.85	0.00006
0.90	0.1841	1.90	0.0287	2.90	0.0019	3.90	0.00005
0.95	0.1711	1.95	0.0256	2.95	0.0016	3.95	0.00004
4.00	0.00003						
4.25	10^{-5}						
4.75	10^{-6}						
5.20	10^{-7}						
5.60	10^{-8}						

Appendix D

Selected Mathematical Formulas

D.1 TRIGONOMETRIC IDENTITIES

$$e^{\pm j\theta} = \cos\theta \pm j\sin\theta$$

$$\cos\theta = \tfrac{1}{2}(e^{j\theta} + e^{-j\theta})$$

$$\sin\theta = \frac{1}{2j}(e^{j\theta} - e^{-j\theta})$$

$$\sin^2\theta + \cos^2\theta = 1$$

$$\cos 2\theta = \cos^2\theta - \sin^2\theta = 2\cos^2\theta - 1 = 1 - 2\sin^2\theta$$

$$\sin 2\theta = 2\cos\theta\sin\theta$$

$$\cos^2\theta = \tfrac{1}{2}(1 + \cos 2\theta)$$

$$\sin^2\theta = \tfrac{1}{2}(1 - \cos 2\theta)$$

$$\cos(\alpha \pm \beta) = \cos\alpha\cos\beta \mp \sin\alpha\sin\beta$$

$$\sin(\alpha \pm \beta) = \sin\alpha\cos\beta \pm \cos\alpha\sin\beta$$

$$\tan(\alpha \pm \beta) = \frac{\tan\alpha \pm \tan\beta}{1 \mp \tan\alpha\tan\beta}$$

$$\cos\alpha\cos\beta = \tfrac{1}{2}\cos(\alpha - \beta) + \tfrac{1}{2}\cos(\alpha + \beta)$$

$$\sin\alpha\sin\beta = \tfrac{1}{2}\cos(\alpha - \beta) - \tfrac{1}{2}\cos(\alpha + \beta)$$

$$\sin\alpha\cos\beta = \tfrac{1}{2}\sin(\alpha - \beta) + \tfrac{1}{2}\sin(\alpha + \beta)$$

$$a\cos x + b\sin x = C\cos(x + \theta) \quad \text{where } C = \sqrt{a^2 + b^2} \quad \text{and} \quad \theta = -\tan^{-1}\frac{b}{a}$$

D.2 SERIES EXPANSIONS AND APPROXIMATIONS

$$(a + b)^n = \sum_{k=0}^{n} \binom{n}{k} a^{n-k} b^k \qquad \text{where } \binom{n}{k} = \frac{n!}{(n-k)!k!}$$

$$(1 + x)^n = \sum_{k=0}^{n} \binom{n}{k} x^k = 1 + nx + \frac{1}{2!}n(n-1)x^2 + \cdots$$

$$e^x = \sum_{k=0}^{\infty} \frac{1}{k!}x^k = 1 + x + \frac{1}{2!}x^2 + \frac{1}{3!}x^3 + \cdots$$

$$a^x = e^{x\ln a} = 1 + x\ln a + \frac{1}{2!}(x\ln a)^2 + \cdots$$

$$\cos x = 1 - \frac{1}{2!}x^2 + \frac{1}{4!}x^4 - \cdots$$

$$\sin x = x - \frac{1}{3!}x^3 + \frac{1}{5!}x^5 - \cdots$$

$$\ln(1 + x) = x - \tfrac{1}{2}x^2 + \tfrac{1}{3}x^3 - \cdots$$

When $|x| \ll 1$,

$$(1+x)^n \approx 1 + nx$$

$$e^x \approx 1 + x$$

$$\cos x \approx 1$$

$$\sin x \approx x$$

$$a^x \approx 1 + x \ln a$$

$$\ln(1+x) \approx x$$

D.3 INTEGRALS

Indefinite Integrals

$$\int \cos ax\, dx = \frac{1}{a} \sin ax$$

$$\int \sin ax\, dx = -\frac{1}{a} \cos ax$$

$$\int \cos ax \cos bx\, dx = \frac{\sin(a-b)x}{2(a-b)} + \frac{\sin(a+b)x}{2(a+b)} \qquad a^2 \neq b^2$$

$$\int \sin ax \sin bx\, dx = \frac{\sin(a-b)x}{2(a-b)} - \frac{\sin(a+b)x}{2(a+b)} \qquad a^2 \neq b^2$$

$$\int \sin ax \cos bx\, dx = -\left[\frac{\cos(a-b)x}{2(a-b)} + \frac{\cos(a+b)x}{2(a+b)}\right] \qquad a^2 \neq b^2$$

$$\int \cos^2 ax\, dx = \frac{x}{2} + \frac{\sin 2ax}{4a}$$

$$\int \sin^2 ax\, dx = \frac{x}{2} - \frac{\sin 2ax}{4a}$$

$$\int e^{ax}\, dx = \frac{1}{a} e^{ax}$$

$$\int e^{ax} \cos bx\, dx = \frac{e^{ax}}{a^2 + b^2}(a \cos bx + b \sin bx)$$

$$\int e^{ax} \sin bx\, dx = \frac{e^{ax}}{a^2 + b^2}(a \sin bx - b \cos bx)$$

$$\int \frac{dx}{a^2 + b^2 x^2} = \frac{1}{ab} \tan^{-1} \frac{bx}{a}$$

$$\int \frac{x^2\, dx}{a^2 + b^2 x^2} = \frac{x}{b^2} - \frac{a}{b^3} \tan^{-1} \frac{bx}{a}$$

Definite Integrals

$$\int_0^\infty x^n e^{-ax}\,dx = \frac{n!}{a^{n+1}} \qquad a>0$$

$$\int_0^\infty e^{-ax^2}\,dx = \frac{1}{2}\sqrt{\frac{\pi}{a}} \qquad a>0$$

$$\int_0^\infty x e^{-ax^2}\,dx = \frac{1}{2a} \qquad a>0$$

$$\int_0^\infty x^{2k} e^{-ax^2}\,dx = \frac{1\cdot 3\cdot \cdots \cdot(2k-1)}{2^{k+1}a^k}\sqrt{\frac{\pi}{a}} \qquad a>0$$

$$\int_0^\infty e^{-ax}\cos bx\,dx = \frac{a}{a^2+b^2} \qquad a>0$$

$$\int_0^\infty e^{-ax}\sin bx\,dx = \frac{b}{a^2+b^2} \qquad a>0$$

$$\int_0^\infty e^{-a^2x^2}\cos bx\,dx = \frac{\sqrt{\pi}}{2a}e^{-b^2/(4a^2)}$$

$$\int_0^\infty \frac{\cos ax}{b^2+x^2}\,dx = \frac{\pi}{2b}e^{-ab} \qquad a>0,\ b>0$$

$$\int_0^\infty \frac{x\sin ax}{b^2+x^2}\,dx = \frac{\pi}{2}e^{-ab} \qquad a>0,\ b>0$$

Integration by parts

$$\int f(x)g'(x)\,dx = f(x)g(x) - \int f'(x)g(x)\,dx$$

Appendix E

Symbols and Abbreviations

E.1 SYMBOLS

Symbol	Meaning
$\approx$	approximately equals
$\longleftrightarrow$	denoting Fourier transform pair
$\oplus$	modulo-2 addition
$\cup$	union of two events
$\cap$	intersection of two events
$\subset$	subset of
$\varnothing$	null event
Σ	summation
Π	product
$\mathscr{F}$	Fourier transform of
$\mathscr{F}^{-1}$	inverse Fourier transform of
$\mathscr{T}$	mapping operator
$\langle \cdot \rangle$	time average
$\bar{x}$	time average of $x(t)$
$\|a\|$	absolute value of a
$E[\cdot]$	expectation of
max	maximum
min	minimum
$\bar{A}$	complement of event A
$A^T(\mathbf{x}^T)$	transpose of matrix A (vector $\mathbf{x}$)
A^{-1}	inverse of matrix A
det A	determinant of matrix A
$H^*(\omega)$	*denotes complex conjugate
$\{x(n)\}$ $(x[n])$	sequence
$x'(\cdot)$	first derivative of x
$x(t) * y(t)$	*denotes convolution
$\hat{m}(t)$	$\hat{}$ denotes Hilbert transform
$\tilde{m}(t)$	$\tilde{}$ denotes staircase approximation
$m_+(\cdot)$	subscript + denotes analytic signal
sgn$(\cdot)$	signum function
$N(\mu; \sigma^2)$	normal random variable with mean μ and variance σ^2

E.2 ABBREVIATIONS

Abbreviations	Meaning
ac	alternating current
A/D	analog-to-digital
ADC	analog-to-digital converter
AM	amplitude modulation
AMI	alternate mark inversion
ASK	amplitude-shift keying
AWGN	additive white gaussian noise
BPF	bandpass filter
b/s	bits per second
BSC	binary symmetric channel
BSF	bandstop filter
CD	compact disk
CNR	carrier-to-noise ratio
CW	continuous wave
dB	decibel
dc	direct current
det	determinant
DM	delta modulation
DMC	discrete memoryless channel
DMS	discrete memoryless source
DSB	double-sideband
DSB-SC	double-sideband suppressed carrier
exp	exponential
FDM	frequency-division multiplexing
FM	frequency modulation
FSK	frequency-shift keying
Gb	gigabit
Hz	hertz
IF	intermediate frequency
ISI	intersymbol interference
kb/s	kilobits per second
kHz	kilohertz
ln	natural logarithm
LPF	low-pass filter
LSB	lower sideband
LTI	linear time-invariant

Abbreviations	Meaning
MAP	maximum a posteriori
Mb	megabit
Mb/s	megabits per second
MHz	megahertz
ms	millisecond
mV	millivolt
μs	microsecond
NB	narrowband
NBFM	narrowband frequency modulation
NRZ	non-return-to-zero
OOK	on-off keying
PAM	pulse amplitude modulation
PCM	pulse code modulation
pdf	probability density function
PM	phase modulation
PSK	phase-shift keying
RF	radio frequency
rms	root mean square
RZ	return-to-zero
s	second
sgn	signum
SNR or S/N	signal-to-noise ratio
$(S/N)_i$	input signal-to-noise ratio
$(S/N)_o$	output signal-to-noise ratio
$(S/N_q)_o$	output signal-to-quantizing-noise ratio
SSB	single-sideband
SSS	strict-sense stationary
TDM	time-division multiplexing
USB	upper sideband
V	volt
VCO	voltage controlled oscillator
var	variance
VSB	vestigial sideband
W	watt
WB	wideband
WSS	wide-sense stationary

Index